H. Sanwein

– privat –

THE MOLECULAR BASIS OF SMELL AND TASTE TRANSDUCTION

The Ciba Foundation is an international scientific and educational charity (Registered Charity No. 313574). It was established in 1947 by the Swiss chemical and pharmaceutical company of CIBA Limited—now Ciba-Geigy Limited. The Foundation operates independently in London under English trust law.

The Ciba Foundation exists to promote international cooperation in biological, medical and chemical research. It organizes about eight international multidisciplinary symposia each year on topics that seem ready for discussion by a small group of research workers. The papers and discussions are published in the Ciba Foundation symposium series. The Foundation also holds many shorter meetings (not published), organized by the Foundation itself or by outside scientific organizations. The staff always welcome suggestions for future meetings.

The Foundation's house at 41 Portland Place, London W1N 4BN, provides facilities for meetings of all kinds. Its Media Resource Service supplies information to journalists on all scientific and technological topics. The library, open five days a week to any graduate in science or medicine, also provides information on scientific meetings throughout the world and answers general enquiries on biomedical and chemical subjects. Scientists from any part of the world may stay in the house during working visits to London.

Ciba Foundation Symposium 179

THE MOLECULAR BASIS OF SMELL AND TASTE TRANSDUCTION

A Wiley–Interscience Publication

1993

JOHN WILEY & SONS

Chichester · New York · Brisbane · Toronto · Singapore

©Ciba Foundation 1993

Published in 1993 by John Wiley & Sons Ltd
Baffins Lane, Chichester
West Sussex PO19 1UD, England

All rights reserved.

No part of this book may be reproduced by any means,
or transmitted, or translated into a machine language
without the written permission of the publisher.

Other Wiley Editorial Offices

John Wiley & Sons, Inc., 605 Third Avenue,
New York, NY 10158-0012, USA

Jacaranda Wiley Ltd, G.P.O. Box 859, Brisbane,
Queensland 4001, Australia

John Wiley & Sons (Canada) Ltd, 22 Worcester Road,
Rexdale, Ontario M9W 1L1, Canada

John Wiley & Sons (SEA) Pte Ltd, 37 Jalan Pemimpin #05-04,
Block B, Union Industrial Building, Singapore 2057

Suggested series entry for library catalogues:
Ciba Foundation Symposia

Ciba Foundation Symposium 179
ix + 287 pages, 58 figures, 9 tables

Library of Congress Cataloging-in-Publication Data

The Molecular basis of smell and taste transduction / Derek Chadwick,
　Joan Marsh, Jamie Goode, editors.
　　　p.　　cm.—(Ciba Foundation symposium; 179)
　Symposium on the Molecular Basis of Smell and Taste Transduction,
held at the Ciba Foundation, London, Feb. 1993.
　Includes bibliographical references and index.
　ISBN 0-471-93946-3
　1. Smell—Molecular aspects—Congresses.　2. Taste—Molecular
aspects—Congresses.　3. Cellular signal transduction—Congresses.
4. Second messengers (Biochemistry)—Congresses.　5. G proteins—
Congresses.　I. Chadwick, Derek.　II. Marsh, Joan.　III. Goode,
Jamie.　IV. Symposium on the Molecular Basis of Smell and Taste
Transduction (1993: London, Eng.)　V. Series.
QP458.M63　1993
591.1'826—dc20　　　　　　　　　　　　　　　　　　　　　　　　93-28783
　　　　　　　　　　　　　　　　　　　　　　　　　　　　　　　　　　CIP

British Library Cataloguing in Publication Data
A catalogue record for this book is
available from the British Library

ISBN 0 471 93946 3

Phototypeset by Dobbie Typesetting Limited, Tavistock, Devon.
Printed and bound in Great Britain by Biddles Ltd, Guildford.

Contents

Symposium on the molecular basis of smell and taste transduction, held at the Ciba Foundation, London 19–21 February 1993

Editors: Derek Chadwick, Joan Marsh (Organizers) and Jamie Goode

This symposium is based on a proposal made by Frank Margolis and Tom Getchell

F. L. Margolis Introduction 1

F. L. Margolis, K. Kudrycki, C. Stein-Izsak, M. Grillo and **R. Akeson** From genotype to olfactory neuron phenotype: the role of the Olf-1-binding site 3
Discussion 20

T. V. Getchell, Z. Su and **M. L. Getchell** Mucous domains: microchemical heterogeneity in the mucociliary complex of the olfactory epithelium 27
Discussion 40

L. B. Buck Receptor diversity and spatial patterning in the mammalian olfactory system 51
Discussion 64

M. M. Wang and **R. R. Reed** Molecular mechanisms of olfactory neuronal gene regulation 68
Discussion 73

M. R. Lerner, M. N. Potenza, G. F. Graminski, T. McClintock, C. K. Jayawickreme and **S. Karne** A new tool for investigating G protein-coupled receptors 76
Discussion 84

General discussion I 88

H. Breer Second messenger signalling in olfaction 97
Discussion 109

S. Firestein and **F. Zufall** Membrane currents and mechanisms of olfactory transduction 115
Discussion 126

D. Lancet, N. Ben-Arie, S. Cohen, U. Gat, R. Gross-Isseroff, S. Horn-Saban, M. Khen, H. Lehrach, M. Natochin, M. North, E. Seidemann and **N. Walker** Olfactory receptors: transduction, diversity, human psychophysics and genome analysis 131
Discussion 141

General Discussion II 147

J. Carlson Molecular genetics of *Drosophila* olfaction 150
Discussion 162

H. Schmale, C. Ahlers, M. Bläker, K. Kock and **A. I. Spielman** Perireceptor events in taste 167
Discussion 180

S. K. McLaughlin, P. J. McKinnon, A. Robichon, N. Spickofsky and **R. F. Margolskee** Gustducin and transducin: a tale of two G proteins 186
Discussion 196

S. C. Kinnamon Role of apical ion channels in sour taste transduction 201
Discussion 210

J. A. DeSimone, Q. Ye and **G. E. Heck** Ion pathways in the taste bud and their significance for transduction 218
Discussion 229

P. Sengupta, H. A. Colbert, B. E. Kimmel, N. Dwyer and **C. I. Bargmann** The cellular and genetic basis of olfactory responses in *Caenorhabditis elegans* 235
Discussion 244

L. M. Bartoshuk Genetic and pathological taste variation: what can we learn from animal models and human disease? 251
Discussion 262

General discussion III 268

F. L. Margolis Summing-up 274

Index of contributors 277

Subject index 279

Participants

B. W. Ache University of Florida, Whitney Laboratory, 9505 Ocean Shore Boulevard, St Augustine, FL 32086, USA

C. I. Bargmann Program in Developmental Biology, University of California, Medical School, 513 Parnassus Avenue, San Francisco, CA 94143-0452, USA

L. M. Bartoshuk Department of Surgery, Yale University School of Medicine, 333 Cedar Street, PO Box 208041, New Haven, CT 06520-8041, USA

H. Breer Institut für Zoophysiologie, Universität Hohenheim (230), Postfach 70 05 62, D-7000 Stuttgart 70, Germany

L. Buck Department of Neurobiology, Harvard Medical School, 220 Longwood Avenue, Boston, MA 02115, USA

J. Caprio Department of Zoology and Physiology, Louisiana State University, Baton Rouge, LA 70803, USA

J. Carlson Department of Biology, Yale University, Kline Biology Tower, PO Box 6666, New Haven, CT 06511-8112, USA

J. A. DeSimone Department of Physiology, Box 551, Medical College of Virginia, Virginia Commonwealth University, Richmond, VA 23298-0551, USA

E. E. Fesenko Institute of Cell Biophysics, Academy of Sciences, 142292 Pushchino, Moscow Region, Russia

S. Firestein Department of Neurobiology, Yale University School of Medicine, 333 Cedar Street, PO Box 208041, New Haven, CT 06520-8041, USA

T. V. Getchell Chandler Medical Centre, University of Kentucky, College of Medicine, 800 Rose Street, Lexington, KY 40536-0084, USA

H. Hatt Physiologisches Institut, Technische Universität München, Biedersteiner Strasse 29, D-8000 München 40, Germany

P. M. Hwang (*Bursar*) Department of Neuroscience, The Johns Hopkins University, School of Medicine, 725 North Wolfe Street, Baltimore, MD 21205, USA

S. C. Kinnamon Department of Anatomy and Neurobiology, Colorado State University, Fort Collins, CO 80523, USA

T. Kurahashi Monell Chemical Senses Center, 3500 Market Street, Philadelphia, PA 19104-3308, USA

D. Lancet Department of Membrane Research and Biophysics, The Weizmann Institute of Science, 76100 Rehovot, Israel

M. Lerner Howard Hughes Medical Institute, Yale University School of Medicine, BCMM 254, PO Box 9812, New Haven, CT 06536-0812, USA

B. Lindemann Physiologisches Institut, Universität des Saarlandes, D-6650 Homburg, Germany

I. Lush Department of Genetics and Biometry, University College London, Wolfson House, 4 Stephenson Way, London NW1 2HE, UK

F. L. Margolis (*Chairman*) Laboratory of Chemosensory Neurobiology, Roche Institute of Molecular Biology, Roche Research Center, 340 Kingsland Street, Nutley, NJ 07110-1199, USA

R. F. Margolskee Roche Institute of Molecular Biology, Roche Research Center, 340 Kingsland Street, Nutley, NJ 07110-1199, USA

P. Pelosi Istituto di Industrie Agrarie, Universitá di Pisa, Via S Michele degli Scalzi 4, I-56100 Pisa, Italy

R. R. Reed Department of Molecular Biology and Genetics, The Johns Hopkins University, School of Medicine, Howard Hughes Medical Institute, 725 North Wolfe Street, Baltimore, MD 21205, USA

G. V. Ronnett Department of Neuroscience, The Johns Hopkins University, School of Medicine, 725 North Wolfe Street, Baltimore, MD 21205, USA

H. Schmale Institut für Zellbiochemie und klinische Neurobiologie, Universität Hamburg, Martinistrasse 52, D-20246 Hamburg, Germany

Participants

O. Siddiqi Molecular Biology Unit, Tata Institute of Fundamental Research, Homi Bhabha Road, Bombay 400 005, India

J. H. Teeter Monell Chemical Senses Center, 3500 Market Street, Philadelphia, PA 19104-3308, USA

J. Van Houten Department of Zoology, Marsh Life Science Building, University of Vermont, Burlington, VT 05405-0086, USA

Introduction

Frank L. Margolis

Laboratory of Chemosensory Neurobiology, Roche Institute of Molecular Biology, 340 Kingsland Street, Nutley, NJ 07110-1199, USA

I would like to start by reviewing the origins of this symposium and indicating some of the broad questions that we hope to answer. Two years ago I was in this room as a discussant at a meeting that was held on the regeneration of sensory cells. The one thing, apart from the science, that I came away with at the end of the meeting was the realization that it was essential that I find some way to come back! After discussions with my friend and collaborator, Tom Getchell, we decided to try and put together an equivalent meeting on the chemical senses, to create a period of intense interaction that would serve as an opportunity to identify the state of the field and the next set of questions to be addressed. So we drafted a proposal which we presented to the Ciba Foundation, resulting in this meeting.

This is actually the second Ciba Foundation symposium on the chemical senses. The first, held nearly a quarter of a century ago in 1969, was called *Taste and Smell in Vertebrates* (Ciba Foundation 1970). I would like to quote from Professor Otto Lowenstein's introduction, where he said 'we have now arrived at the point where what are sometimes considered to be the Cinderellas of the senses, taste and smell, are to be discussed. You may know only too well how anyone teaching sensory physiology usually enjoys himself until he comes to taste and smell, and then he tends to run out of teachable material. But I have a feeling that we are nearing a breakthrough in this field, and therefore it is very timely that we have this symposium now.'

Finally, nearly 25 years later, we have reached the point where we are seeing the breakthrough predicted by Professor Lowenstein in 1969. Over the last few years, the contemporary techniques of molecular biology and patch-clamp recording have led to major advances. Consequently, we now know a great deal about the mechanisms, molecules and interactions involved in chemosensory transduction. As this meeting proceeds over the next few days, I think we will see how true this is.

I want to state what I believe the questions are that we are going to try to answer or, at least, discuss in this symposium. One of the things that became apparent to me in rereading the proposal that Tom Getchell and I had written was that several of the issues we raised as open questions two years ago have already been at least partly answered. Nevertheless, there are several questions that I hope we will address over the next few days:

1) In terms of chemosensory transduction—what can we learn about receptor–ligand interactions and second messenger cascades?

2) Perireceptor events—what do we know about stimulus arrival, interaction and elimination?

3) What are the molecular properties of the chemosensory cells themselves, in both vertebrates and invertebrates?

4) Does novel function require expression of genes coding for novel components of the transduction pathway?

5) With regard to initiation and transmission of information—how and what do the sensory cells tell the central nervous system?

6) In terms of the genetic basis of dysfunction—what can we learn from animal models, human disease and reverse genetics?

Hopefully, at the end we will be in a position where we can make predictions, draw some conclusions and try to answer the question of what additional tools, techniques and conceptual insights are required to study the mechanisms of chemodetection.

Reference

Ciba Foundation 1970 Taste and smell in vertebrates. Churchill, London

From genotype to olfactory neuron phenotype: the role of the Olf-1-binding site

Frank L. Margolis, Katarzyna Kudrycki, Cathy Stein-Izsak*, Mary Grillo and Richard Akeson[†]

Roche Institute of Molecular Biology, Roche Research Center, Department of Neurosciences, Nutley, NJ 07110, USA

Abstract. The highly organized pattern of gene expression leading to the determination of cellular phenotype derives from the interplay between genetic and epigenetic factors. This is mediated in part by distinctive DNA sequence motifs present in the regulatory regions of various genes and the transcription factors with which they interact. The phenotype of olfactory neurons is determined in part by the selective expression of novel isoforms of several genes involved in chemosensory transduction. To characterize the mechanisms determining olfactory neuron phenotype we have been studying the olfactory marker protein (OMP), the first olfactory-specific protein to be isolated and cloned. The temporal and spatial expression of *OMP* is regulated stringently and is highly restricted to mature olfactory neurons in all vertebrates from amphibians to humans. Identification of the specific elements responsible for regulating the expression of the *OMP* gene will elucidate the mechanisms leading to the determination of olfactory neuron phenotype. Using a combined *in vivo* (transgenic mice) and *in vitro* (electrophoretic mobility shift assays and DNase I footprinting) approach, we have identified and characterized a novel genomic motif that binds an olfactory tissue nuclear protein(s) that we designate Olf-1. We propose that Olf-1 is a novel olfactory-specific *trans*-acting factor responsible for directing the expression of genes containing the Olf-1 motif in olfactory neurons. Thus it may play a role in regulating the expression of genes associated with neuronal turnover and olfactory transduction.

1993 The molecular basis of smell and taste transduction. Wiley, Chichester (Ciba Foundation Symposium 179) p 3–26

Determination of cellular phenotype in the mammalian central nervous system (CNS) must involve exceptionally complex patterns of transcriptional regulation. The CNS consists of billions of cells that show distinctive morphological,

---

*Present address: Organon Inc., West Orange, NJ 07052, USA.
[†]We regret to report the death of Dr Richard Akeson.

biochemical, biophysical and functional properties. In addition, patterns of gene expression in these cells change continuously during cell development (He & Rosenfeld 1991, McKay 1989). The mature peripheral olfactory system offers several advantages for studying neuron-specific gene expression. It manifests an unusual ability to generate new receptor neurons from precursor cells to replace olfactory neurons lost because of environmental, surgical or chemical damage (Constanzo & Graziadei 1983, Graziadei & Monti-Graziadei 1978, Samanen & Forbes 1984, Verhaagen et al 1990). Significantly, the mature olfactory neurons are the major site of expression of several novel gene products associated with chemosensory signal transduction (Bakalyar & Reed 1990, Buck & Axel 1991, Dhallan et al 1990, Jones & Reed 1989, Ludwig et al 1990). The first olfactory neuron-specific protein to be isolated (Margolis 1972) and cloned (Danciger et al 1989, Rogers et al 1987) was the olfactory marker protein (OMP). OMP expression is highly restricted to mature olfactory neurons (Margolis 1980, Margolis 1988) in species as diverse as salamanders (Krishna et al 1992) and humans (Chuah & Zheng 1987, Nakashima et al 1984). This 19 kDa cytoplasmic protein of unknown physiological function appears in the olfactory tissues at the beginning of the last trimester of gestation and is characteristic of olfactory neurons in their final stages of differentiation (Schwob et al 1992, Verhaagen et al 1990).

Since OMP expression is tightly controlled in a spatial and temporal manner, identification of the genomic elements responsible for this regulation should offer insight into the mechanisms regulating expression of various gene products associated with olfactory chemosensory transduction. To understand the mechanism of *OMP* transcriptional regulation, we initiated a study to identify genomic motifs within the *OMP* promoter responsible for age- and tissue-specific expression of this protein. Analyses of regulatory elements within neuronal genes have been conducted largely in transgenic animals as permanent differentiated cell lines are rarely available. Initial studies with transgenic mice demonstrated that 5.2 kb of the *OMP* promoter contained the information required for *OMP*-like tissue-specific reporter gene expression (Danciger 1989). Largent et al (1993) have confirmed this observation. Subsequent experiments in our laboratory indicated that as little as 0.3 kb upstream of the *OMP* gene translational start site is sufficient for correct cell-specific expression of a reporter gene *in vivo* (Grillo et al 1992). We have identified several elements within the *OMP* gene promoter that may be responsible for tissue- and developmentally specific expression of this protein. Two of these contain a similar novel sequence motif that interacts selectively with nuclear proteins from olfactory neuroepithelia of several species. The protein(s) that is involved in this olfactory-specific binding was designated Olf-1. We propose that Olf-1 represents an olfactory neuron-specific *trans*-acting factor or a complex of factors involved in regulating the expression of genes associated with olfactory transduction and neuronal regeneration.

Identification of the Olf-1-binding site

Two elements that interact with olfactory-specific factor(s) have been identified within the rat *OMP* promoter by electrophoretic mobility shift assay (EMSA) and DNase I footprinting analyses. Initial experiments indicated that a fragment of the *OMP* promoter, located between nucleotides -239 and -124 (fragment F-2) upstream of the transcription start site, formed a complex with component(s) of the nuclear protein extract prepared from rat olfactory epithelium. The gene fragment involved in this binding was analysed by EMSA of several DNA fragments derived from this *OMP* promoter region. The results of these experiments are summarized in Fig. 1A. A bandshift identical to that observed in the initial experiments was present when probe F-7, F-8 or F-9 was used. Formation of this complex was eliminated completely by prior digestion of the F-2 fragment with either *Ava*II or *Hph*I restriction endonucleases (fragments F-3 to F-6). These results demonstrate that the DNA sequence that binds component(s) of the olfactory neuroepithelium nuclear protein extract is located within the 36 bp *Nla*III–*Mnl*I restriction endonuclease fragment (F-9) of the *OMP* promoter.

The specificity of this DNA–protein interaction was evaluated by EMSAs performed with nuclear protein extracts prepared from a variety of tissues and in the presence of specific and non-specific competitors (Fig. 1B). Preincubation of the olfactory epithelium nuclear protein extract with a 20-fold molar excess of the unlabelled F-2 fragment competitively inhibited binding with the ^{32}P-labelled F-8 probe. In contrast, neither the fragment F-1 nor a fragment of β-lactamase cDNA competed with F-8. The specific bandshift observed in the presence of a nuclear protein extract prepared from olfactory neuroepithelium was not detected with protein extracts prepared from cerebellum, cerebral hemispheres, thymus, lungs or liver (Fig. 1B). Thus, this binding must involve specific interactions between a nuclear protein(s) from olfactory neuroepithelium and a specific nucleotide sequence motif. A more slowly migrating complex of lower intensity was also formed between *OMP* gene fragments containing this binding site and nuclear proteins of the cerebellum (Fig. 1B). However, more detailed evaluation of this interaction indicated that it is unrelated to the olfactory-specific complex (Kudrycki et al 1993).

The nucleotide sequence of the binding site was characterized by DNase I footprinting experiments. Nucleotides -181 to -163 of the sense strand and nucleotides -186 to -166 of the antisense strand were protected from DNase I digestion in the presence of nuclear protein extracts from rat olfactory epithelia. Slightly larger footprints of the same region were obtained with mouse olfactory nuclear protein extract (Fig. 2). We have named the olfactory-specific factor Olf-1 (Kudrycki et al 1993). Over 2.3 kb of the 5' flanking region of the rat *OMP* gene (Danciger et al 1989, C. Stein-Izsak, unpublished results) were searched for similarities with the protected sequence and several regions with

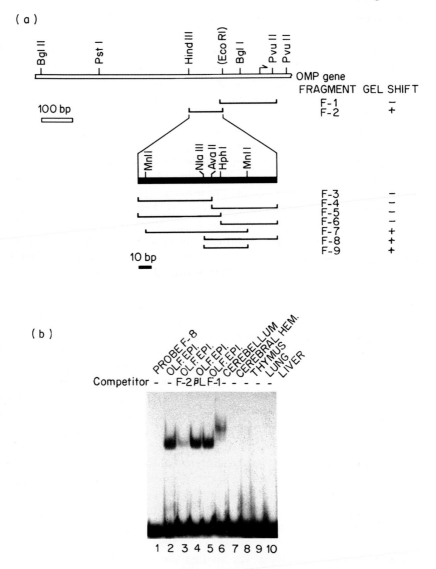

FIG. 1. Identification of a region within the *OMP* promoter that specifically binds the olfactory-specific nuclear factor Olf-1. (a) Summary of the results of the initial EMSA experiments. The restriction enzyme map of a portion of the *OMP* gene is shown at the top. Only selected restriction sites are shown. The *Eco*RI site was created in fragment F-2 using the polymerase chain reaction and is shown in parentheses. The arrow represents the transcription initiation site. DNA fragments that were used as probes and/or competitors in EMSAs are identified as brackets. Shaded bar represents an *OMP* gene region that was first identified as binding to Olf-1 (nucleotides −239 to −124). Binding activity of the radiolabelled DNA fragments F-1 to F-9 in the presence of nuclear proteins

some degree of sequence similarity were identified. One such region, located approximately 700 bp upstream of the *OMP* gene transcription initiation site, contains the sequence 5′-CTCCCAGGGGAGG-3′ in which nine out of 13 nucleotides are complementary to the motif 5′-GGTCCCCAAGGAG-3′ that forms the core of the footprint described above (Table 1). This region was analysed by DNase I footprinting and a prominent footprint that includes the 5′-CTCCCAGGGGAG-3′ sequence was observed between nucleotides −708 to −683 on the sense strand and nucleotides −710 to −691 on the antisense strand. Further analysis indicated that both sequence motifs bind the same component that is present only in olfactory nuclear extracts. Therefore, these sites are referred to as the proximal (sense strand, nucleotides −181 to −163) and distal (sense strand, nucleotides −708 to −683) Olf-1-binding sites.

An additional, unrelated footprint was observed between nucleotides −588 to −563 on the sense strand and nucleotides −590 to −565 on the antisense strand; it is referred to as the upstream binding element (UBE). Comparison of the nucleotide sequences of all three footprints with sequences of *cis*-acting elements present in various data banks provided no exact matches.

Tissue specificity of the Olf-1 and upstream binding element motifs

The footprinted regions of the *OMP* promoter were tested for their ability to interact with nuclear proteins from a number of neuronal and non-neuronal tissues by DNase I footprinting and EMSA (Fig. 3). The footprints of the proximal and distal Olf-1-binding sites were observed only in the presence of nuclear extracts from olfactory neuroepithelia. Furthermore, synthetic double-stranded oligonucleotides A and B, representing the proximal and distal sites, respectively (Table 1), bound only nuclear proteins derived from olfactory neuroepithelium of mouse and rat, not those from other tissues (Fig. 3B). This is the first example of molecular analysis of an interaction between the promoter region of an olfactory neuron-specific gene with putative tissue-specific regulatory nuclear protein(s) from olfactory neuroepithelium.

In contrast to the tissue specificity of the Olf-1-binding site, the UBE interacted with nuclear protein extracts from every tissue tested. The boundaries of the

---

from olfactory neuroepithelium is shown on the right: Olf-1-specific shift (+); no Olf-1 binding (−). EMSA analysis of one of these fragments is shown in panel (b). The F-8 (*Nla*III–*Eco*RI) fragment was analysed in the presence of 5 μg of nuclear protein-enriched extract from olfactory neuroepithelium (lanes 2–5) and from other neural (lanes 6 and 7) and non-neural tissues (lanes 8–10). The specificity of the DNA–protein interaction was determined in the presence of at least a 10-fold molar excess of competitors that are specific (F-2, lane 3) or non-specific (F-1, lane 4; βL, fragment of the β-lactamase coding region, lane 5) for Olf-1 binding. Migration of the free probe is also shown (probe). For details on the preparation of probes and competitors see Kudrycki et al (1993).

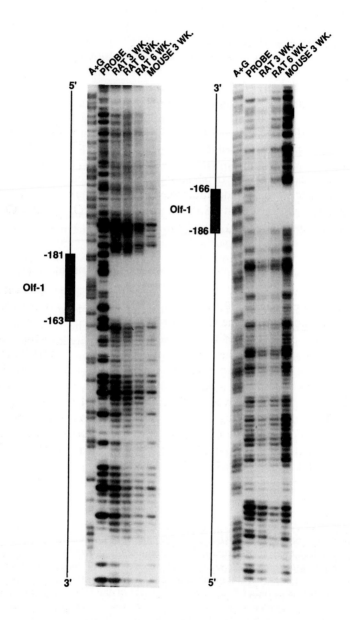

protected areas differed slightly when extracts from various tissues were used in DNase I footprinting (data not shown). Furthermore, the complexes formed between the UBE and nuclear proteins from various tissues migrated with different rates. Coupled with other results from our lab, this indicates that in contrast to the restricted distribution of Olf-1, the factor(s) binding to the UBE of the *OMP* promoter are ubiquitous and have distinct properties.

Analysis of the Olf-1-binding site

The consensus sequence of the proximal and distal Olf-1-binding sites has been defined by analysis of Olf-1 binding using double-stranded synthetic oligonucleotides containing nucleotide substitutions within the Olf-1 sites and using sequences resembling the Olf-1-binding site that occur elsewhere within the *OMP* gene. The 13-nucleotide imperfect inverse repeats at the centre of the Olf-1-specific footprints consist of the sequence (G/C)(G/C)TCCCC(A/T)-(A/G)GGAG (A and B, Table 1). Six double-stranded oligonucleotides spanning *OMP* gene regions that exhibit partial homology to this motif (oligonucleotides C, D, E, F, G and H, Table 1) and other synthetic oligonucleotides (not shown) with substitutions within the Olf-1-binding sequences were examined by EMSA for their ability to bind Olf-1 and to compete for its binding with oligonucleotides which contain the proximal Olf-1-binding site. Five oligonucleotides carrying sites partially homologous to the Olf-1 motif (oligonucleotides C, D, E, F and G) were incapable of either binding to or competing for the olfactory-specific factor (data not shown). In all these oligonucleotides, at least two purines within the GGAG segment of the Olf-1-binding site are replaced by pyrimidines. Purine substitutions within this segment (oligonucleotide H) weaken but do not eliminate the Olf-1 binding. This result was confirmed by DNase I footprinting of the *OMP* gene fragment containing the sequence of oligonucleotide H, which showed weak protection of the Olf-1-like motif (data not shown). Extensive detailed analyses (Kudrycki et al 1993) have led us to define the consensus sequence for Olf-1 binding as TCCCC(A/T)NGGAG.

FIG. 2. (*opposite*) DNase I footprint analysis of the Olf-1-binding site. The *OMP* 5' flanking region was analysed by DNase I footprinting using 25 µg of nuclear protein extract from olfactory neuroepithelia of three- and six-week-old rats and 20 µg of nuclear protein extract obtained from olfactory neuroepithelia of three-week-old mice. Shaded bars outline regions protected by interaction with Olf-1. Positions of binding sites relative to the transcription initiation start site in the *OMP* gene are indicated by numbers. Probes treated with DNase I in the absence of any protein extract are shown in lanes labelled probe. Chemically cleaved probes (A+G reactions) were used as sequence markers (A+G). Olf-1 binding to its proximal binding site on antisense (left) and sense (right) strands.

TABLE 1 Olf-1-binding activity of various oligonucleotides present in the rat *OMP* gene

Oligonucleotide	Position of oligo[a]	Sequence[b]		Olf-1-binding activity[c]
A (Proximal Olf-1 site)	−190 to −158	5′-CCATGCTCTGG TCCCCAAGGAG CCTGTCACCCT-3′		+++
B (Distal Olf-1 site)	−683 to −713	5′-GATCCTCCACC TCCCCTGGGAG ATGTGAGGC-3′		+++
C	−734 to −705	5′-AGATCTCCAGCGTCC TCCCCGG**CCTC** ACAT-3′		−
D	−105 to −76	5′-TCATGTG TCCCCTG**TTCT** GACAACTGGGTG-3′		−
E	−409 to −438	5′-GGTTGCTTCC TCCCCACC**TCA** TTCTCTCGA-3′		−
F	−994 to −965	5′-ATGTGCCTCTGG TCCCCGCCGTG TGTGTGT-3′		−
G	+2162 to +2193	5′-AAAGACTGTATGCCC TCCCCTCTGTG GTGTGG-3		−
H	+2140 to +2109	5′-AGACACAACTG TCCCCATAGGG CAAATGGAGT-3′		+
	Consensus sequence of the Olf-1-binding motif:	TCCCC$\frac{A}{T}$NGGAG		

[a]Numbers indicate positions of the sequence within the rat *OMP* gene that corresponds to the indicated strand of the double-stranded oligonucleotides used in EMSA experiments.
[b]Bold capital letters indicate residues which differ from the consensus sequence of Olf-1-binding motif.
[c]+++, strong Olf-1-specific binding; +, weak Olf-1-specific binding; −, no binding.

Biological responses of the Olf-1-binding activity

In olfactory tissue, OMP is expressed exclusively in mature olfactory receptor neurons (for review see Margolis 1980) and appears one week after precursor cell mitosis (Miragall & Monti-Graziadei 1982). The thickness of the cell layer expressing OMP increases progressively from a single layer one day after birth until 3.5 weeks of age, when most of the cells of the neuroepithelium express OMP, except for those in the basal cell region (Verhaagen et al 1989). To identify a possible correlation between the Olf-1-binding activity and the expression of OMP at different stages of olfactory neuron maturation, we have measured Olf-1-binding activity of nuclear protein extracts prepared from olfactory neuroepithelium of rats of various ages (Fig. 4 and data not shown). On the basis of both DNase I footprinting titrations and EMSA, Olf-1 activity is somewhat higher in the tissues of animals before weaning. In contrast, concomitant age-dependent decreases in binding activity were not observed when oligonucleotides containing the UBE motif, or the binding sites for NF-$\varkappa$B, AP-1 or Oct-1, were used as EMSA probes with these extracts. To confirm the finding that Olf-1-binding activity was higher in olfactory tissues enriched in immature olfactory neurons, we also measured it in actively regenerating olfactory tissues of animals that had undergone bilateral olfactory bulbectomy. In two separate experiments, olfactory neuroepithelia of four-week old surgically treated rats were allowed to regenerate for four or 14 days and then were used for preparation of nuclear protein extracts. The binding activity of Olf-1 in these extracts was tested by EMSA and compared with the binding activity of nuclear protein extracts prepared from olfactory tissues of age-matched controls. The Olf-1-binding activity of a nuclear protein extract from rats sacrificed four days after bulbectomy was greater than that of the extract obtained from untreated age-matched control animals or from rats whose neuroepithelium was allowed to regenerate for two weeks (Fig. 5). These observations are consistent with, but do not prove that, expression of Olf-1 precedes that of OMP.

Evidence for biological function of the region of the *OMP* gene containing the Olf-1 motif

To begin assessment of the biological significance of the Olf-1 motifs, we have examined the ability of a fragment of the rat *OMP* gene spanning nucleotides -239 to $+55$ (where the translation start site is located at position $+56$, Danciger et al 1989) to drive expression of *lacZ* in transgenic mice. In three strains of mice expressing the *OMP–lacZ* transgene, high levels of exogenous β-galactosidase activity were observed in olfactory neuroepithelium, but not in cortex or cerebellum (Table 2). Histochemical staining for β-galactosidase activity demonstrates selective cellular expression in sites that express OMP,

(a)

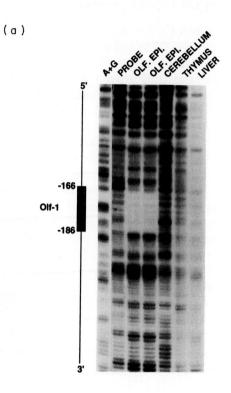

(b)

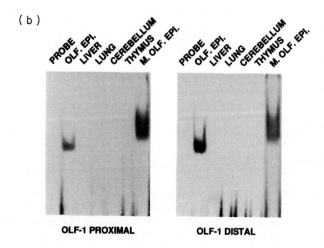

OLF-1 PROXIMAL   OLF-1 DISTAL

Olf-1-binding site

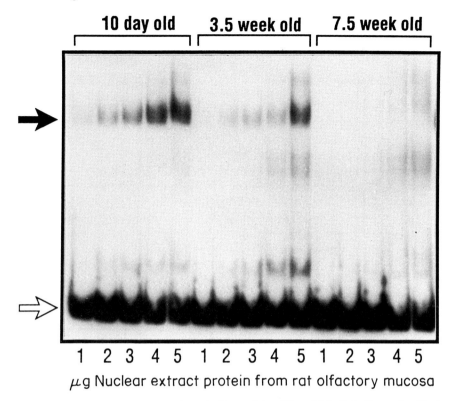

FIG. 4. Age dependence of the Olf-1-binding activity. ^{32}P-end-labelled oligonucleotide A, which contains the proximal Olf-1-binding site (see Table 1), was analysed by EMSA in the presence of increasing amounts of nuclear protein-enriched extracts from olfactory neuroepithelia of 10-day-old, 3.5-week-old and 7.5-week-old rats. Open arrow indicates mobility of free oligonucleotide probe and closed arrow mobility of Olf-1–oligonucleotide complex.

FIG. 3. *(opposite)* Tissue specificity of the Olf-1 and the UBE sequence motifs. (A) DNase I footprint analyses of ^{32}P-labelled sense strand probe containing the proximal Olf-1-binding site in the presence of 25 µg of nuclear protein extract from neuroepithelia (olf. epi.), cerebellum and thymus, and 19 µg of liver extract, of three-week postnatal rat. Position of the binding site relative to the transcription initiation site in the *OMP* gene is indicated by numbers. Chemically cleaved footprinting probes (A + G reactions) were used as sequence markers (A + G). Probe treated with DNase I in absence of protein extract is also shown (probe). (B) Tissue-specific binding of nuclear proteins with ^{32}P-end-labelled oligonucleotides, which contain the proximal Olf-1-binding site or the distal Olf-1-binding site, were examined by EMSA in the presence of 2.6 µg of nuclear protein-enriched extracts from three-week-old rat olfactory neuroepithelia (olf. epi.), 9.2 µg liver, 1.9 µg lung, 7.5 µg cerebellum and 7 µg thymus extracts, and 4 µg of protein extract from olfactory neuroepithelia of three-week-old mice (m. olf. epi.). Binding reactions performed in the absence of protein extract are shown in lanes labelled probe.

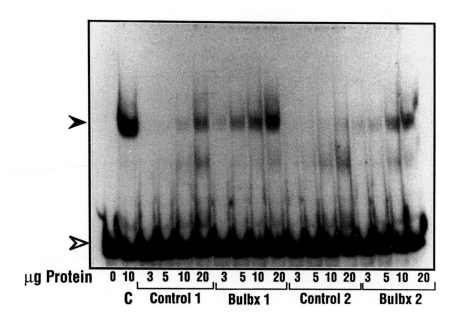

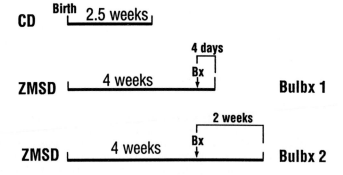

TABLE 2 *LacZ* activity in tissues of *OMP-lacZ* transgenic mice

Strain	Olfactory mucosa	Cortex	Cerebellum
H-*OMP-lacZ*-1	15.1	<1	<1
	16.9	<1	1.7
	8.0	<1	<1
H-*OMP-lacZ*-3	95.6	<1	<1
	70.2	<1	<1
	103.1	<1	<1
H-*OMP-lacZ*-6	27.7	<1	<1
	29.7	<1	<1
	37.5	<1	<1

Picograms of *lacZ* activity detected per milligram of tissue are presented as means of duplicate assays for individual 9-week-old male mice. The observed *lacZ* activity in control mice is routinely below 1 pg per mg of tissue.

i.e. olfactory, vomeronasal and septal organ neuroepithelium (R. Smeyne & E. Walters, unpublished observations). Since the Olf-1 motif is located within the transgene, approximately 180 nucleotides upstream of the *OMP* gene transcription initiation site, these results are consistent with the hypothesis that this motif is involved in olfactory-specific regulation of the *OMP* gene. This pattern of expression was not observed when a reporter gene was placed under the control of a smaller fragment of the *OMP* promoter from which the Olf-1 motif was absent (Grillo et al 1992), demonstrating the requirement for the presence of the Olf-1 motif for olfactory neuron expression. More detailed analysis of the biological role of the Olf-1 motif is currently under way, both *in vivo* and *in vitro*.

Discussion

We have identified several sequence motifs within 800 bp of the 5' flanking region of the *OMP* gene that may be involved in tissue- and developmentally specific transcriptional regulation of *OMP* expression (Fig. 6). Two of the identified sequences are located approximately 0.5 kb apart and share a common

FIG. 5. (*opposite*) EMSA analysis of Olf-1-binding activity in olfactory neuroepithelium of bulbectomized rats. Rats of the ZMSD strain were bilaterally olfactory bulbectomized at four weeks of age, killed four days or two weeks after surgery and their olfactory neuroepithelia were used to prepare nuclear protein-enriched extracts. A ^{32}P-end-labelled fragment of the *OMP* promoter (nucleotides -241 to -113), amplified using PCR, that contains the proximal Olf-1-binding site was incubated with increasing amounts of protein extracts from bulbectomized animals (Bulbx) and from age-matched littermate controls (Control). Migration of free probe is shown (0 μg protein). For comparison, an extract of olfactory neuroepithelium from 2.5-week postnatal rats of the CD strain was also included (C).

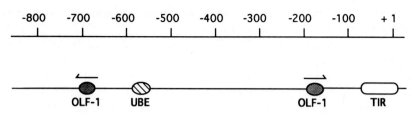

FIG. 6. Summary diagram of the locations of sequence motifs involved in binding of nuclear proteins relative to the *OMP* transcription initiation start (+1). The shaded ovals with right and left pointing arrows represent proximal and distal Olf-1-binding sites, respectively. The striped oval represents the UBE. The open oval represents regions located in proximity to the transcription initiation start site (TIR).

motif that binds a novel, olfactory-specific factor(s), Olf-1. Another region involved in binding of nuclear proteins, interacts with components of nuclear extracts from all tissues that we have tested. In addition, several areas adjacent to the transcription initiation start site are protected from DNase I digestion in the presence of nuclear proteins of the olfactory neuroepithelium (Kudrycki et al 1993).

One possible role of Olf-1 is to act as a part of the mechanism that restricts olfactory gene expression to the receptor neurons of the olfactory epithelium. Both the proximal and distal Olf-1-binding sites reside within 0.8 kb of the *OMP* gene 5′ flanking region that contains elements responsible for olfactory neuron-specific *OMP* expression since, like the 5.2 kb upstream fragment (Danciger et al 1989), it is capable of directing the expression of a reporter gene to olfactory neurons in transgenic mice (Grillo et al 1992). Furthermore, our results indicate that an *OMP* gene fragment that contains 239 bp upstream of the transcription initiation site, including the proximal Olf-1-binding site, is sufficient to confer olfactory neuron-specific expression of a reporter gene in transgenic mice (Table 2). More detailed studies are needed to evaluate the roles of these various motifs in regulating temporal and quantitative aspects of olfactory gene expression in addition to the spatial information obtained to date. Interestingly, EMSA experiments with nuclear extracts prepared from olfactory neuroepithelium at different stages of maturation (Figs. 4 and 5) suggest that Olf-1-binding activity is higher in tissues enriched in immature olfactory neurons. Thus, the expression of Olf-1 may itself be developmentally regulated with a time course that differs from that seen for OMP. It is conceivable that Olf-1 activity is modified during different stages of olfactory neuronal differentiation by interactions with other transcription factors or by post-translational modifications (Berk 1989, He & Rosenfeld 1991, Maguire et al 1991, Maniatis et al 1987, Ptashne 1986), resulting in suppression of *OMP* expression in immature cells and activation in fully differentiated olfactory neurons.

The sequences that we have identified within the 5' flanking region of the *OMP* promoter are likely to function as regulatory elements responsible for tissue-specific expression of *OMP* and other genes within the olfactory receptor neurons. Selective gene expression in neuronal cells can be achieved by more than one mechanism. For example, a silencer can impose tissue-specific expression on a relatively neutral core promoter, as in the rat brain type II Na^+ channel gene (Kraner et al 1992, Mori et al 1992), or the core promoter can exhibit substantial tissue specificity, as was observed for the *GAP-43* gene (Nedivi et al 1992). Both positively and negatively acting promoter elements have been described for an increasing number of neuronal genes, including those for neural cell adhesion molecule (Hirsch et al 1990), type II Na^+ channel (Kraner et al 1992), β-amyloid precursor protein (Lahiri & Robakis 1991), synapsin I (Howland et al 1991) and GAP-43 (Nedivi et al 1992).

Immunochemical studies have demonstrated the presence of OMP in olfactory tissues of virtually all vertebrates (for review see Margolis 1980, 1988) and immunocytological localization to mature olfactory neurons in species as diverse as amphibians (Krishna et al 1992) and humans (Chuah & Zheng 1987, Nakashima et al 1984). Our recent studies, demonstrating Olf-1 activity by EMSA in species as diverse as pig and fish (H. Baumeister & O. Buiakova, unpublished observations), in addition to the rodent data presented here, indicate that both Olf-1 and OMP are similarly conserved phylogenetically. This provides further support to the theory that Olf-1 may also be involved in the regulation of transcription of other olfactory neuronal genes. Thus, although the consensus motif (TCCCC(A/G)NGGAG) of the proximal and distal Olf-1-binding sites reported here is unique when compared with sequences of other *cis*-acting elements from various computer databases, a sequence 90% identical to the Olf-1 motif is present in the 5' flanking region of an olfactory neuronal gene involved in chemosensory transduction, the olfactory cyclic nucleotide-activated channel (R. Reed, personal communication), suggesting that Olf-1 may be of general importance in regulating expression of genes participating in olfactory neuronal transduction events.

Summary

The high level of tissue- and developmentally-specific control of *OMP* gene expression implies a complex pattern of regulation in which both positive and negative regulatory elements may be involved. Results described in this report suggest that the sequence motif interacting with the newly defined olfactory-specific binding factor(s), Olf-1, plays a role in establishing the specific pattern of *OMP* gene expression. It is probable that the *cis*- and *trans*-acting factors involved in DNA–Olf-1 complex formation also participate in regulating transcriptional activation of other olfactory neuron-specific genes. These initial observations provide an opportunity for studying the mechanisms governing chemosensory gene transcription and responsible for generating the olfactory neuron phenotype.

Acknowledgements

We thank R. Wurzburger for oligonucleotide syntheses, and C. Bocchiaro and A. Phillips for technical assistance in generating and maintaining transgenic mice.

References

Bakalyar HA, Reed RR 1990 Identification of a specialized adenylyl cyclase that may mediate odorant detection. Science 250:1403-1406

Berk AJ 1989 Regulation of eukaryotic transcription factors by post-translational modification. Biochim Biophys Acta 1009:103-109

Buck L, Axel R 1991 A novel multigene family may encode odorant receptors: a molecular basis for odor recognition. Cell 65:175-187

Chuah MI, Zheng DR 1987 Olfactory marker protein is present in olfactory receptor cells of human fetuses. Neuroscience 23:363-370

Constanzo RM, Graziadei PPC 1983 A quantitative analysis of changes in the olfactory epithelium following bulbectomy in the hamster. J Comp Neurol 215:370-381

Danciger E, Mettling C, Vidal M, Morris R, Margolis F 1989 Olfactory marker protein gene: its structure and olfactory neuron-specific expression in transgenic mice. Proc Natl Acad Sci USA 86:8565-8569

Dhallan RS, Yau KW, Schrader KA, Reed RR 1990 Primary structure and functional expression of a cyclic nucleotide-activated channel from olfactory neurons. Nature 347:184-187

Graziadei PPC, Monti-Graziadei GA 1978 The olfactory system: a model for the study of neurogenesis and axon regeneration in mammals. In: Cotman CW (ed) Neuronal plasticity. Raven Press, New York, p 131-153

Grillo M, Morris R, Akeson R et al 1992 Transgenic analysis of OMP promoter using two reporter genes. Soc Neurosci Abstr 18:239

He X, Rosenfeld MG 1991 Mechanisms of complex transcriptional regulation: implications for brain development. Neuron 7:183-196

Hirsch M-R, Gaugler L, Deagostini-Bazin H, Bally-Cuif L, Goridis C 1990 Identification of positive and negative regulatory elements governing cell-type specific expression of the neural cell adhesion molecule gene. Mol Cell Biol 10:1959-1968

Howland DS, Hemmendinger LM, Carroll PD, Estes PS, Melloni RHJ, DeGennaro LJ 1991 Positive- and negative-acting promoter sequences regulate cell type-specific expression of the rat synapsin I gene. Mol Brain Res 11:345-353

Jones DT, Reed RR 1989 G_{olf}: an olfactory neuron specific-G protein involved in odorant signal transduction. Science 244:790-795

Kraner SD, Chong JA, Tsay H-J, Mandel G 1992 Silencing the type II sodium channel gene: a model for neural-specific gene regulation. Neuron 9:37-44

Krishna NSR, Getchell TV, Margolis FL, Getchell ML 1992 Amphibian olfactory receptor neurons express olfactory marker protein. Brain Res 593:295-298

Kudrycki K, Stein-Izsak C, Behn C, Grillo M, Akeson R, Margolis FL 1993 Olf-1 binding site: characterization of an olfactory specific promoter motif. Mol Cell Biol 13:3002-3014

Lahiri DK, Robakis NK 1991 The promoter activity of the gene encoding alzheimer β-amyloid precursor protein (APP) is regulated by two blocks of upstream sequences. Mol Brain Res 9:253-257

Largent BL, Sosnowski RG, Reed RR 1993 Directed expression of an oncogene to the olfactory neuronal lineage in transgenic mice. J Neurosci 13:300–312

Ludwig J, Margalit T, Eismann E, Lancet D, Kaupp UB 1990 Primary structure of cAMP-gated channel from bovine olfactory epithelium. FEBS (Fed Eur Biochem Soc) Lett 270:24–29

Maguire HF, Hoeffler JP, Siddiqui A 1991 HBV X protein alters the DNA binding specificity of CREB and ATF-2 by protein–protein interactions. Science 252: 842–844

Maniatis T, Goodbourn S, Fischer JA 1987 Regulation of inducible and tissue-specific gene expression. Science 236:1237–1245

Margolis FL 1972 A brain protein unique to the olfactory bulb. Proc Natl Acad Sci USA 69:1221–1224

Margolis FL 1980 A marker protein for the olfactory chemoreceptor neuron. In: Bradshaw RA, Schneider DM (eds) Proteins of the nervous system. Raven Press, New York, p 59–84

Margolis FL 1988 Molecular cloning of olfactory specific gene products. In: Margolis FL, Getchell TV (eds) Molecular neurobiology of the olfactory system. Plenum, New York, p 237–265

McKay RDG 1989 The origins of cellular diversity in the mammalian central nervous system. Cell 58:815–821

Miragall F, Monti-Graziadei GA 1982 Experimental studies on the olfactory marker protein. II. Appearance of the olfactory marker protein during differentiation of the olfactory sensory neuron of mouse: an immunohistochemical and autoradiographic study. Brain Res 239:245–250

Mori N, Schoenherr C, Vandenbergh DJ, Anderson DJ 1992 A common silencer element in the SCG10 and type II Na^+ channel genes binds a factor present in nonneuronal cells but not in neuronal cells. Neuron 9:45–54

Nakashima T, Kimmelman GP, Snow JP 1984 Structure of human fetal and adult olfactory neuroepithelium. Arch Otolaryngol 110:641–646

Nedivi E, Basi GS, Akey IV, Skene JHP 1992 A neural-specific GAP-43 core promoter located between unusual DNA elements that interact to regulate its activity. J Neurosci 12:691–704

Ptashne M 1986 Gene regulation by proteins acting nearby and at a distance. Nature 322:697–701

Rogers KE, Dasgupta P, Gubler U, Grillo M, Khew-Goodall Y-S, Margolis FL 1987 Molecular cloning and sequencing of a cDNA for olfactory marker protein. Proc Natl Acad Sci USA 84:1704–1708

Samanen DW, Forbes WB 1984 Replication and differentiation of olfactory receptor neurons following axotomy in the adult hamster: a morphometric analysis of postnatal neurogenesis. J Comp Neurol 225:201–211

Schwob JE, Szumowski KEM, Stasky AA 1992 Olfactory sensory neurons are trophically dependent on the olfactory bulb for their prolonged survival. J Neurosci 12: 3896–3919

Verhaagen J, Oestreicher AB, Gispen WH, Margolis FL 1989 The expression of the growth associated protein B50/GAP43 in the olfactory system of neonatal and adult rats. J Neurosci 9:683–691

Verhaagen J, Oestreicher AB, Grillo M, Khew-Goodall Y-S, Gispen WH, Margolis FL 1990 Neuroplasticity in the olfactory system: differential effects of central and peripheral lesions of the primary olfactory pathway on the expression of B50/GAP43 and the olfactory marker protein. J Neurosci Res 26:31–44

DISCUSSION

Buck: Frank, can you take the upstream element that you have identified, place it in front of a different sequence and see expression of that sequence in olfactory neurons? What happens if you remove it from the rest of the *OMP* 5′ flanking sequence?

Margolis: We are now generating transgenic mice carrying a mutated Olf-1-binding element in an *OMP* promoter–*lacZ* construct. We are trying to make *OMP* promoter constructs in which we replace the proximal Olf-1-binding site with a copy of the distal Olf-1-binding site, to see if it still works. We have generated a whole variety of additional constructs, but so far we have no answers.

The obvious thing to do next is to take a tandem repeat of the Olf-1-binding site and incorporate it into a general promoter, put it into mice and see what happens. It would be really nice if there was a real cell line that we could differentiate in culture, so we could do these things more easily. Doing all of these experiments with transgenic animals is insane—if I had realized at the beginning how much work was involved, I never would have started! For example, we are now in the process of seeing the effect of taking the *OMP* promoter—*lacZ* construct and replacing the SV40 polyadenylation signal sequence with the 3′ fragment of the *OMP* gene that contains the endogenous poly(A) signal sequence domain. We are also testing whether the downstream Olf-1-like binding site is really an active regulator. I suspect that it is, but I really don't know. We need cell lines.

Hwang: Have you considered knocking out *OMP* to discover its function?

Margolis: We are trying to do this. The construct is almost ready for injection.

Getchell: What do you anticipate might be the results of this experiment? What if it doesn't have any effect on the animals?

Margolis: That would be very upsetting! Although it would be nice to be able to anticipate some logical result, we can't. We are doing the knockout experiments partly out of a sense of frustration: OMP doesn't look like anything else, so it's hard for me to make any kind of rational guess as to what it might be doing. We plan to knock out the *OMP* gene, look and see whether there are major changes in the morphology, neuronal turnover or the ability of these animals to respond to odours. In a sense, this approach is the contemporary version of classical neural lesion or endocrine ablation studies, except that now we can do single gene ablations. Hopefully, this strategy will give us some clue as to the role *OMP* plays.

Reed: Does the olfactory epithelium go through a wave of development that diminishes in adult animals?

Margolis: Verhaagen in my lab (Verhaagen et al 1989, 1990) and subsequently Schwob et al (1992) confirmed that in neonates there is a large population of immature olfactory neurons that express B50/GAP-43 and only a fraction of neurons that express OMP. This was demonstrated by both immunocytochemistry and *in situ* hybridization. There is then a progressive shift over the first few postnatal weeks to the mature pattern of many OMP-positive neurons and very few B50/GAP-43-positive neurons. This progression is largely replicated after lesions. The molecular mechanisms responsible for these events are unknown.

Siddiqi: My colleagues Dr Swati Sathe, Dr Neelam Shirsat and V. C. Jayaram are studying mutations that affect the firing pattern of taste neurons in *Drosophila*. In the labellar hair of flies there is one neuron which has all acceptor sites for sugars; there are two neurons called L1 and L2 which respond to salts. L1 has two acceptor sites, one highly specific for Na^+ and the other relatively non-specific. Mutations in the gene *gust E* eliminate the Na^+-specific response of L1. Dr Sathe determined the site of expression of *gust E* by mosaic mapping experiments. To our surprise, she found that the gene is not expressed in the sensory neuron. As far as we can tell from the mosaic data, it appears to be expressed in the ventral region of the brain. So here is a gene which changes the receptor phenotype of a neuron and yet it is not expressed in the sensory neuron but in the brain (Siddiqi et al 1989, Sathe 1992).

Firestein: Is it not expressed in cells that are the direct target of the sensory neuron either?

Siddiqi: It could be expressed in the direct target of the sensory neuron. I'm inclined to think that even in *Drosophila*, where it is commonly assumed that all development proceeds stereotypically, the receptor specificity of the gustatory neurons and perhaps also of the olfactory neurons is one of the last things to be specified. It could be open to instruction.

Caprio: How are we defining an immature olfactory neuron: on the basis of synaptic events or on the basis of morphology of receptors? What do we know about the specificity of olfactory receptor neurons that haven't synapsed to the bulb?

Margolis: The only hint that I am aware of is from some results reported by Gesteland et al (1982) that suggest the olfactory neuroepithelial sheet has relatively low specificity, as though the immature neurons respond to a large variety of odours and that subsequently the specificity seems to become more narrowly tuned and selective.

Caprio: But what are the criteria you use to decide whether a neuron is immature or mature?

Margolis: The description of a neuron as immature is indicated by its position in the epithelium and by the expression of B50/GAP-43, which is a molecular marker for immature neurons. The expression of B50/GAP-43 declines as synaptic connection takes place. I don't know anything about the signal transduction properties of immature as opposed to mature neurons.

Getchell: Different definitions of maturation of receptor neurons have been used to reflect the different perspectives of investigators. For example, the criterion that Frank Margolis has used is the expression of *OMP*, which appears to be greatly enhanced when synaptic contacts are made with the olfactory bulb. The timing of this event coincides with when olfactory receptor neurons in mammals begin to become odorant selective, rather than simply detecting a wide range of compounds. A way to investigate the transition from genotype to phenotype would be with the experiment that Frank has described. He investigated the relative time of onset of the expression of *OMP* in olfactory, septal and vomeronasal systems, because these three sensory systems located within the nasal cavity mature at different times in the developing rat.

Caprio: Peter Zippel, working in my lab, looked at the effects of cutting the olfactory nerve in goldfish. We knew from Peter's earlier work that goldfish with transected olfactory nerves lose their ability to perform an olfactory discrimination task in which they were previously trained (Zippel et al 1988, Hudson et al 1990). However, 8–10 days after bilateral axotomy, the fish regain this ability, presumably because of the renewal of olfactory receptor neurons and the reestablishment of specific neural connections within the olfactory bulb. During our experiments, electroolfactogram (EOG) recordings were made from the olfactory organs of goldfish over 24 d, during which the olfactory receptor renewal process occurred and through the time that prior experiments indicated that olfactory discrimination behaviour recovered. We were amazed to find that in unilaterally axotomized goldfish the sensitivity and specificity of both control and experimental olfactory organs were similar at any time from day 1 to day 24. For example, the most stimulatory amino acid to olfactory receptors of goldfish was *L*-arginine. The sensitivity to *L*-arginine remained throughout the days after axotomy. Concurrent scanning electron microscope studies of the olfactory mucosae of these fish indicated that both microvillus and ciliated olfactory receptor neurons were degenerating post axotomy, but a number of ciliated receptors were evident even at day 1 post axotomy. Thus, either a portion of the ciliated olfactory receptor neurons did not die following olfactory nerve transection or the chemical specificity of the olfactory receptor neurons in the goldfish is determined prior to the emergence of the dendritic knob of young receptor neurons.

Getchell: Do fish olfactory receptor neurons express *OMP*?

Caprio: Riddle & Oakley (1992) recently published immunocytochemical evidence for OMP-like immunoreactivity in the rainbow trout, *Oncorhynchus mykiss*.

Lancet: I would like to mention some results that T. Margalit, in my lab, has produced to do with the developmental expression of various olfactory-specific genes (Margalit & Lancet 1993). From reverse transcriptase PCR (polymerase chain reaction) assays, it seems that genes that code for olfactory-specific components, including olfactory G proteins, olfactory adenylate cyclase, olfactory receptors, olfactory ion channels and several olfactory biotransformation enzymes, are active at different times during embryogenesis. To tie in with what Frank Margolis has said, although there may be only one DNA element, in this case the Olf-1-binding element, there may be several versions of it operating. Consequently, there could be a whole orchestration of events taking place. That is, the same Olf-1 could give rise to several different expression sequences.

A second point is that the onset of expression for two gene sets, namely those coding for the channels and the receptors, occurs at around E18/E19, which is a very important age in rat embryogenesis. This is when sensory neuronal specificity appears to become sharper and the majority of synapses are formed. At this stage, the receptors begin to show massive expression, although I'm not ruling out the possibility that some are expressed earlier. Although this isn't the only way to interpret the data, such a coincidence may tell us that when the synapses form a signal is generated that tells the neurons to begin to express receptors: this is a retrograde-type mechanism.

Buck: We see a small number of cells expressing odorant receptors at E12 in the mouse, which is equivalent in embryogenesis to about E14–E15 in the rat. Perhaps what you are seeing is an increase in the number of cells that are expressing receptors rather than the initiation of receptor expression.

Lancet: There are several ways of assessing the ontogeny of different genes. One criterion is to look at the half maximal response; alternatively, you can measure the time at which the first cell in the tissue expresses the gene. Of course, this might be much earlier. All the data need to be taken into consideration to understand the onset of specificity. We think specificity may arise when many receptors are expressed and that the non-specific responses that are there to begin with may have to do with non-receptor responses that people have been addressing for many years—direct odorant effects on transduction components, membrane potential and so forth. It is possible, therefore, that most of the cells must express receptors before specific responses to odorants are observed.

Buck: Has anyone used antibodies or done *in situ* hybridization with any of these other genes, such as those coding for the cAMP-gated channel or G_{olf}, to look at patterns of expression in early development?

Lancet: Mania-Farnell & Farbman (1990), using antibodies, have seen expression of adenylate cyclase and G proteins beginning more or less at the same time (E15) as we see in the rat.

Reed: As Doron Lancet said, the number of cells in the epithelium expressing a particular gene is important. But potentially much more important are the levels of expression of that gene. As antibodies have a very limited dynamic range, the dynamic ranges detected by PCR are going to be significantly broader.

The *OMP* data, which are the best on the patterns of expression, are not inconsistent with an elevation of expression as a function of a variety of specific environmental factors. That is, *Olf-1* could either be the gene that turns on high levels of expression of *OMP* about the time that synaptic connection is made, or it may be the first event which causes a small level of *OMP* expression and then there's a subsequent event, which could be transcriptional or translational, that raises expression 10- or 100-fold in each individual cell. It may be true for receptors as well, that there is a low and broad expression that then increases in response to some environmental control.

Lancet: Frank, have you investigated the effect of using different lengths of the 5' region of the *OMP* gene to drive the *lac-Z* reporter in transgenic mice?

Margolis: Our preliminary data indicate that when we use longer upstream fragments of *OMP* than I showed here we get mice with fewer cells expressing *lacZ* but those that do express it give robust expression.

We don't really understand what is going on here. It's clear that there are multiple events taking place but we don't yet understand how these events are interacting. The nature of the reporter gene and small variations in the nature of the construct seem to have major effects on the level of expression that we see in the transgenic mice.

Reed: Do you mean that some transgenic mouse lines can be scored as expressing low levels of *lacZ* activity but that the few cells which are making *lacZ* produce it at a high level?

Margolis: Yes, it's the same kind of issue as Doron Lancet was referring to in his ontogeny study: you can obtain apparently different results from the same transgenic mouse line depending on whether you count numbers of cells producing *lacZ* or the levels of *lacZ* produced by the total tissue or by individual cells. Thus, how we score the data is crucial, as there are several options.

Reed: A similar concern exists for measurements by single cell recording, EOGs, PCR, *in situ* hybridization or *lacZ* staining.

Lancet: EOG is a population property, as it records the sum of electrophysiological responses of many cells. So is gene expression as detected by PCR. On the other hand, single-cell recordings are more comparable to single-cell expression assays.

Margolis: An additional problem has been noticed by Eric Walters, a postdoc in my lab, who is analysing the *lacZ* transgenic mice. There seems to be a variation in the rostrocaudal level of expression across the epithelium, as though there's some kind of positional information as well (unpublished observations).

Reed: Jeremy Nathan observed this in the retina of mice carrying the rhodopsin promoter fused to *lacZ*. He saw dramatic gradients of expression in terms of the fraction of cells that respond across the eye, in an impressive spatial map.

Lancet: Linda Buck's beautiful data on the different receptors expressed at different positions in the olfactory epithelium may be related to this (this volume: Buck 1993). Perhaps different subclasses of receptors have different ontogenies.

Firestein: Do bulbectomized mice that regenerate their epithelium express *OMP*? In these animals, the regenerating cells presumably send the axons back to the brain, but they no longer have a target.

Margolis: Initially, the cells do express *OMP*. One of the characteristics of the olfactory epithelium of bulbectomized animals is that their apparent turnover is increased as though they are undergoing mitosis and they begin to undergo maturation. The cells appear to be trying to find their target, which is no longer there, so they die: it's an abortive process (for review see Farbman 1992).

Firestein: So if these animals live for a long time, you end up with a steady-state level of OMP, but it's lower than you would find in any normal animal.

Margolis: Yes.

Hatt: In collaboration with Professor I. Parnas' group at the Hebrew University in Israel, we are working on another system that might help to answer some questions here. We are looking at the influence of the presynaptic terminal in crustacean muscle on the postsynaptic transmitter-activated channels. Cutting an excitatory motor neuron of a lobster muscle (complete denervation), we found that the kinetic behaviour of postsynaptic transmitter channels (glutamate activated) is completely changed about two weeks after denervation. We concluded that there must be some sort of chemical messages from the presynaptic side which influence the channels on the post-synaptic membrane.

Margolis: The same phenomenon occurs in the olfactory system. If you remove the presynaptic input to the targets in the bulb, you see major biochemical changes and major transcriptional changes in genes in the target neurons in the olfactory bulb (Margolis et al 1991). For instance, tyrosine hydrolase and substance P levels change. So in this system as well, there's information going both ways. The question is: what is the information and what is its molecular basis?

Caprio: There is a relatively simple experiment one can do to answer some of these questions concerning the chemical specificity of young olfactory receptor neurons. Assuming that one accepts the EOG or integrated multiunit activity as an indicator of the specificity of a population of olfactory receptors, you simply determine the olfactory specificity of an adult animal, transect the olfactory nerve and look to see whether there are changes in chemical specificity both during the time of receptor renewal and subsequent to synaptic connections within the olfactory bulb. It appears that in goldfish the olfactory receptor specificity is established quite early during the renewal process.

Getchell: Peter Simmons reported similar experiments in the early 1980s on salamander, investigating the time course of changes in EOGs (Simmons & Getchell 1981a) and single units (Simmons & Getchell 1981b) prior and subsequent to olfactory nerve section. These showed similar results to yours: the specificity was there early on.

Schmale: Does anyone know anything about the death or degeneration of these neurons: is there a regulated pathway of cell death, such as apoptosis, which requires new gene expression?

Margolis: For a long time it was thought that cell death was a determined process—that every 30 days the cells died and they were replaced. Now it seems more likely that these cells die in response to damage; that it's not a programmed cell death. However, there may be some combination of these two mechanisms (Farbman 1992).

References

Buck LB 1993 Receptor diversity and spatial patterning in the mammalian olfactory system. In: The molecular basis of smell and taste transduction. Wiley, Chichester (Ciba Found Symp 179) p 51–67

Farbman AI 1992 Cell biology of olfaction. Cambridge University Press, Cambridge

Gesteland RC, Yancey RA, Farbman AI 1982 Development of olfactory receptor neuron selectivity in the rat fetus. Neuroscience 7:3127–3136

Hudson R, Distel H, Zippel HP 1990 Perceptual performance in peripherally reduced olfactory systems. In: Schild D (ed) Chemosensory information processing. Springer-Verlag, Berlin, p 259–269

Mania-Farnell B, Farbman AI 1990 Immunohistochemical localization of guanine-binding proteins in rat olfactory epithelium during development. Dev Brain Res 51:103–112

Margalit T, Lancet D 1993 Expression of olfactory receptor and transduction genes during rat development. Dev Brain Res 73:7–16

Margolis FL, Verhaagen J, Biffo S, Huang F, Grillo M 1991 Regulation of gene expression in the olfactory neuroepithelium: a neurogenetic matrix. Prog Brain Res 89:97–122

Riddle DR, Oakley B 1992 Immunocytochemical identification of primary olfactory afferents in rainbow trout. J Comp Neurol 324:575–589

Sathe S 1992 Genetic and physiological investigation of salt reception in *Drosophila melanogaster*. PhD thesis, Bombay University, Bombay, India

Schwob JE, Szumowski KEM, Stasky AA 1992 Olfactory sensory neurons are trophically dependent on the olfactory bulb for their prolonged survival. J Neurosci 12:3896–3919

Siddiqi O, Joshi S, Arora K, Rodrigues V 1989 Genetic investigation of salt reception in *Drosophila melanogaster*. Genome 31:646–651

Simmons PA, Getchell TV 1981a Neurogenesis in olfactory epithelium: loss and recovery of transepithelial voltage transients following olfactory nerve section. J Neurophysiol 45:516–528

Simmons PA, Getchell TV 1981b Physiological activity of newly differentiated olfactory receptor neurons correlated with morphological recovery from olfactory nerve section in the salamander. J Neurophysiol 45:529–549

Verhaagen J, Oestreicher AB, Gispen WH, Margolis FL 1989 The expression of the growth associated protein B50/GAP43 in the olfactory system of neonatal and adult rats. J Neurosci 9:683–691

Verhaagen J, Oestreicher AB, Grillo M, Khew-Goodall Y-S, Gispen WH, Margolis FL 1990 Neuroplasticity in the olfactory system: differential effects of central and peripheral lesions of the primary olfactory pathway on the expression of B50/GAP43 and the olfactory marker protein. J Neurosci Res 26:31–44

Zippel HP, Meyer DL, Knaust M 1988 Peripheral and central post-lesion plasticity in the olfactory system of the goldfish: behaviour and morphology. In: Flohr H (ed) Post-lesion neural plasticity. Springer-Verlag, Berlin, p 577–591

Mucous domains: microchemical heterogeneity in the mucociliary complex of the olfactory epithelium

Thomas V. Getchell*†‡, Zhaoyu Su* and Marilyn L. Getchell†‡

*Department of Physiology and Biophysics, †Division of Otolaryngology—Head and Neck Surgery, Department of Surgery, ‡Sanders-Brown Center on Aging, University of Kentucky College of Medicine, Lexington, KY 40536, USA

>*Abstract.* Access to and clearance of odorants from binding sites on olfactory cilia are regulated by a complex interplay of molecular, physical and cellular factors. These perireceptor events occur primarily in the mucociliary complex. The use of gold-labelled lectinoprobes, one from *Limax flavus* (LFA) which is specific for terminal sialic acid residues, and one from *Datura stramonium* (DSA) specific for *N*-acetylglucosamine residues, demonstrated intricate patterns of binding in mucous domains of the olfactory mucus and ectodomains of the glycocalyx of olfactory cilia. In electron micrographs of Lowicryl-embedded salamander olfactory mucosa, the mucus consisted of an electron-dense domain that lay superficial to an electron-lucent domain; the interface between the two was irregular. A significantly higher density of binding sites for both lectins was present in the superficial than in the deeper domain. The two domains were not homogeneous: there were small electron-lucent domains (hsL) within the superficial electron-dense domain (hsD) that bound a 4.8-fold lower density of gold-labelled DSA than the surrounding matrix, and the olfactory cilia, which project into hsD, were surrounded by an electron-lucent sheath that appeared to be continuous with the deeper domain. Ectodomains of the glycocalyx associated with olfactory cilia exhibited a higher density of binding sites for both LFA and DSA than did either microvilli of sustentacular cells or respiratory cilia. Specificity of the lectinoprobes was confirmed by inhibition of binding with specific sugars or enzymic removal of specific sugar residues. These results demonstrated microchemical heterogeneity of the non-homogeneous mucous domains in olfactory mucus and in the attendant glycocalyx of olfactory cilia based on the differential localization of sialic acid and *N*-acetylglucosamine sugar residues.
>
>*1993 The molecular basis of smell and taste transduction. Wiley, Chichester (Ciba Foundation Symposium 179) p 27–50*

The access to and clearance of odorants from binding sites on olfactory cilia are regulated through a complex interplay of molecular, physical and cellular factors. These perireceptor events occur primarily in the mucosensory compartment of the olfactory mucosa. Five interrelated aspects of the olfactory mucus require investigation in order for us to gain further insight into these events.

Chemical composition of the olfactory mucus

The synthesis, glycosylation and distribution of mucous glycoproteins are being studied using primarily biochemical, molecular biological and cell biological methods. Snyder et al (1991) have identified and characterized a novel glycoprotein called olfactomedin that is synthesized by secretory cells in the olfactory epithelium and is distributed within the olfactory mucus of the frog. Foster et al (1992) have localized the site of terminal sialylation in the post-translational processing of mucous glycoconjugates to the *trans* cisternae of the Golgi apparatus in sustentacular cells of the salamander olfactory mucosa. They have described quantitatively the distribution of sialylated glycoconjugates in the olfactory mucus and their localization in the glycocalyx of the olfactory cilia.

Physical state

The viscosity and other rheological properties of olfactory mucus that derive from sol–gel transformations resulting from the hydration of mucous glycoproteins are being studied using the methods of physical chemistry and polymer biophysics. Except for the theoretical studies (e.g. Getchell et al 1984, Getchell & Getchell 1990, Pevsner & Snyder 1990), where estimates of the viscosity of olfactory mucus have been utilized to calculate the diffusion times of odorants, the fundamental physicochemical properties of olfactory mucus have not been investigated.

Organization

The microstructural and microchemical organization of olfactory mucus is being investigated using primarily cell biological methods. Reese (1965), in an *in vivo* description of olfactory mucus in the frog, reported that 'a very thin layer of watery mucus covered a thicker viscid layer on which the cilia floated, and that it was the movement of this thin layer which moved the carbon particles' that he had placed in the nasal cavity. This important observation established that the organization and physical properties of olfactory mucus (Fig. 1) were unlike those of other mucus surfaces such as that associated with the tracheal mucosa of mammals (Lopez-Vidriero 1989) and that olfactory mucus could be divided into a superficial watery, sol-like layer (hs) and a deeper, mucoid, gel-like layer (hd) for theoretical studies (Getchell et al 1984). More recently, Menco & Farbman (1992) demonstrated the ultrastructural complexity of olfactory mucus, and Foster et al (1991, 1992) demonstrated its microchemical complexity.

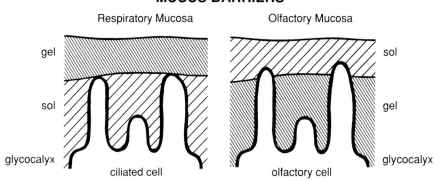

FIG. 1. Differences in physical organization of the mucus barriers associated with the respiratory and olfactory mucosae. Respiratory mucosa modified from Lopez-Vidriero 1989.

Regulation

The odorant, pharmacological and neural regulation of the secretion of olfactory mucus have been studied using neuropharmacological and cell biological techniques. We have identified several of the neurotransmitters and neuromodulators and types of nerve terminals mediating the regulation of secretion from sustentacular cells and the acinar cells of Bowman's glands in salamanders and frogs (e.g. Getchell & Getchell 1984, Zielinski et al 1989, Getchell et al 1989), as well as the adrenergic and peptidergic regulation of secretion from Bowman's glands in humans (Chen et al 1993).

Ligand translocation

The mechanisms by which the physico-chemical properties of olfactory mucus influence the diffusion and/or transport of odorants and pheromones to receptor molecules are being studied using the methods of physical chemistry and biochemistry. Although limited experimental progress is being made on the physicochemical mechanisms, substantial progress is being achieved using molecular biological techniques. This includes the identification of and the cloning of the genes for odorant transporters such as odorant-binding protein of olfactory mucus (Pevsner et al 1985, Pevsner & Snyder 1990), the putative pheromone transporter vomeromodulin of vomeronasal mucus (Khew-Goodall et al 1991, Rama Krishna et al 1993) and insect pheromone transporters (Vogt et al 1989, Steinbrecht et al 1992).

From this brief introduction, it can be seen that the investigation of olfactory mucus is rich in intriguing questions that, when addressed with contemporary

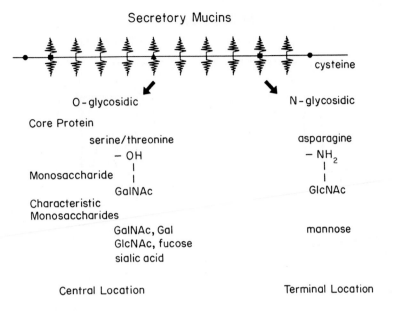

FIG. 2. Secretory mucins consist of a protein core, including cysteine (circles) and serine/threonine (triangle) residues. These form O-glycosidic linkages with GalNAc to attach oligosaccharide side chains composed of five characteristic monosaccharides to the central portion of the mucin molecule. Asparagine (square) residues form N-glycosidic linkages with GlcNAc to attach oligosaccharide side chains with characteristic mannose residues to a terminal location in the mucin molecule.

methodologies, will yield fundamental insights that will contribute substantially to knowledge of perireceptor events that modulate sensory transduction.

Secretory glycoconjugates

Mucous glycoproteins are a distinct chemical category of glycoconjugates, which are high molecular weight (250–1000 kDa) macromolecules that consist of a protein backbone from which several hundred oligosaccharide side chains of variable length may project (Allen 1983). The hydration of mucous glycoproteins subsequent to their secretion, resulting in sol–gel transformations, confers on epithelial mucus its characteristic rheological properties, including viscosity and elasticity. Mucous glycoproteins are of two types: soluble or secretory mucins and membrane-bound mucins (Strous & Dekker 1992). Secretory mucins (Fig. 2) are characterized by a protein core to which oligosaccharide side chains are attached by O-glycosidic linkages between the hydroxyl groups of the amino acids serine and threonine, which occur clustered in the central region of the protein, and the monosaccharide N-acetylgalactosamine at the reducing end of the sugar chain. The termini of the protein backbone contain a typical distribution of amino acids, including asparagine, to which oligosaccharides may be attached by N-glycosidic linkages, and cysteine, whose sulphydryl groups form intra- and intermolecular disulphide bonds. The formation of intermolecular disulphide bonds links the glycoprotein molecules into oligomers; the breaking and forming of these disulphide bonds underlie sol–gel transformations. Membrane-bound mucins appear to lack the intermolecular disulphide bonds and, additionally, possess a region of hydrophobic amino acids that anchors these mucins to cell membranes. Mucous glycoproteins include bronchial, cervical, gastric, respiratory and submaxillary mucins. The side chains of the O-linked oligosaccharides contain primarily galactose, N-acetylgalactosamine, N-acetylglucosamine, fucose and sialic acid; side chains of the N-linked oligosaccharides are characterized by the presence of mannose. Elongation of the O-linked oligosaccharide chains proceeds by the sequential addition of sugars by glycosyltransferases in the Golgi apparatus of the secretory cell (Paulson & Colley 1989). The sequence of sugars in the oligosaccharide side chains is genetically determined and is characteristic, not only of the specific type of glycoprotein, but also the cellular source of the mucus. The physicochemical properties of mucus depend largely on its glycoprotein content, which is determined by the regulated expression and enzymic activity of the glycosyltransferases and the hydration of these macromolecules.

Microstructural organization of olfactory mucus

Mucus is the first biological fluid that air-borne compounds encounter when they are inhaled into the nasal cavity. The mucus layer functions not only as a physical, chemical and biological barrier, but also as a selective barrier by which certain compounds are sequestered for uptake, binding and translocation through the mucus. The early observations of Reese (1965), together with theoretical (Getchell et al 1984, Getchell & Getchell 1990), ultrastructural (Menco & Farbman 1992, Menco 1992) and cell biological (e.g. Foster et al 1992) studies, have shown that olfactory mucus has a complex structural and microchemical organization.

As shown in Fig. 3, which is an electron micrograph of the salamander's mucosensory compartment, the cellular component of the olfactory epithelium is covered by a layer of mucus that includes the mucociliary complex. This complex consists of a sensory component, comprising the olfactory cilia together with their attendant glycocalyx, and a mucoid component, comprising olfactory mucus. We have chosen this terminology because of the intimate association between perireceptor events that occur primarily in mucus with the transductional events that occur in olfactory cilia. In Lowicryl-embedded tissue from salamanders, the average thickness of the mucosensory compartment, h, is 10.3 ± 1.3 μm ($n = 56$). The mucoid component clearly consists of

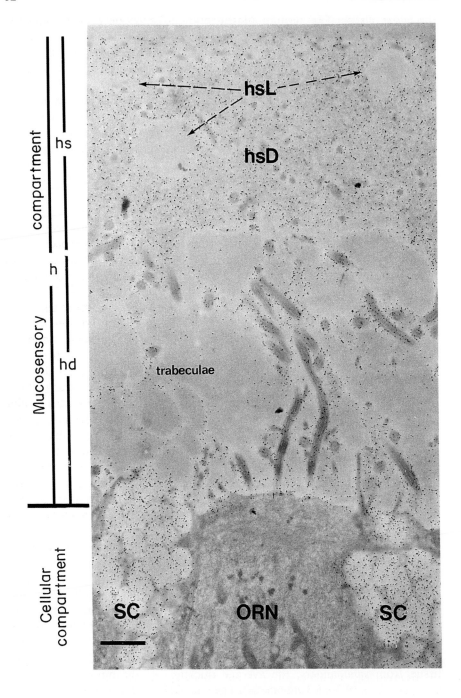

two layers: a superficial electron-dense domain (hs) and a deeper electron-lucent domain (hd). This observation is consistent with the *in vivo* observation of Reese (1965), the theoretical designations by Getchell et al (1984) and the cell biological studies of Foster et al (1992). The interface between the two domains is clearly irregular: the thickness of the superficial layer ranges from 2.6 to 7.5 μm (mean = 4.8 ± 1.0 μm, $n = 28$) and that of the deeper layer ranges from 3.7 to 7.7 μm (mean = 5.7 ± 1.2 μm). The electron density within the two domains is not homogeneous. The superficial electron-dense domain consists of a large electron-dense matrix (hsD) in which smaller electron-lucent domains (hsL) are located. These quantitative ultrastructural studies demonstrate a complex microstructural organization of olfactory mucus.

Microchemical organization of olfactory mucus

The sugar residues in the oligosaccharide side chains of secretory glycoconjugates in olfactory mucus make them particularly amenable to *in situ* binding studies with lectins (Sharon & Lis 1989, Foster et al 1991, 1992). Five monosaccharides are commonly identified in *O*-linked secretory mucins: galactose, *N*-acetylgalactosamine (GalNAc), *N*-acetylglucosamine (GlcNAc), L-fucose and sialic acid. Foster et al (1991, 1992) used sugar-specific lectins to identify and localize the distribution of sugars in the mucosensory compartments and the two primary sources of secretory glycoconjugates in the cellular compartment, the epithelial sustentacular cells and the acinar cells of Bowman's glands in the lamina propria. We selected two lectins for detailed quantitative ultrastructural studies in the salamander's mucosensory compartment. These lectins exhibit high binding specificity to sugars that are present in relatively high concentrations in secretory glycoconjugates of the mucosensory compartment and/or the secretory cells, on the basis of our results with histofluorescence (Foster et al 1991). We utilized an indirect fetuin-gold-labelling technique with the lectin isolated from the slug, *Limax flavus* agglutinin (LFA), which has a high specific affinity for the monosaccharide sialic acid, and a direct gold-labelling technique with the lectin isolated from the jimson weed, *Datura stramonium* agglutinin (DSA), which has a high specific affinity for the monosaccharide GlcNAc. As shown in Fig. 3, the 15 nm diameter gold particles, representing binding sites of the gold microsphere-labelled DSA lectinoprobe, are unevenly

---

FIG. 3. (*opposite*) DSA-gold labelling of salamander olfactory mucosa (OM). The surface of the OM consists of a cellular compartment containing olfactory receptor neurons (ORNs) and sustentacular cells (SC), and a mucosensory compartment. In the mucosensory compartment, an electron-dense domain (hs) lies superficial to an electron-lucent domain (hd); density of gold particles is significantly higher in hs than in hd. Within the electron-dense matrix (hsD) of hs are small-area electron lucent domains (hsL). Trabeculae project from the epithelial surface through hd. Bar = 1 μm.

TABLE 1 Localization and density of *N*-acetylglucosamine and sialic acid sugar residues in mucous domains

Mucous domains	Sugar localized (gold particles/μm^2)	
	N-acetylglucosamine[a]	Sialic acid[b]
hs	25.1 ± 1.9	17.2 ± 1.5
hsD	27.0 ± 1.6	19.0 ± 2.4
hsL	5.7 ± 0.4	5.8 ± 1.7
hd	11.3 ± 1.4	6.1 ± 2.0

[a]Detected with *Datura stramonium* agglutinin.
[b]Detected with *Limax flavus* agglutinin.
hs, superficial mucous domain; hsD, electron-dense matrix of hs; hsL, small electron-lucent domains within hs; hd, deep mucous domain (±SEM).

distributed in mucus domains of the mucosensory compartment and in the apical region of cells in the cellular compartment. In general, more gold particles are localized in the superficial than in the deeper domain (Table 1). Careful inspection of the electron micrographs (Figs. 3 and 4a) clearly demonstrates that the distribution of gold particles in the superficial domain is not homogeneous. Within the superficial electron-dense domain, there are small electron-lucent sub-domains (hsL) with areas ranging from 0.02 to 1.9 μm^2, which are distributed within the electron-dense matrix (hsD). The distribution of gold particles, reflecting the distribution of GlcNAc residues, differs in these two areas, with fewer gold particles in hsL than in hsD (Table 1). Similarly, the distribution of gold particles in the deeper domain is not homogeneous (Figs. 3 and 4b). Trabeculae, containing secretory glycoconjugates with a high density of GlcNAc residues that bind to the gold-labelled lectinoprobe, extend from the epithelial surface through the deeper domain to the interface with the superficial domain.

Similar results were obtained when the binding of the gold microsphere-labelled sialic acid lectinoprobe LFA was used to investigate the distribution of sialic acid residues. There was a higher density of gold particles in the superficial domain than in the deeper domain, and hsL with fewer gold particles are also distributed within hsD (Table 1). To confirm the specificity of lectin binding in the tissue sections, we incubated lectins with the appropriate monosaccharide prior to staining, resulting in the virtual absence of gold particle binding. Neuraminidase digestion of the tissue sections to remove terminal sialic acid residues also resulted in loss of binding of the LFA–fetuin–gold complex.

The results of these studies demonstrate the differential electron density of the superficial and deep layers of olfactory mucus that presumably reflects different sol–gel states of secretory mucins within these domains. These domains are clearly not homogeneous, nor is the border between the two sharply separated. The microchemical heterogeneity within these domains and within hsL, hsD and the trabeculae reflects major differences in the sugar composition of oligosaccharide side chains in secretory mucins within these domains.

Microchemical organization of ciliary glycocalices

The glycocalyx is an essential element of the mucous barrier in epithelial cells and in the mucociliary complex of olfactory receptor neurons (Fig. 1). The glycocalyx consists of the ectodomains of integral membrane proteins, such as transmembrane mucins, receptor molecules and channel molecules, and of membrane-bound glycolipids together with any adsorbed molecules such as secretory mucins and immunoglobulins (Schnitzer 1988, Silberberg 1989, Foster et al 1992, Strous & Dekker 1992). We have initiated quantitative ultrastructural studies on the binding of gold-labelled lectinoprobes specific for sialic acid and GlcNAc monosaccharide residues in order to examine the microchemical organization of the glycocalyx of olfactory cilia (Foster et al 1992, Su et al 1993). There is a higher density of gold particles representing the binding of the DSA lectinoprobe to GlcNAc residues (Fig. 5a) and of the LFA lectinoprobe to sialic acid residues (Fig. 5c) in the glycocalyx of olfactory cilia than in the corresponding glycocalices of tracheal (Fig. 5b) and respiratory (Fig. 5d) epithelial cells. There was a 2.5-fold greater density of LFA-fetuin–gold particles associated with the cilia of olfactory receptor neurons ($6.1 \pm 1.6 \,\mu m^{-2}$) than with the microvilli of sustentacular cells ($2.5 \pm 1.1 \,\mu m^{-2}$) in the mucociliary complex (*cf.* Table 3 in Foster et al 1992). The results of these quantitative studies indicate differences in the microchemical composition of the glycocalyx associated with olfactory cilia compared to that of tracheal and respiratory epithelial cell cilia and sustentacular cell microvilli. The results demonstrate the characteristic presence of GlcNAc and sialic acid monosaccharide residues in the glycocalyx of olfactory cilia. These monosaccharide residues may be associated with the ectodomains of integral membrane proteins and/or molecules that have become adsorbed to the surface of the ciliary membrane, such as secretory mucins. For example, in biochemical studies, Khan et al (1992) reported the presence of high levels of GlcNAc and sialic acid residues in preparations of isolated olfactory cilia compared with the levels of these monosaccharides found in endoplasmic reticular fractions of cerebellar neurons, suggesting the presence of inositol 1,4,5-trisphosphate receptors associated with Ca^{2+} conductances in olfactory cilia. In cell biological studies, we (Foster et al 1992, Su et al 1993) demonstrated the presence of high levels of GlcNAc and sialic acid residues associated with the glycocalyx of olfactory cilia (Fig. 5a and c), the mucoid component of the mucociliary complex (Table 1) and secretory vesicles (Figs. 3 and 4) of sustentacular cells and the secretory granules of the acinar cells of Bowman's glands. Further studies are now required for a microchemical dissection of the molecular organization of the olfactory

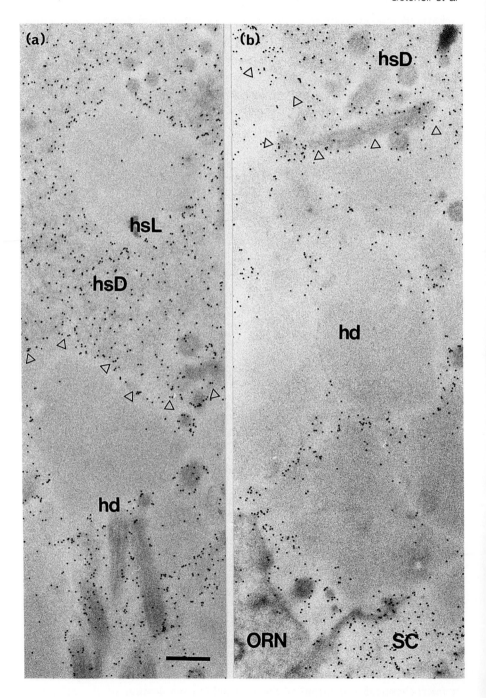

ciliary glycocalyx to investigate the function of particular mono- and oligo-saccharides in receptor and perireceptor events in olfactory transduction.

Summary

The results of our studies and those from other laboratories indicate that the physical and microchemical properties of the olfactory mucous barrier are integral determinants of perireceptor processes associated with olfactory transduction. Further insight into the role that the microstructural and microchemical organization of the olfactory mucus and the ciliary glycocalyx play in sensory transduction requires the articulation of research strategies to establish novel hypotheses and to test them empirically using contemporary methodologies. Areas for further study encompass: (a) elucidation of the three-dimensional organization of the mucosensory compartment, including the course of the trabeculae and the disposition of islets of high and low densities of glycoconjugates rich in particular monosaccharide residues; (b) the functions of the specific monosaccharide residues in the secretory glycoconjugates of olfactory mucus; (c) the regulation of gene expression for the various glycosyltransferases in specific cell types in the olfactory mucosa and the regulation of their enzymic activity during development as well as under different physiological and pathological conditions; (d) the differential distribution of secretory mucins into microchemically defined mucous domains as a result of autonomic, peptidergic and odorant regulation of the secretory cycle; (e) identification of the physiological, pharmacological and environmental factors that regulate sol–gel transformations associated with the changing rheological properties of olfactory mucus, and the effects of certain diseases on these factors; (f) regulation of the expression and activity of odorant and pheromone transporters; (g) the functional role in sensory transduction of specific monosaccharides in the glycocalyx of olfactory cilia; and (h) the genetic and physiological regulation of the composition of the glycocalyx of olfactory cilia. New insights into perireceptor and receptor events will be achieved with the conceptual and experimental implementations of these strategies.

Acknowledgements

The authors thank Dr Nikk Katzman for technical assistance. This research was supported by National Institutes of Health grant NIDCD-00159 (TVG) and National Science Foundation grant BNS-8821074 (MLG).

FIG. 4. (*opposite*) High power magnifications of areas indicated in Fig. 2. (a) Electron-lucent area (hsL) exhibits lower density of DSA–gold labelling than surrounding matrix (hsD). The irregular interface between hs and hd is indicated by triangles. (b) Trabeculae with a high density of gold particles project through hd. Bar = 0.5 μm.

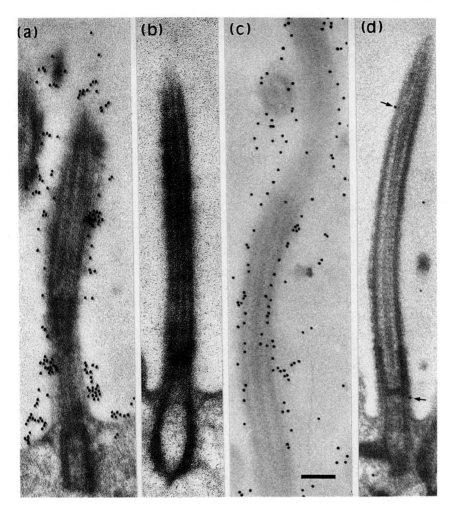

FIG. 5. Lectinoprobe labelling of ciliary glycocalices. (a, b) LFA-fetuin-gold label; (c, d) DSA-gold label. Glycocalices of the cilia of olfactory receptor neurons (ORNs, a) contain sialic acid residues as indicated by gold particle labelling, while those of respiratory epithelial cells (b) do not. Glycocalices of ORNs (c) exhibit a higher density of N-acetylglucosamine residues as indicated by gold particle labelling than do those of tracheal cilia (d, arrows). Bar = 0.2 μm.

References

Allen A 1983 Mucus—a protective secretion of complexity. Trends Biochem Sci 8:169–173
Chen Y, Getchell TV, Sparks DL, Getchell ML 1993 Patterns of adrenergic and peptigergic innervation in human olfactory mucosa: age-related trends. J Comp Neurol 334:104–116

Foster JD, Getchell ML, Getchell TV 1991 Identification of sugar residues in secretory glycoconjugates of olfactory mucosae using lectin histochemistry. Anat Rec 229:525–544

Foster JD, Getchell ML, Getchell TV 1992 Ultrastructural localization of sialylated glyconconjugates in cells of the salamander olfactory mucosa using lectin cytochemistry. Cell Tissue Res 267:113–124

Getchell ML, Getchell TV 1984 β-Adrenergic regulation of the secretory granule content of acinar cells in olfactory glands of the salamander. J Comp Physiol A Sens Neural Behav Physiol 155:435–443

Getchell TV, Getchell ML 1990 Regulatory factors in the vertebrate olfactory mucosa. Chem Senses 15:223–231

Getchell TV, Margolis FL, Getchell ML 1984 Perireceptor and receptor events in vertebrate olfaction. Prog Neurobiol (Oxf) 23:317–345

Getchell ML, Bouvet J-F, Finger TE, Holley A, Getchell TV 1989 Peptidergic regulation of secretory activity in amphibian olfactory mucosa: immunohistochemistry, neural stimulation, and pharmacology. Cell Tissue Res 256:381–389

Khan AA, Steiner JP, Snyder SH 1992 Plasma membrane inositol 1,4,5-triphosphate receptor of lymphocytes: selective enrichment in sialic acid and unique binding specificity. Proc Natl Acad Sci USA 89:2849–2853

Khew-Goodall Y, Grillo M, Getchell ML, Danho W, Getchell TV, Margolis FL 1991 Vomeromodulin, a putative pheromone transporter: cloning, characterization, and cellular localization of a novel glycoprotein of lateral nasal gland. FASEB (Fed Am Soc Exp Biol) J 5:2976–2982

Lopez-Vidriero MT 1989 Mucus as a natural barrier. Respiration 55(suppl 1):28–32

Menco BPhM 1992 Lectins bind differentially to cilia and microvilli of major and minor cell populations in olfactory and nasal respiratory epithelia. Microsc Res Tech 23:181–199

Menco BPhM, Farbman AI 1992 Ultrastructural evidence for multiple mucous domains in frog olfactory epithelium. Cell Tissue Res 270:47–56

Paulson JC, Colley KJ 1989 Glycosyltransferases. Structure, localization, and control of cell type-specific glycosylation. J Biol Chem 264:17615–17618

Pevsner J, Snyder SH 1990 Odorant-binding protein: odorant transport function in the vertebrate nasal epithelium. Chem Senses 15:217–222

Pevsner J, Trifiletti RR, Stritmatter SM, Snyder SH 1985 Isolation and characterization of an olfactory receptor protein for odorant pyrazines. Proc Natl Acad Sci USA 82:3050–3054

Rama Krishna NS, Getchell ML, Margolis FL, Getchell TV 1993 Vomeromodulin gene expression in the nasal mucosa of rats during postnatal development. Chem Senses 18, in press

Reese TS 1965 Olfactory cilia in the frog. J Cell Biol 25:209–230

Schnitzer JE 1988 Glycocalyx electrostatic potential profile analysis: ion, pH, steric, and charge effects. Yale J Biol Med 61:427–446

Sharon N, Lis H 1989 Lectins as cell recognition molecules. Science 246:227–234

Silberberg A 1989 Mucus glycoprotein, its biophysical and gel-forming properties. Symp Soc Exp Biol 43:43–63

Snyder DA, Rivers AM, Yokoe H, Menco BPhM, Anholt RRH 1991 Olfactomedin: purification, characterization, and localization of a novel olfactory glycoprotein. Biochemistry 30:9143–9153

Steinbrecht RA, Ozaki M, Ziegelberger G 1992 Immunocytochemical localization of pheromone-binding protein in moth antennae. Cell Tissue Res 270:287–302

Strous GJ, Dekker J 1992 Mucin-type glycoproteins. Crit Rev Biochem Mol Biol 27:57–92

Su Z, Getchell ML, Getchell TV 1993 Sub-cellular localization of *N*-acetylglucosamine-containing glycoconjugates in the salamander olfactory mucosa. Chem Senses 18, in press

Vogt RG, Köhne AC, Dubnau JT, Prestwich GD 1989 Expression of pheromone binding proteins during antennal development in the gypsy moth *Lymantria dispar*. J Neurosci 9:3332–3346

Zielinski BS, Getchell ML, Wenokur RL, Getchell TV 1989 Ultrastructural localization and identification of adrenergic and cholinergic nerve terminals in the olfactory mucosa. Anat Rec 225:232–245

DISCUSSION

DeSimone: Can you tell us anything about the stages of mobilization of the secretory apparatus in mucus-secreting cells following stimulation by an odorant or other stimulus type?

Getchell: Using electron microscopy, we demonstrated granule depletion in the acinar cells and also secretory vesicle release in the sustentacular cells (Zielinski et al 1988). We identified five stages in the secretory cycle within sustentacular cells. We have not begun to address this at the microchemical level, but these qualitative results should provide the basis for future experiments.

Lindemann: What is know about the electrolyte composition of the olfactory mucus?

Getchell: Earlier studies from our laboratory indicated that the overall electrolyte composition of olfactory mucus appeared to be much more like that of other mucus than the extracellular fluid around neurons, either in the retina or in the brain (Getchell & Getchell 1984, Joshi et al 1987). The extracellular Ca^{2+} concentration appeared to be somewhat higher than we had anticipated, but again, it was more in line with that measured in other secretory mucus than in neural tissue. More recently, Khayari et al (1991) reported that the electrolyte composition of olfactory mucus appeared to be closer to that associated with neuronal tissue than that of secretory material. This issue needs to be readdressed using more contemporary techniques. In the experiments that we reported (Joshi et al 1987), we were always very hesitant about giving absolute concentrations of electrolytes in olfactory mucus. Of more importance in our experiments was the systematic change in the concentration of Na^+ and K^+ with either odorant stimulation or the application of cholinergic and adrenergic agonists.

Buck: Have you looked at where odorant-binding protein (OBP) might be localized in the mucus?

Getchell: No, this needs to be done. It would be a way of testing whether or not OBP is involved specifically in the transport of odorants. Jonathan Pevsner, in Solomon Snyder's lab, has done elegant binding and localization studies on binding proteins in the lateral nasal glands (Pevsner & Snyder 1990), but a direct test of this hypothesis is needed.

Pelosi: We haven't done any immunocytochemistry to locate OBP in mucus. The available data were obtained with slices of tissue that were stained with fluorescent antibodies after fixation. The results of these immunocytochemical experiments indicated that OBPs are highly concentrated in the tubular acinar glands of the nasal respiratory epithelium in the cow (Avanzini et al 1987, Pevsner et al 1986), in the lateral nasal glands of the rat (Pevsner et al 1988a, Dear et al 1991) and in the Bowman's glands of the frog (Lee et al 1987). It would be nice to know how OBPs are distributed in the nasal mucus and in particular whether they are concentrated around the olfactory receptor proteins.

Getchell: How would you do this experiment and what would the appropriate controls be?

Buck: You would use anti-OBP antibodies and your immunogold technique.

Hwang: Olfactory mucus is a very complex mixture of substances and it could be more than just OBP that's involved in odorant transport.

We cloned a small soluble glycoprotein, which had previously been cloned, from a rat circumvallate papilla cDNA library. It shows homology to the human lipoprotein A that is involved in cholesterol transport in the serum. The cDNA is detected in the von Ebner's gland by *in situ* hybridization (P. Hwang, unpublished data).

Interestingly, this glycoprotein is highly expressed by the Sertoli cells in the rat testis and is thought to be involved in the maturation of the germ cells. If this protein is secreted by the von Ebner's gland, it may be involved in the transport of cholesterol-like molecules to or from taste cells, or in the maturation of taste receptor cells.

Schmale: There is an apolipoprotein D that belongs to the same lipocalin superfamily as the von Ebner's gland protein that we have identified (Drayna et al 1987). Is the protein you are talking about related to apolipoprotein D?

Hwang: Our soluble glycoprotein has homology to apolipoprotein A.

Carlson: Has anyone fractionated protein from mucus extracts on 2-D gels? Is a small number of predominant species observed?

Margolis: Bob Anholt's lab has actually run some gels with mucus extracts (Snyder et al 1991).

Lancet: When you run olfactory epithelium (or even olfactory cilia) on 2-D gels, you are bound to see some sub-patterns that derive from the mucus component. When we originally looked for receptors in olfactory cilia preparations by trying to find membrane-associated glycoproteins, we often saw proteins that weren't integral membrane proteins, as well as membrane-attached mucus proteins, in the same gels. The original glycoprotein gp95 we looked at may have a role similar to olfactomedin or to other mucus-related proteins. But we have never run pure mucus on a gel.

Getchell: We've tried, but we have always been concerned that the mucus we used wasn't pure. We have done some preliminary experiments where we stimulated the animals with β-adrenergic agonists to stimulate mucus so as to increase the volume produced (Getchell & Getchell 1984).

Khan et al (1992) have identified and characterized channel molecules in which there appears to be an N-glycosylated protein rich in GlcNAc monosaccharides. These channel molecules seem to be associated with Ca^{2+} conductance for the release of inositol trisphosphate ($InsP_3$) as a second messenger in olfactory transduction.

Ronnett: The basic observation was that there was an $InsP_3$-like receptor on the surface of lymphocytes that was thought to gate Ca^{2+}. This was thought to occur by a slow mechanism whereby intracellular Ca^{2+} could be elevated for sustained periods, which might be relevant to lymphocyte activation. The sugar composition of those receptors is thought to be somewhat different from the classical $InsP_3$ receptor. They bind $InsP_3$, but these receptors also bind other inositol phosphates in different rank order from the classical cerebellar receptor.

Getchell: Khan et al (1992) used a DSA lectinoprobe on gels to identify the sugar residues in the isolates from cerebellum as well as olfactory mucosa; they were quite different. The LFA lectinoprobe was also used to identify O-glycosylated sugar residues. Certainly, the presence of GlcNAc residues was proposed to be associated with a slow Ca^{2+} release.

Fesenko: What is the viscosity or microviscosity of the different layers of mucus?

Getchell: In terms of physical measurements, it's not really known what the true viscosity of olfactory mucus is; we only have the qualitative observations that many of us have made, based upon Professor Reese's report in 1965. This is something that needs to be measured under control as well as experimental conditions.

Lindemann: Let us come back to the layered structure of the mucus. I understood you as saying that in respiratory tissue, the ciliary beat prevents extensive association of mucus molecules. Where these short cilia beat, the mucus remains fluid, while above, it becomes gel-like. In contrast, in the olfactory mucus, the mucus density will be higher where there is binding of mucus molecules to the rather stationary olfactory cilia. Where it rises up between these cilia the mucus is more fluid. Is this correct?

Getchell: Yes, this is what the observations indicate.

Schmale: Tom, you hypothesized that stimulation of the olfactory receptors leads to a change in the composition of this mucus. For this to happen, not only would new mucus have to be produced, but old mucus would have to be degraded. I find it hard to envisage that this rather rigid system, with all those disulphide bonds, could change so rapidly. At least, you would expect to find disulphide-bond-degrading enzymes or some other similar factor present for this to occur.

Getchell: That is an excellent observation; experiments need to be performed along these lines. We need a better understanding of the hydration of these macromolecules.

In terms of the enzymic composition of olfactory mucus, there are some immunohistochemical observations that indicate that there are esterases present in the olfactory mucus (reviewed in Getchell & Getchell 1977), although there are no detailed studies of the enzymic composition of mucus.

DeSimone: The Bowman's gland is a mixed gland that contains both serous and mucous cell types. Various factors, including different odorants, might result in stimulation of one cell type more than the other. This is certainly one means by which the viscosity and electrolyte composition of mucus could be regulated. In the salivary system, stimulus control over the quantity and quality of the secretion has been observed. Like the Bowman's gland, the submaxillary gland is also mixed, containing both serous and mucous cells. Acids are such potent stimuli of the serous component that the volume so stimulated has been used as an index of sourness (Makhlouf & Blum 1972). Although both branches of the autonomic nervous system stimulate salivation, the serous component is mobilized to a greater degree by parasympathetic stimulation. Is there evidence for this type of differential activation in the olfactory system?

Getchell: Yes; it is true in the salamander olfactory mucosa and in the mammalian olfactory mucosa, where there are serous and mucous components. Barbara Zielinski, in our lab, has looked very carefully at the types of synaptic vesicles that make contact on the acinar cells and has shown a regional distribution of synaptic contacts (Zielinski et al 1989). For example, cholinergic synapses are always found between the cell membranes, deep within the acinar cells, whereas the adrenergic synapses are always found along the periphery. Presumably, there is typical autonomic as well as peptidergic innervation of these cells that regulates changes in the composition of olfactory mucus. But, again, the details of the regulatory mechanism are not known, nor is the enzymology of olfactory mucus.

Margolis: However, it is thought that the nature of the stimuli that olfactory and vomeronasal cell receptor sites interact with is significantly different. Is anything known about the differences in composition of the vomeronasal compared with the olfactory mucus?

Getchell: To my knowledge, no. We have not completed the analogous experiments in vomeronasal mucosa with the variety of lectinoprobes similar to those we have completed with olfactory mucosa (Foster at al 1991, 1992). Some results are just beginning to emerge showing that the immunocytochemical properties of the secretory cells of the vomeronasal glands appear to be different from those of the olfactory glands (Chen et al 1992, Rama Krishna et al 1992a, b). We are not certain at this time if these results indicate that there are major microchemical differences in the composition of the mucus between these two sensory systems.

Hwang: In the lateral nasal gland, where OBP is synthesized in the rat, β-adrenergic agonists such as isoproterenol and muscarinic agonists like carbachol both stimulate secretion. If you cut sections of the lateral nasal gland after you stimulate with these agonists, you can see that the serous glands have released all their contents.

We looked at the level of expression of the *OBP* mRNA, after stimulation with muscarinic or adrenergic agonists (Pevsner et al 1988a, P. Hwang, unpublished results). In both cases the level of mRNA decreased. We had expected it to go up on secretory stimulation. There may be other more essential secretory proteins that need to be synthesized first.

Getchell: Was it the same mRNA whose expression was regulated by both adrenergic and cholinergic stimulation?

Hwang: Yes; we used the same *OBP* cDNA as a probe for our Northern analyses.

Margolis: There are now supposedly two genes involved in OBP synthesis; *OBP-1* and *OBP-2*.

Hwang: We used Jonathan Pevsner's original clone of *OBP*, the first one to be reported (Pevsner et al 1988b).

Fesenko: Are components such as OBP necessary for solubilizing hydrophobic stimuli? Is OBP present in fish, which detect water-soluble stimuli?

Pelosi: To my knowledge, nobody has yet found OBPs in aquatic animals, although I am not sure anyone has really looked for them. In some preliminary experiments we failed to measure any binding of [^{3}H]2-isobutyl-3-methoxypyrazine to crude extracts of nasal mucosa from several species of fish (Pelosi et al 1982, Baldaccini et al 1986), but this does not exclude the presence of OBPs with different binding specificities in those animals. Soluble proteins that bind chemical stimuli are known to be present in chemoreception systems using water-soluble stimuli. Examples are the von Ebner's gland proteins that are similar to OBPs in their amino acid sequences and are probably involved in taste perception (Schmale et al 1990) and the soluble binding proteins of bacterial chemotaxis (Koshland 1981). These observations could suggest that the function of OBPs is not simply to solubilize and concentrate odorants, but might be related to a more specific role in odour perception and recognition.

Margolis: Other than the binding activity, do we know anything else about the function of OBP or related proteins?

Hatt: We looked at the response of moth olfactory neurons to pheromone in culture. We found that the neurons responded to pheromone in a dose-dependent way: co-application of pheromone-binding protein (PBP) was not necessary. Our preliminary results on adding PBP, extracted from antennae, or bovine serum albumin (BSA), are consistent with the possibility that PBP and BSA scavenge the pheromone and hence prevent adaptation due to overstimulation by pheromone (Stengl et al 1992).

Getchell: Stuart, didn't you do a similar experiment a number of years ago with tissue slice preparations of salamander olfactory mucosa, in which the sensitivity of olfactory receptor cells from which you were doing patch-clamp recordings *in situ* appeared to be greater than that of isolated receptor cells?

Firestein: I was never able to determine a real dose–response curve in those tissue slices (unpublished results). Also, the slices were thoroughly washed in a Ringer's bath, so I'm not sure how much mucus would be left on them.

Lindemann: It may be possible to record odorant responses from single receptor cells *in situ* without impaling the mucosa and with the air–mucus interface preserved. As a step in this direction, we recorded action potentials from olfactory cells of the frog. The cells remained in the mucosa while one sensory cilium of (ideally) one cell was used as a capacitive electrode (Frings & Lindemann 1990, 1991). The remaining bunch of cilia was free to detect odorants. In some cells we have seen sensitivities to cineole in the nanomolar and picomolar range. In these experiments the odorant was dissolved in the saline superfusing the mucosal surface. The method may be modified such that the air–mucus border is retained and the odorants are applied in the gas phase.

Firestein: That's certainly more sensitive than anything I have seen in isolated cells.

Fesenko: Is OBP necessary for odour discrimination?

Firestein: No. Isolated cells without OBP can discriminate between different odours (this volume: Firestein & Zufall 1993).

Schmale: A lipid-binding protein like albumin will do the same job: you don't need OBP or PBP specifically, just any protein that dissolves the lipophilic ligand so that accessibility to the membrane is increased.

Hatt: Marc van den Berg in my lab did exactly those experiments in *Antheraea polyphemus* (van den Berg & Ziegelberger 1991). He could see no significant differences between the effects of PBP and BSA, measuring the electro-olfactogram (EOG) responses to pheromone. In mixtures with pheromone, both substances decreased the threshold pheromone concentration by a factor of 100.

Margolis: Doron, didn't you show that lipophilic molecules at the air–water interface dissolve preferentially into the aqueous phase anyway?

Lancet: I didn't show any data—it's all in the physical chemistry textbooks! If you calculate the partition coefficient correctly (that is, molar in air versus molar in water), you find that there is a significant enrichment in the water phase for most odorants. The most hydrophobic odorants have a partition coefficient, molar:molar, of about one. It's quite clear that you don't really need a protein to bring odorants into the aqueous subphases. As Linda Buck alluded, it is very difficult to distinguish between the processes of bringing odorants to the receptors and taking odorants from the receptors. We should look at OBP, PBP and related proteins as buffering or carrier proteins; they are like calcium-binding proteins in cytoplasm. At equilibrium, they regulate the concentration of odorants. It is possible that, at high odorant concentrations, more odorant is dissolved than would be if there had not been a binding protein. You don't really need to bring the ligand to the receptor, and no matter what, the ligand needs to dissociate from the carrier prior to binding to the receptor. We need a model, kinetic and thermodynamic, of what happens in this system of air, odorant, water, membrane, carrier protein and receptor protein. If this model incorporated all the parameters, such as the kinetic constants and binding constants, we could find out more accurately what the carrier proteins actually do.

DeSimone: This is important: it is not sufficient to just talk about the binding of a protein, we must also consider the parameters with which it binds. As you pointed out, if OBP is going to function as a carrier, binding is just one part of the process—unbinding is just as important, as we know in the case of haemoglobin. If a protein binds something too tightly, it can't deliver it; consequently, it isn't useful physiologically.

Getchell: These calculations are very important. Using the theoretical model we established (Getchell et al 1984), we and Professor Snyder's group have calculated the rate of diffusion through mucus of OBP to which an odorant was bound (Pevsner & Snyder 1990; Getchell & Getchell 1990). We compared this rate with the onset latencies of single olfactory receptor neuron recordings and we then tied this data in with the biophysical model that John DeSimone and I had worked on (Getchell et al 1980). The conclusion was that if one invokes a carrier function for OBP and assumes diffusion of OBP plus a bound odorant, such as amyl acetate or a musk, the diffusion times for that complex are far too long by tens of seconds compared with the onset latency that one records electrophysiologically from single olfactory neurons (Getchell et al 1980).

DeSimone: If the mucus is already charged with binding protein, let's say, and you introduce the odorant, you have a charging time—a capacitance. You are going to get transport only after you reach steady state. This charging process could take so long as to be totally unphysiological.

Lancet: But this may well be the function of OBP and similar proteins. You need very little odorant to stimulate the receptors. Perhaps if there is a very strong pulse of odorant, then OBP is a buffer; it absorbs most of the odorant, leaving whatever little is necessary to go to the receptors. As Tom Getchell said, it cannot serve as a kinetic carrier.

Pelosi: I like this idea, but on the other hand, results of experiments performed with insects have shown that in the presence of PBP the sensitivity of moth antennae to the specific pheromone increases (van den Berg & Ziegelberger 1991).

Lancet: Karl-Ernst Kaissling also got similar results. I can't explain them!

Firestein: What is the time span of these measurements? We find that receptors are not only responsive to the concentration of odorant at any given moment, but they integrate over time. Really, instead of 'concentration', you should talk about 'flux': the number of molecules per second. So if OBP or PBP keeps the odorant or pheromone around longer, you could get responses at what appear to be lower concentrations, depending on the time period of your recordings.

Lancet: Are you alluding to a slow-release mechanism?

Firestein: Yes; a sort of integration of the concentration over time, so a lower concentration for a longer time would elicit the same response as a high concentration presented for a short time.

Getchell: The argument against that is from the basic electrophysiological data, which show a very sharp termination of the signal—the membrane current—or the inactivation of the spike electrogenesis (Getchell 1986).

Lancet: The integrating time may be in the range of 10–1000 ms.

Getchell: I would agree with that; certainly not longer than a second.

Breer: One possibility that is frequently mentioned when discussing the role of PBP is that these binding proteins transport odorants in a manner similar to the way that myoglobin transports oxygen in muscles; a process that is called facilitated diffusion (Gayeski et al 1985). In this case, OBP would be increasing the speed of transport of odorants to the receptors.

Getchell: Jonathan Pevsner and Sol Snyder (1990) proposed a model for OBP function in the facilitated diffusion of odorants based upon the example of oxygen transport by haemoglobin; this is another model that needs to be tested empirically.

Breer: Your criticism concerning a carrier function of OBP depends on the distance that you think an odour has to travel to the receptor. We don't know how near the surface of the mucus the cilia are located.

Getchell: It's true that the distance could be short, but in the Stokes–Einstein equation used to calculate the diffusion coefficient (Getchell et al 1984), the diffusion time is more a function of η, the viscosity of mucus, than of the diffusion distance.

Firestein: But if a significant proportion of the cilium is in the top, more sol-like layer of the mucus, viscosity is not a major consideration.

Getchell: It is true that viscosity could be of lesser importance if the receptors are located in the apical region of the cilia. However, some biophysical calculations that John DeSimone's associates and our group did together suggested that the receptor sites are at the base of the cilia (Getchell et al 1980).

Lerner: I don't think it's safe to consider PBP and OBP together for a number of reasons. First, there is no evolutionary relationship between them. Second, there is a 300-fold difference between the concentration of PBP in the sensilla of the moth and the concentration of OBP in the rat olfactory mucosa: this might suggest they are not doing the same thing. Another point is that in the insect system, a further possible function of PBP is to protect incoming odours from degrading enzymes, which are pretty avid, to at least get the odour in.

Reed: Is it possible that olfactory mucus has an important physical protecting function?

DeSimone: Mucus has a protective role in the lower gastrointestinal tract, where it is secreted in copious amounts. A good example is in the pH–mucus barrier of the gastric mucosa that protects the stomach against its own peptic secretions (Flemstrom & Turnberg 1984). I think it is likely that tissue protection is also a function served by olfactory mucus.

Getchell: In cell physiology it's called a 'mucus barrier': it's a biological barrier in that a variety of immunological factors, such as immunoglobulin A and lysozyme, are localized in the mucus; it's a chemical barrier in terms of its microchemical composition, buffering capacity, enzymes and its regulation of pH; and it's also a physical barrier in that it protects the epithelial surface from dehydration.

Reed: Regarding the pH regulation of olfactory mucus: is it clear from calculations that you can actually get pH buffering of, say inhaled ammonia?

DeSimone: You can certainly get it in the oral cavity. Acids, for instance, cause rather specific reflexes that give rise to copious secretion of the serous component of saliva, which contains a high concentration of bicarbonate and neutralizes the acid. I don't know whether this has been studied in the olfactory mucus, but I would be amazed if a similar reflex doesn't exist there also.

Getchell: In the olfactory system, there are descriptive observations that suggest a secretomotor reflex also exists there. Observing the mucus macroscopically, it appears to change in composition, becoming more fluid with certain types of stimulation. But strict measurements of its viscosity, elasticity and other properties have not been made.

Margolis: Is there any way of getting enough olfactory mucus to do some quantitative measurements of its ability to buffer any of these components?

Firestein: We need Gabriele Ronnett's olfactory neuron cell line, which turned out to be mucus secreting!

Lancet: We have to realize that the olfactory system is very unusual. We have receptor mechanisms, transduction channels, disposed just $2\,\mu$m from the air that we breathe. Without mucus protection, these mechanisms would be dead in no time. There is little precedence for this system: almost all the other receptor systems are protected, rather than just being exposed directly to the air. It would be interesting to ask specific questions about how the olfactory mucociliary system is specially suited for a receptor–air interface.

Caprio: It would also be worth doing a comparison with fish, where this receptor–air interface doesn't exist. Fish don't have Bowman's glands either.

References

Avanzini F, Bignetti E, Bordi C et al 1987 Immunocytochemical localization of pyrazine-binding protein in the cow nasal mucosa. Cell Tissue Res 247:461–464

Baldaccini NE, Gagliardo A, Pelosi P, Topazzini A 1986 Occurrence of a pyrazine binding protein in the nasal mucosa of some vertebrates. Comp Biochem Physiol B Comp Biochem 84:249–253

Chen Y, Getchell ML, Ding X, Getchell TV 1992 Immunolocalization of two cytochrome P450 isozymes in rat nasal chemosensory tissue. NeuroReport 3:749–752

Dear TN, Campbell K, Rabbitts TH 1991 Molecular cloning of putative odorant-binding and odorant metabolizing proteins. Biochemistry 30:10376–10382

Drayna DT, McLean JW, Wion KL, Trent JM, Drabkin HA, Lawn RM 1987 Human apolipoprotein D gene: gene sequence, chromosome localization and homology to the $\alpha 2\mu$-globulin superfamily. DNA 6:199–204

Firestein S, Zufall F 1993 Membrane currents and mechanisms of olfactory transduction. In: The molecular basis of smell and taste transduction. Wiley, Chichester (Ciba Found Symp 179) p 115–130

Flemstrom G, Turnberg LA 1984 Gastroduodenal defense mechanisms. Clin Gastroenterol 13:327–354

Foster JD, Getchell ML, Getchell TV 1991 Identification of sugar residues in secretory glycoconjugates of olfactory mucosae using lectin histochemistry. Anat Rec 229:525–544

Foster JD, Getchell ML, Getchell TV 1992 Ultrastructural localization of sialylated glycoconjugates in cells of the salamander olfactory mucosa using lectin cytochemistry. Cell Tissue Res 267:113–124

Frings S, Lindemann B 1990 Single unit recording from olfactory cilia. Biophys J 57:1091–1094

Frings S, Lindemann B 1991 Current recording from sensory cilia of olfactory receptor cells *in situ*. I. The neuronal response to cyclic nucleotides. J Gen Physiol 97:1–16

Gayeski TEJ, Connet RJ, Honig CR 1985 Oxygen transport in rest–work transition illustrates new functions for myoglobin. Am J Physiol 248:914–921

Getchell ML, Getchell TV 1984 β-Adrenergic regulation of the secretory granule content of acinar cells in olfactory glands of the salamander. J Comp Physiol A Sens Neural Behav Physiol 155:435–443

Getchell TV 1986 Functional properties of vertebrate olfactory receptor neurons. Physiol Rev 66:772–780

Getchell TV, Getchell ML 1977 Early events in vertebrate olfaction. Chem Senses Flavor 2:313–326

Getchell TV, Getchell ML 1990 Regulatory factors in the vertebrate olfactory mucosa. Chem Senses 15:223–231

Getchell TV, Heck GL, DeSimone JA, Price S 1980 The location of olfactory receptor sites: inferences from latency measurements. Biophys J 29:397–412

Getchell TV, Margolis FL, Getchell ML 1984 Perireceptor and receptor events in vertebrate olfaction. Prog Neurobiol (Oxf) 23:317–345

Joshi H, Getchell ML, Zielinski B, Getchell TV 1987 Spectrophotometric determination of cation concentrations in olfactory mucus. Neurosci Lett 82:321–326

Khan AA, Steiner JP, Snyder SH 1992 Plasma membrane inositol 1,4,5-trisphosphate receptor of lymphocytes: selective enrichment in sialic acid and unique binding specificity. Proc Natl Acad Sci USA 89:2849–2853

Khayari A, Math F, Trotier D 1991 Odorant-evoked potassium changes in frog olfactory epithelium. Brain Res 539:1–5

Koshland DE 1981 Biochemistry of sensing and adaptation in a simple bacterial system. Annu Rev Biochem 50:765–782

Lee HK, Wells RG, Reed RR 1987 Isolation of an olfactory cDNA: similarity to retinol binding protein suggests a role in olfaction. Science 253:1053–1056

Makhlouf GM, Blum AL 1972 Kinetics of the taste response to chemical stimulation: a theory of acid taste in man. Gastroenterology 63:67–75

Pelosi P, Baldaccini NE, Pisanelli AM 1982 Identification of a specific olfactory receptor for 2-isobutyl-3-methoxypyrazine. Biochem J 201:245–248

Pevsner J, Snyder SH 1990 Odorant-binding protein: odorant transport function in the vertebrate nasal epithelium. Chem Senses 15:217–222

Pevsner J, Sklar PB, Snyder SH 1986 Odorant-binding protein:localization to nasal glands and secretions. Proc Natl Acad Sci USA 83:4942–4946

Pevsner J, Hwang PM, Sklar PB, Venable JC, Snyder SH 1988a Odorant-binding protein and its mRNA are localized to lateral nasal gland implying a carrier function. Proc Natl Acad Sci USA 85:2383–2387

Pevsner J, Reed RR, Feinstein PG, Snyder SH 1988b Molecular cloning of odorant-binding protein: member of a ligand carrier family. Science 241:336–339

Rama Krishna NS, Getchell ML, Getchell TV 1992a Differential distribution of γ-glutamyl cycle molecules in the vomeronasal organ of rats. NeuroReport 3:551–554

Rama Krishna NS, Getchell ML, Tate SS, Margolis FL, Getchell TV 1992b Glutathione and γ-glutamyl transpeptidase are differentially distributed in the olfactory mucosa of rats. Cell Tissue Res 270:475–484

Reese TS 1965 Olfactory cilia in the frog. J Cell Biol 25:209–230

Schmale H, Holtgreve-Grez H, Christianse H 1990 Possible role for salivary gland protein in taste reception indicated by homology to lipophilic-ligand carrier proteins. Nature 343:366–369

Snyder DA, Rivers AM, Yokoe H, Menco BPhM, Anholt RRH 1991 Olfactomedin: purification, characterization, and localization of a novel olfactory glycoprotein. Biochemistry 30:9143–9153

Stengl M, Zufall F, Hatt H, Hildenbrand JG 1992 Olfactory receptor neurons from antennae of developing male *Manduca sexta* respond to components of the species-specific sex pheromone *in vitro*. J Neurosci 12:2523–2531

van den Berg MJ, Ziegelberger G 1991 On the function of the pheromone binding protein in the olfactory hairs of *Antheraea polyphemus*. J Insect Physiol 37:79–85

Zielinski BS, Getchell ML, Getchell TV 1988 Ultrastructural characteristics of sustentacular cells in control and odorant-treated olfactory mucosae of the salamander. Anat Rec 221:769–779

Zielinski BS, Getchell ML, Wenokur RL, Getchell TV 1989 Ultrastructural localization and identification of adrenergic and cholinergic nerve terminals in the olfactory mucosa. Anat Rec 225:232–245

Receptor diversity and spatial patterning in the mammalian olfactory system

Linda B. Buck

Department of Neurobiology, Harvard Medical School, 220 Longwood Avenue, Boston, MA 02115, USA

Abstract. In order to gain insight into the mechanisms underlying olfactory perception in mammals, we have performed experiments to identify and characterize the basic receptive elements of the olfactory system, the odorant receptors. We have identified a novel multigene family that encodes odorant receptors on olfactory sensory neurons in the nasal cavity. The tremendous size and diversity of this family indicate that perceptual acuity in the olfactory system relies heavily on the differential binding properties of hundreds of different receptor types. In order to determine how the information supplied by such a large collection of diverse receptors might be organized, we have examined the patterns of expression of different odorant receptor genes in the olfactory epithelium. We have observed distinct topographical patterns of odorant receptor RNAs that indicate that the olfactory epithelium is divided into a series of expression zones. These zones are likely to provide for a broad organization of sensory information in the nasal cavity which is maintained in the axonal projection to the olfactory bulb.

1993 The molecular basis of smell and taste transduction. Wiley, Chichester (Ciba Foundation Symposium 179) p 51–67

The fine discriminatory power of the mammalian olfactory system is likely to derive from information processing events that occur at several distinct anatomical sites: the olfactory epithelium of the nasal cavity, where odours are first sensed by olfactory sensory neurons; the olfactory bulb, where information received from the sensory neurons is thought to be processed; and the cortex, where information received from the olfactory bulb is thought to be further refined to allow for the discrimination of thousands of different odours (Shepherd 1990). Odorous molecules that enter the nasal cavity are thought to bind to specific odorant receptors on the lumenal cilia of olfactory sensory neurons (Rhein & Cagan 1983). This binding event induces a cascade of transductive events (reviewed in Breer et al 1992, Reed 1992) that culminate in the transmission of action potentials along olfactory neuron axons that synapse in the olfactory bulb. Electrophysiological studies indicate that different olfactory neurons recognize different odorants or sets of odorants (Sicard & Holley 1984, this volume: Firestein & Zufall 1993). The ability of different neurons to recognize different odorants presumably derives from the existence of a variety of odorant receptors that are differentially expressed by olfactory neurons.

Our work has focused on the mechanisms by which sensory information is organized in the olfactory epithelium. In our initial studies, the goal was to identify and characterize the receptive elements of the olfactory sensory neuron, the odorant receptors (Buck & Axel 1991). In more recent studies, we have used odorant receptor genes to investigate how the olfactory system organizes incoming sensory information in the olfactory epithelium (Ressler et al 1993). In the present article, critical aspects of these two studies will be described and their implications discussed.

Results and discussion

Identification and characterization of a multigene family encoding odorant receptors

The identification of a novel multigene family encoding odorant receptors. Our initial studies focused on the identification of genes encoding odorant receptors (Buck & Axel 1991). The experimental strategy that we devised to search for these genes was based on two ideas. One was that odorant receptors, because they recognize structurally diverse odorants, must themselves be variable in structure and would therefore be encoded by a multigene family. The second was that odorant receptors would be members of a large superfamily of receptors that transduce signals via interactions with G proteins. This assumption was based on previous observations that had implicated G proteins in olfactory signal transduction (reviewed in Breer et al 1992, Reed 1992). Members of the G protein-coupled receptor superfamily exhibit seven hydrophobic domains that are believed to serve as membrane spanning regions and are commonly referred to as seven-transmembrane-domain receptors (Strader et al 1989, Dohlman et al 1991). We first performed polymerase chain reactions (PCRs) with a large number of different degenerate primers to amplify members of the seven-transmembrane-domain superfamily that are expressed in the olfactory epithelium. We then used restriction enzymes to determine whether any of the PCR products contained multiple DNA species, consistent with the amplification of multiple members of a multigene family. This approach allowed us to identify a single PCR product that contained a series of related DNAs that encoded proteins that appeared to be members of the seven-transmembrane-domain superfamily. These proteins shared sequence motifs with one another that were not present in other superfamily members, thus suggesting that the olfactory proteins were representatives of a novel receptor family.

Using the PCR product as probe, we isolated a series of homologous olfactory cDNA clones. In Northern blots with RNA from a variety of tissues, a mix of these cDNAs hybridized only to olfactory epithelium RNA, thus suggesting that this receptor family might be expressed uniquely in the olfactory epithelium. An elevated frequency of homologous clones in a cDNA library prepared from an enriched population of olfactory neurons further suggested that the olfactory receptor family might be expressed exclusively in olfactory neurons (Buck & Axel 1991).

Extensive sequence diversity and subfamilies within the odorant receptor family. Sequence analyses of the coding regions of a number of cDNAs confirmed that they encode a series of proteins that belong to the G protein-coupled receptor superfamily (Buck & Axel 1991). Each of the olfactory proteins displays seven hydrophobic domains and several small sequence motifs commonly seen in members of the superfamily. However, the olfactory receptors share a number of novel sequence motifs that are not seen in other members of the superfamily.

An alignment of the olfactory receptors reveals several intriguing features of the olfactory receptor family that may be important to its role in olfactory perception (Buck & Axel 1991). First, there is tremendous sequence diversity in this receptor family. Interestingly, extensive diversity is seen in several of the transmembrane domains, regions which have been implicated in ligand binding in some other G protein-coupled receptors (Strader et al 1989, Dohlman et al 1991). This diversity in potential ligand-binding regions is consistent with the ability of the olfactory system to recognize a variety of structurally diverse odorous ligands.

A second striking feature is the presence of subfamilies within the family (Buck & Axel 1991). Although the cDNAs that we sequenced encode a diverse set of receptors, some of these receptors are almost identical to each other. Southern blotting experiments with individual cDNAs revealed that, in fact, most members of the olfactory multigene family belong to subfamilies of highly related genes. The subfamilies appear to range in size from one to about 20 genes. These observations led us to propose that different subfamilies might recognize different structural classes of odorants while the individual members of a subfamily might detect subtle differences between structurally related odorants (Buck & Axel 1991).

All of the aforementioned characteristics of the olfactory multigene family that we identified are consistent with the conclusion that this family encodes odorant receptors on olfactory sensory neurons (Buck & Axel 1991). This conclusion is supported by recent studies performed by Heinz Breer and his colleagues who have examined the ability of these molecules to confer odorant responsiveness on heterologous cells following transfection of receptor cDNAs (Raming et al 1993).

Sequence analyses of additional members of the rat olfactory multigene family have further emphasized the extensive diversity of this family (Levy et al 1991). In addition, experiments performed in a number of laboratories have demonstrated that families of genes homologous to the rat olfactory multigene family are present in a variety of different species, including human (Parmentier et al 1992, Selbie et al 1992, this volume: Lancet et al 1993, this volume: Wang & Reed 1993), dog (Parmentier et al 1992), mouse (Nef et al 1992, Ressler et al 1993), salamander (A. Jesurum, D. Chicaraishi & J. Kauer, unpublished results), catfish (Ngai et al 1993a) and zebrafish (S. Korshing & H. Baier, unpublished results).

Size and organization of the odorant receptor multigene family. Southern blotting experiments with receptor gene probes suggested to us that the olfactory multigene family might be quite large. This suspicion was confirmed when we screened genomic libraries with a mixture of coding region segments from the cloned cDNAs. Even with a limited set of probes, we obtained evidence for at least several hundred family members in the rat genome (Buck & Axel 1991). Experiments performed with more complex probes indicate that there may be as many as 500–1000 genes in the olfactory multigene families of mouse (E. Liman & L. Buck, unpublished results) and rat (R. Reed, personal communiction). This family is one of the largest gene families known.

One intriguing issue raised by the size of the olfactory multigene family concerns the mechanisms that regulate gene expression in this system. How does an olfactory neuron express selectively one or a few members of such an enormous gene family? Selective receptor expression could depend upon familiar strategies utilizing transcription factors or, as in some other multigene families (Borst 1986, Alt et al 1987), might involve a more novel mechanism such as gene rearrangement or gene conversion. Preliminary studies on the structures of several members of the multigene family in rat have shown that the coding regions of these genes are not interrupted by introns, but that one or more introns are present in the 5' untranslated regions (L. Buck & R. Axel, unpublished results). We have obtained no evidence for the somatic generation of sequence diversity by gene rearrangement, gene conversion, or somatic mutation. However, it is possible that selective gene expression in this system requires genomic alterations such as gene rearrangement or gene conversion. On the other hand, it is possible that gene expression in this system does not involve alterations in genomic structure, but relies instead on transcription, or enhancer, factor-based gene regulatory mechanisms.

Spatial patterns of odorant receptor gene expression in the olfactory epithelium

The remarkable size and diversity of the odorant receptor multigene family indicate that odour discrimination relies upon the differential binding properties of hundreds of different types of odorant receptor. How does the olfactory system organize the information provided by such a large collection of diverse receptors? Does the olfactory system use physical space within the nervous system to organize or 'encode' information? Are there topographical maps or spatial codes for odours in the olfactory epithelium or olfactory bulb? If so, neurons that express the same receptor type might all be clustered in one region of the epithelium or might all form synapses in a discrete set of glomeruli in the olfactory bulb. On the other hand, other strategies for information processing are also possible (Buck 1992) and neurons that express the same odorant receptor gene could be distributed randomly in the olfactory epithelium or form synapses in a distributed fashion in the olfactory bulb.

We have begun to address these questions by first asking if there is any evidence for spatial maps in the olfactory epithelium. In these experiments we have used *in situ* hybridization techniques to analyse the patterns of expression of a large number of different odorant receptor genes in the olfactory epithelium.

Neurons that express the same receptor gene are not clustered in the epithelium.
For these analyses, we decided to use the mouse rather than the rat, because of the availability of isogenic mouse strains, which would circumvent potential variability due to genetic polymorphisms. In preliminary experiments, we amplified a series of coding region segments of mouse odorant receptor genes from genomic DNA and cloned the PCR products (Ressler et al 1993). Sequence analyses of the PCR clones confirmed that they encode a variety of proteins that share novel sequence motifs with proteins encoded by the rat odorant receptor multigene family (Fig. 1A). In Southern blotting experiments, most of the PCR clones hybridize to distinct sets of restriction fragments in mouse liver DNA, indicating that they belong to different subfamilies within the mouse olfactory multigene family (Fig. 1B).

We next performed a series of *in situ* hybridization studies in which we hybridized ^{35}S-labelled antisense RNAs prepared from three different clones (K4, K7 and K18 (Fig. 1)) to coronal and horizontal sections that spanned the mouse olfactory epithelium (Ressler et al 1993). The *OMP* (olfactory marker protein) gene, which is expressed in all mature olfactory neurons (Danciger et al 1989), was used as a positive control.

We found that each of the subfamily probes hybridizes to cells that are centrally located within the olfactory epithelium, thus indicating that odorant receptor genes are expressed in olfactory neurons, but not in olfactory stem cells or supporting cells. This is consistent with the observation that antibodies against conserved regions of the rat odorant receptor family bind only to olfactory sensory neurons (R. Reed, N. Levy, H. Bakalyar & A. Cunningham, unpublished results). Each of the probes hybridizes to neurons in several different areas of the nasal cavity. Furthermore, within these areas, the hybridized

A

```
         II
I9    LFLSNLSFADLCFSSVTMPKLIQNMQSQVPSIPYAGCLAQIYFFLFGDLGNFLLVAMAYDRYVAICFPLHYMS
K4    YFLSSLSFIDFCQSTVVIPKMIVSFLTEMNIISYSECMAQLVFLTFGIAGCYTLAAMAYDRYVAICNPLYNV
K7    FFLSHAIVDIAYACNIVPQMEVNLLDPVKPISYAGCMTQTFLFLTFAITECLLIVMSYDRYVAICHPLRYSA
K18                                                       RYVAICKPLTYKV

                              III
                                                                    V
I9    IMSPKLCVSIVLSWVLTTFHAMLFTLLMARLSFCEDSVIPHYFCDMSTLLKVACSDTHDNELAIFILGGPIVV
K4    TMSYQIYSSLISGVYIFAVICSSFNTGFMLRTQFCNLDVINHYFCDLLPLENLASSNTYINEILLFFATLNSF
K7    IMSWRVCSTMAVTSWIGVLLSLIHLVLLPLPFCVSQKVNHFFCEITAILKLACADTHLNETMVLAGAVSVLV
K18   IMSPKICLIIFSSYLMGFASAMAHTGCMIRLSFCDSNIINHYMCDIFPLIPLSCSSTYVNELMSSVVVGSAII

             VI                                                  VII
I9    LPFLLIIVSYARIVSSIFKVPSSQSIHKAFSTCGSHLSVSLFYGTVIGLYLCPSANNSTVKETVMSLMYTVT
K4    VPVLTIITSYIFIIVTLLSIHSREGKFKAFSTCSTHISAVAIFYGSGAFTYLQPSSLNSMGQAKVSSVFVTVV
K7    GPFSSIVVSVACILGAILKIQSEEGQRKAFSTCSSHICVVGLFYGTAIVMYVGPRHGSPKEQKKYLLLFHSLFN
K18   LCCLILILISYAMLFNIIHMSSGKGWSKALGTCGSHIITVSLFYGSGLLAYVKPSSAKTVGQGKFFSVFYLLV
```

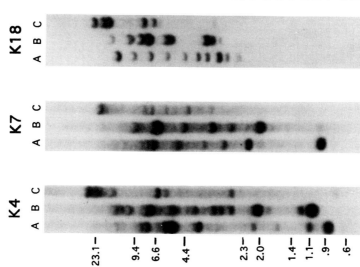

FIG. 1. Sequence and Southern blot analyses of mouse odorant receptor clones (Reproduced with permission from Ressler et al 1993). (A) The protein sequences encoded by three mouse odorant receptor clones (K4, K7 and K18) are aligned with a member of the rat odorant receptor family (19). Shaded residues are those present in at least 60% of rat odorant receptors previously analysed (Buck & Axel 1991). The putative transmembrane domains are indicated by bars and Roman numerals. (B) 5 μg of mouse liver DNA were digested with EcoRI (A), HindIII (B), or BamHI (C), subjected to electrophoresis in 0.75% agarose, blotted onto nylon membranes and hybridized with the ^{32}P-labelled probes indicated. The size markers indicate kilobase pairs.

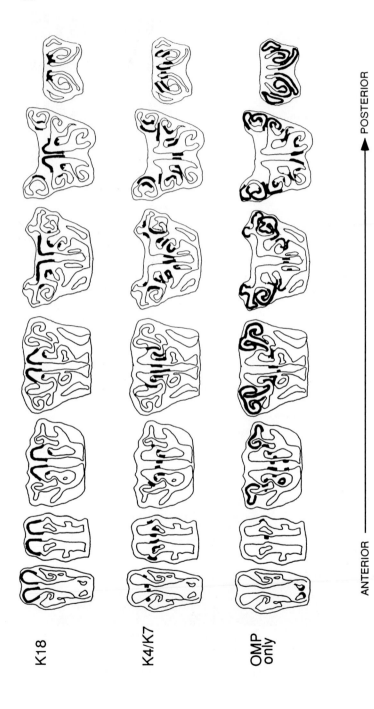

FIG. 2. Schematic diagrams of the mouse nasal cavity showing regional patterns of odorant receptor gene expression. Tracings of coronal sections from regions spaced approximately 300 μm apart along the anterior–posterior axis of the nasal cavity are shown. The shaded areas of the olfactory epithelium indicate regions that hybridize to the K18 receptor probe, the K4 and K7 receptor probes, or to the OMP probe, but not to the K4, K7, or K18 probe.

cells are broadly distributed. Each hybridized neuron is surrounded by many neurons that have not hybridized. Thus, neurons that express the same odorant receptor gene are not clustered in the epithelium. This argues against the hypothesis that spatial codes for odours consist of patches of neurons that recognize the same odours.

Expression zones in the olfactory epithelium. In our initial experiments, we noticed that each receptor probe hybridized to neurons in many (but not all) areas of the nasal cavity. Furthermore, the patterns of hybridized neurons in the two nasal cavities appeared to be bilaterally symmetrical. Bilaterally symmetrical patterns of hybridization have also been noted by Nef et al (1992) and by Strotmann et al (1992). To define the hybridization patterns more precisely, we made tracings from projected photographs of a large number of hybridized coronal sections from different animals. The schematic diagrams in Fig. 2 were constructed on the basis of these data. From these coronal schematics, we prepared schematic diagrams of predicted medial and lateral views of the hybridization patterns (Fig. 3).

The patterns of hybridization that we observed exhibit a number of striking features that emphasize that different odorant receptor genes are expressed in highly specified topographical patterns in the olfactory epithelium (Ressler et al 1993):

1. The expression of each receptor subfamily is limited to distinct regions of the olfactory epithelium. These regions exhibit bilateral symmetry in the two nasal cavities and are virtually identical in different individuals, regardless of sex.

2. The K4 and K7 subfamilies are always expressed together in the same regions while the K18 subfamily is expressed in different regions. In fact, the K4/K7 and K18 regions appear to be mutually exclusive except at shared boundaries where a small degree of overlap is observed. 93–99% of cells that hybridize to each probe lie within the zonal boundaries.

3. The K4/K7 and K18 expression regions show a consistent dorsal–ventral or medial–lateral relationship, with the K18 regions always located dorsal or medial to the K4/K7 regions. Interestingly, regions that hybridize to the OMP probe, but none of these receptor probes are always located ventral or lateral to the K4/K7 regions.

4. Each subfamily appears to be expressed in several separate bands that are located in different regions of the nasal cavity (e.g. the septum and individual turbinates). These bands, which extend along the anterior–posterior axis of the nasal cavity, are apparent in the coronal schematics (Fig. 2), but are most obvious in the predicted lateral and medial views in Fig. 3.

Our measurements indicate that the K4/K7 and K18 expression zones each cover about a quarter of the olfactory epithelium. If the olfactory epithelium is divided into non-overlapping expression zones of approximately equal size,

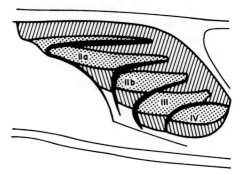
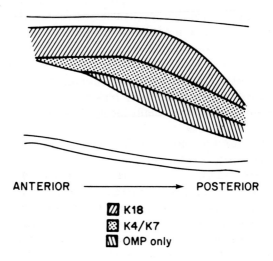

FIG. 3. Schematic diagrams showing the zonal expression of odorant receptor genes in medial and lateral views of the nasal cavity. Predicted views of the medial and lateral nasal cavity based on reconstructions of coronal sections are shown. A lateral view of the nasal cavity shows that cells that hybridize to the K18 probe are located on the roof, the dorsal lateral wall, and the dorsal surface of each turbinate (II–IV). Cells that hybridize to the K4 and K7 probes lie just ventral to the K18+ regions on each of these structures. Regions that hybridize to the OMP probe, but not to these receptor probes are ventral to the K4/K7+ regions. The medial view shows bands of expression extending from anterior to posterior along the nasal septum. (Reproduced with permission from Ressler et al 1993).

there could be as few as four such zones. To investigate this possibility, we performed limited *in situ* hybridization studies with a large set of additional odorant receptor probes that hybridize to distinct sets of bands on Southern blots of EcoRI digested mouse liver DNA (Ressler et al 1993). Many of these probes give patterns of hybridization typical of those seen with K4 and K7 or with K18, including some that appear to recognize only a single gene in Southern blots. However, some hybridize to regions that hybridized to the OMP probe, but not to K4, K7, or K18 probes. The patterns of hybridization that we have observed thus far indicate that there are at least four expression zones that are organized along the dorsal–ventral and medial–lateral axes.

Expression of receptor genes within a zone. The topographical patterns of expression we have observed indicate that the olfactory epithelium is divided into a limited series of expression zones. Although the expression of each receptor subfamily is confined to neurons in a single zone, we have not observed clustering or any other obvious spatial ordering of hybridized neurons within an expression zone. Quantitative measurements of the densities of hybridizing neurons further indicate that there is a similar density of cells that hybridize to each probe throughout the expression zone. Interestingly, in the catfish, odorant receptor genes are also expressed in neurons that are broadly distributed; however, there is no zonal patterning evident in the catfish (Ngai et al 1993b). On the basis of the reported density of olfactory neurons in the mouse olfactory epithelium (Mackay-Sim & Kittel 1991) and the densities of hybridized neurons in our studies, we estimate that the K4 and K18 subfamilies are expressed in about 0.7% of olfactory neurons in their respective zones, while the K7 subfamily is expressed in about 0.3%.

Our analyses indicate that the zonal assignment of each odorant receptor gene is regulated strictly. However, within each zone, neurons that express a particular receptor gene are broadly distributed. It thus appears that when an olfactory neuron or its progenitor chooses which odorant receptor gene(s) to express, it is restricted to a single zonal gene set, but may choose a receptor gene (or set of genes) to express from among the members of the set via a stochastic mechanism (Ressler et al 1993).

What is the functional significance of the expression zones? Our results indicate that neurons that express a given odorant receptor gene (and therefore must recognize the same odorants) are confined to a single expression zone. The restriction of multiple members of the same subfamily to the same zone further indicates that neurons that express related odorant receptor genes that are likely to recognize related odorants are also located in the same zone. These findings suggest that the expression zones might provide for a broad organization of incoming sensory information in the olfactory epithelium.

Numerous studies of the axonal projection from the olfactory epithelium to the olfactory bulb indicate that this projection, like the receptor zones we have observed, is organized along the dorsal–ventral and medial–lateral axes (Astic & Saucier 1986, Astic et al 1987, Clancy et al 1985, Saucier & Astic 1986, Schwarting & Crandall 1991, Schwob & Gottlieb 1986). In some studies, retrograde tracers were deposited in different bulbar regions and the patterns of back-labelled neurons in the epithelium were examined (Astic & Saucier 1986, Astic et al 1987, Saucier & Astic 1986, T. Schoenfeld, personal communication). Comparisons of our zonal patterns with the patterns of labelled neurons seen in those studies suggest that neurons that belong to the same odorant receptor expression zone project axons to the same broad region along the dorsal–ventral axis of the bulb (Ressler et al 1993). Thus the zonal organization of sensory information in the epithelium may be maintained in the transmission of this information to the bulb. It is possible that, within each bulbar region, a further refinement of this patterning occurs such that neurons that express the same odorant receptor gene synapse within the same set of glomeruli. By analogy with the visual system (Shatz 1990), such a sharpening of the axonal projection might conceivably arise via ligand-dependent, activity-dependent mechanisms (Buck 1992, Ressler et al 1993).

Summary

The mammalian olfactory system is capable of discriminating a vast array of structurally diverse odours. We have identified a novel multigene family whose unusual size and diversity suggest that odour discrimination may rely heavily on the existence of many hundreds of different types of odorant receptors which are differentially expressed by olfactory sensory neurons in the nasal cavity. We have found that the members of this family are segregated in their expression into a series of distinct and highly specified zones within the olfactory epithelium. Our experiments suggest that the odorant receptor expression zones may provide for an initial organization of sensory information in the nasal cavity which is maintained in the transmission of this information to the olfactory bulb of the brain.

References

Alt F, Blackwell T, Yancopoulos G 1987 Development of the primary antibody repertoire. Science 238:1079–1087

Astic L, Saucier D 1986 Anatomical mapping of the neuroepithelial projection to the olfactory bulb in the rat. Brain Res Bull 16:445–454

Astic L, Saucier D, Holley A 1987 Topographical relationships between olfactory receptor cells and glomerular foci in the rat olfactory bulb. Brain Res 424:144–152

Borst P 1986 Discontinuous transcription and antigenic variation in trypanosomes. Annu Rev Biochem 55:701–732

Breer H, Boekhoff I, Krieger J, Raming K, Strotmann J, Tareilus E 1992 Molecular mechanisms of olfactory signal transduction. In: Corey DP, Roper SD (eds) Sensory Transduction. The Rockefeller University Press, New York, p 93–108

Buck L, Axel R 1991 A novel multigene family may encode odorant receptors: a molecular basis for odor recognition. Cell 65:175–187

Buck LB 1992 The olfactory multigene family. Curr Opin Genet Dev 2:467–473

Clancy AN, Schoenfeld TA, Macrides F 1985 Topographic organization of peripheral input to the hamster olfactory bulb. Chem Senses 10:399–400

Danciger E, Mettling C, Vidal M, Morris R, Margolis F 1989 Olfactory marker protein gene: its structure and olfactory neuron-specific expression in transgenic mice. Proc Natl Acad Sci USA 86:8565–8569

Dohlman HG, Thorner J, Caron MG, Lefkowitz RJ 1991 Model systems for the study of seven-transmembrane-segment receptors. Annu Rev Biochem 60:653–688

Firestein S, Zufall F 1993 Membrane currents and mechanisms in olfactory transduction. In: The molecular basis of smell and taste transduction. Wiley, Chichester (Ciba Found Symp 179) p 115–130

Lancet D, Ben-Arie N, Cohen S et al 1993 Olfactory receptors: transduction, diversity, human psychophysics and genome analysis. In: The molecular basis of smell and taste transduction. Wiley, Chichester (Ciba Found Symp 179) p 131–146

Levy NS, Bakalyar HA, Reed RR 1991 Signal transduction in olfactory neurons. J Steroid Biochem Mol Biol 39:633–637

Mackay-Sim A, Kittel PW 1991 On the life span of olfactory receptor neurons. Eur J Neurosci 3:209–215

Nef P, Hermans-Borgmeyer I, Artieres-Pin H, Beasley L, Dionne VE, Heinemann SF 1992 Spatial pattern of receptor expression in the olfactory epithelium. Proc Natl Acad Sci USA 89:8948–8952

Ngai J, Dowling MM, Buck L, Axel R, Chess A 1993a The family of genes encoding odorant receptors in the channel catfish. Cell 72:657–666

Ngai J, Chess A, Dowling MM, Necles N, Macagno ER, Axel R 1993b Coding of olfactory information: topography of odorant receptor expression in the catfish epithelium. Cell 72:667–680

Parmentier M, Libert F, Schurmans F et al 1992 Expression of members of the putative olfactory receptor gene family in mammalian germ cells. Nature 355:453–455

Raming K, Krieger J, Strotmann J et al 1993 Cloning and expression of odorant receptors. Nature 361:353–356

Reed RR 1992 Signaling pathways in odorant detection. Neuron 8:205–209

Ressler KJ, Sullivan SL, Buck LB 1993 A zonal organization of odorant receptor gene expression in the olfactory epithelium. Cell 73:597–609

Rhein LD, Cagan RH 1983 Biochemical studies of olfaction: binding specificity of odorants to a cilia preparation from rainbow trout olfactory rosettes. J Neurochem 41:569–577

Saucier D, Astic L 1986 Analysis of the topographical organization of olfactory epithelium projections in the rat. Brain Res Bull 16:455–462

Schwarting GA, Crandall JE 1991 Subsets of olfactory and vomeronasal sensory epithelial cells and axons revealed by monoclonal antibodies to carbohydrate antigens. Brain Res 547:239–248

Schwob JE, Gottlieb DI 1986 The primary olfactory projection has two chemically distinct zones. J Neurosci 6:3393–3404

Selbie LA, Townsend-Nicholson A, Iismaa TP, Shine J 1992 Novel G protein-coupled receptors: a gene family of putative human olfactory receptor sequences. Brain Res 13:159–163

Shatz CJ 1990 Impulse activity and the patterning of connections during CNS development. Neuron 5:745–756

Shepherd GM 1990 The synaptic organization of the brain. Oxford University Press, New York

Sicard G, Holley A 1984 Receptor cell responses to odorants: similarities and differences among odorants. Brain Res 292:283–296

Strader CD, Sigal IS, Dixon RAF 1989 Structural basis of β-adrenergic receptor function. FASEB (Fed Am Soc Exp Biol) J 3:1825–1832

Strotmann J, Wanner I, Krieger J, Raming K, Breer H 1992 Expression of odorant receptors in spatially restricted subsets of chemosensory neurones. NeuroReport 3:1053–1056

Wang MM, Reed RR 1993 Molecular mechanisms of olfactory neuronal gene regulation. In: The molecular basis of smell and taste transduction. Wiley, Chichester (Ciba Found Symp 179) p 68–75

DISCUSSION

Getchell: Did you observe any receptor gene expression zones in the vomeronasal organ?

Buck: No. Only a few of the clones that we used hybridized to the vomeronasal organ and only a few hybridized neurons were seen.

Getchell: Is there any overlap of receptor gene expression zones that might suggest a common ligand stimulation of receptors between the vomeronasal and olfactory zones?

Buck: We don't know yet, because we have not yet identified receptor genes that we are convinced are expressed in the vomeronasal organ. We are currently looking for such genes, however.

Firestein: Linda, in some of the sections towards the rostral end of the nasal cavity, it looked like there were only two zones that took up almost the entire section.

Buck: Yes. There is another zone there, but it is very small in that posterior region of the nasal cavity.

Firestein: So in certain anterior–posterior regions of the olfactory epithelium, are whole families of receptors virtually excluded?

Buck: No, except at the anterior end of the olfactory epithelium, where the zone defined by clone K18 predominates and at the posterior end, where the zone which contains M50$^+$ cells predominates.

Kinnamon: I don't know if your *in situ* hybridization techniques have the resolution for you to look at this, but do you ever see any evidence that the same cell expresses two or more different receptors?

Buck: No. We can't address that question with the techniques that we've used. However, we do have evidence that the K4 and K7 subfamilies are not generally expressed in the same cells. This conclusion is based on determinations of the number of cells in adjacent sections that hybridize to the individual or a mix of probes.

Reed: That argument is valid only if the receptors are always expressed in the same cells as opposed to overlapping randomly.

Buck: These results indicate that there is not a large overlap in the populations of cells that express members of the K4 and K7 subfamilies. They don't address whether a single neuron expresses one, or more than one, receptor gene.

Reed: Using a statistical argument; suppose every cell expresses 10% of all the receptors that are expressed within that zone and there are a couple of hundred individual receptor genes expressed within that zone. If 1% of the cells hybridized with the K7 receptor and 1% were recognized by the K4 probe with a 10% overlap, you would see only 1.8% of the cells in the epithelium hybridizng with a mixed probe instead of the 2% expected for distinct cell populations identified by each receptor probe. Isn't that what your argument is based on?

Buck: No. What I'm saying is that a large proportion of neurons that express members of the K4 subfamily do not also express members of the K7 subfamily. We can't say anything about the number of different receptor genes that are expressed in a single cell. One would have to do single-cell PCR experiments to address that question.

Reed: I agree. It's essential to remember that there's a real distinction between how many genes are expressed in each cell and whether the same cell always express a particular set of receptors in a coordinated pattern.

Buck: I am biased. I think that we will find that each cell expresses just a single receptor gene, but this may not be the case. For example, each neuron might express an entire receptor subfamily.

Reed: You suggested that there is a zone in the nasal cavity in which the receptors all project back to the olfactory bulb. Have you considered the possibility that it's the other way around—that the neurons are directed by the bulb as to which repertoire of receptors to express in their region and that what you are seeing is a zonal representation of the topographic pattern of projection with the bulb?

Buck: Yes, that's a possibility. The expression zones could be intrinsic to the olfactory epithelium or, as you noted, they could be imposed by the bulb during or following synapse formation.

Firestein: Couldn't you test this? If you bulbectomize the rats, wait for the olfactory neurons to grow back and then do these *in situ* hybridizations, you could look to see whether similar patterns (or no patterns at all) arise.

Reed: A topographic map between part of the epithelium and a place on the bulb already exists. I don't know whether this has any significance for olfactory processing. If you have two sheets of cells, they will tend to connect so that the cells near the top of one sheet will tend to project to near the top of the target cell sheet and the ones near the bottom of one sheet will tend to project to near the bottom of the other. You could cut the connection as many times as you want and you wouldn't be able to discriminate between a retrograde signal that tells the first set of cells which receptor repertoire is expressed and some other complicated signal.

Firestein: Anatomically, that's not the case—there's a tremendous mixing of axons close to bulb.

Buck: Yes, there is evidence for axonal reassortment in the bulb (Daston et al 1990, Greer & Kaliszewski 1992). In addition, it is clear from the retrograde labelling studies (Saucier & Astic 1986, Astic et al 1987) that when a retrograde label is placed in a single bulb region, there are rather sharp boundaries in the epithelium between areas that contain back-labelled neurons and those that do not. Furthermore, the regional patterns of back-labelled neurons bear an uncanny resemblance to the odorant receptor expression zones that we have observed.

Bargmann: That depends on when receptors are expressed relative to when the projections are made to the olfactory bulb. If you sever the olfactory tract, and new receptors are expressed before the regenerating neuron gets back to the target, then the targets probably aren't instructing the neuron as to which receptor should be expressed.

Buck: We are trying to do this experiment by asking whether zonal patterning is evident in the developing olfactory epithelium before olfactory axons have reached the bulb or the presumptive bulb.

Lancet: We looked at rat development to see whether olfactory receptor genes are expressed early or just at and after the time that the sensory axons reach their targets in the olfactory bulb. The answer, with some reservations, seems to be the latter—that olfactory neurons begin to express the receptors when they reach their target (Margalit & Lancet 1993). It is extremely important to see whether this developmental sequence is recapitulated after bulbectomy.

Breer: Our *in situ* hybridization experiments studying various developmental stages show that in rats, receptor-expressing cells can be found as early as E14. Several receptor subtypes were found to be expressed in cells of the olfactory pit at this stage.

Linda, you say there are four expression zones. Isn't it possible that there may be other overlapping zones?

Buck: Yes, we have evidence for four zones, but there could be more.

Breer: From the results of our *in situ* hybridization studies on the olfactory epithelium of rats, we can confirm most of the principles you have found in mice. We have seen two main expression zones so far; most of the receptor subtypes are widely distributed in either of these areas.

However, in addition we have found a receptor subtype (OR37) which shows a unique spatial distribution; it is segregated in a restricted region on the tip of endoturbinate II and ectoturbinate 3 (Strotmann et al 1992). So, the main expression zones may be superimposed by additional sets of expression patterns which make the chemotopic organization of the olfactory epithelium much more complex.

Buck: Yes. For example, we have a number of clones that give patterns that resemble the K4/K7 patterns. We are now hybridizing these to see if they show exactly the same patterns or patterns that are slightly different. There could conceivably be a series of expression patterns that are largely, but not completely, overlapping.

Caprio: Riddle & Oakley (1992) showed that the olfactory nerve in the rainbow trout projects non-topographically to the olfactory bulb, that the axons re-sort at the nerve–bulb interface and culminate in nine discrete terminal fields within the glomerular layer in the olfactory bulb. These nine terminal fields occurred in rainbow trout of different sizes and ages and re-appeared subsequent to degeneration of the olfactory receptors following olfactory nerve transection. Furthermore, widely distributed olfactory receptor neurons, characterized by specific lectin-binding properties, projected to restricted regions of the glomerular layer of the olfactory bulb (Riddle & Oakley 1993).

Firestein: The same thing has been shown in rat where, 500 μm from the bulb, these axons which have come out of the epithelium as parallel fibres just take off in all sorts of different directions and then sort themselves in the bulb.

Lancet: This was shown by Charles Greer, in an experiment in which he did very elegant electron microscopic tracing of rat olfactory axons: he mapped their spatial distribution, following them through sections taken at 0.1 μm intervals (Greer & Kaliszewski 1992). He showed that the axons behave as if there is a signal just as the axons enter the bulb, so they go to their appropriate targets. This indicates that there is recognition at the bulb, rather than axons going randomly into the bulb and then being told what to express in a retrograde way.

References

Astic L, Saucier D, Holley A 1987 Topographical relationships between olfactory receptor cells and glomerular foci in the rat olfactory bulb. Brain Res 424:144–152

Daston MM, Adamek GD, Gesteland RC 1990 Ultrastructural organization of receptor cell axons in frog olfactory nerve. Brain Res 537:69–75

Greer CA, Kaliszewski C 1992 Topographic courses of individual axons within the olfactory nerve. Soc Neurosci Abstr 18:1199

Margalit T, Lancet D 1993 Expression of olfactory receptor and transduction genes during rat development. Dev Brain Res 73:7–116

Riddle DR, Oakley B 1992 Immunocytochemical identification of primary olfactory afferents in rainbow trout. J Comp Neurol 324:575–589

Riddle DR, Oakley B 1993 Lectin identification of olfactory receptor neuron subclasses with segregrated central projections. J Neurosci 13:3018–3033

Saucier D, Astic L 1986 Analysis of the topographical organization of olfactory epithelium projections in the rat. Brain Res Bull 16:455–462

Strotmann J, Wanner I, Krieger J, Raming K, Breer H 1992 Expression of odorant receptors in spatially restricted subsets of chemosensory neurones. NeuroReport 3:1053–1056

Molecular mechanisms of olfactory neuronal gene regulation

Michael M. Wang and Randall R. Reed

Howard Hughes Medical Institute, Department of Molecular Biology and Genetics, 818 PCTB, Johns Hopkins School of Medicine, 725 N Wolfe St, Baltimore, MD 21205, USA

> *Abstract.* The mammalian olfactory system utilizes a biochemical cascade mediated by specialized proteins to detect odorants with high sensitivity and specificity. The recent identification of olfactory neuron-specific components for each step in the signalling cascade suggests that expression of these proteins is coordinately controlled by *cis*-acting regulatory sequences and *trans*-acting transcriptional activators. We have used molecular genetic methods to characterize sequences encoding tissue-specific DNA-binding sites in the genes for components of the odorant transduction pathway and have identified a putative transcription factor, Olf-1, that functions at these sites to regulate gene expression.
>
> *1993 The molecular basis of smell and taste transduction. Wiley, Chichester (Ciba Foundation Symposium 179) p 68–75*

The mammalian olfactory system is remarkable for its ability to detect odorants at concentrations as low as a few parts per billion in air. The capacity to discriminate between more than 10 000 different odorants arises from specialization at the anatomical, cellular, biochemical and genetic level, as well as in the patterns of neuronal connections. The cells responsible for odorant detection, the olfactory neurons, are located within a neuroepithelium and project a dozen or more cilia into the mucus that bathes the epithelium. These cilia are the presumed sites of the initial events in odorant signal transduction.

Odorant ligands are thought to bind integral membrane receptors in the plasma membrane of the cilium. These receptors then initiate a second messenger cascade, leading to activation of G proteins and their interaction with the membrane-bound adenylate cyclase (Pace et al 1985). Increases in cAMP levels result in the opening of a cyclic nucleotide-responsive cation channel, depolarization of the cell and initiation of an action potential. Olfactory-specific or olfactory-enriched forms of many of the participants in this cascade have now been identified: G_{olf} (Jones & Reed 1989), type III adenylate cyclase (Bakalyar & Reed 1990) and a cyclic nucleotide-gated ion channel (Dhallan et al 1990).

Olfactory neuronal gene regulation

The olfactory neurons are equally remarkable for their ability to undergo continual replacement throughout adult life. The receptor neurons arise from neuroblast-like precursor cells that lie near the basal lamina that defines the olfactory epithelial layer. As the neuronal progenitors mature, they migrate into the more apical regions of the epithelium, begin to extend dendritic processes that are the site of attachment of the sensory cilia to the luminal surface and send axonal projections to the olfactory bulb. The gene products responsible for the differentiated properties of the mature neurons must be induced in a coordinated fashion during the differentiation. The genes known to be induced during this process include those responsible for components of the signal transduction cascade: G_{olf}, the type III adenylate cyclase and the olfactory cyclic nucleotide-activated channel (OcNC). Additional genes that are expressed specifically in mature olfactory neurons but are of unknown function have been identified using molecular genetic methods (Rogers et al 1987, R. Reed, unpublished observations). The proteins these genes encode may be involved in the signal transduction cascade or, alternatively, may play a role in the establishment of the mature neuronal phenotype.

Mechanisms for coordinated regulation of olfactory neuron-specific genes

The regulation of gene expression that leads to the terminally differentiated phenotype in mammalian cells is often mediated by tissue-specific transcription factors. These regulatory factors control target genes through interactions with specific DNA sequences present in the region of the transcriptional start sites. For example, during muscle differentiation, the *MyoD* gene is thought to play a central role in the induction of muscle-specific gene products (Davis et al 1987). It is one member of the myogenic helix-loop-helix family of transcription factors, which have been shown to trigger differentiation of myoblasts into mature muscle cells through sequence-specific interactions. It seemed likely that a transcription factor analogous to MyoD might direct the expression of olfactory neuronal-specific products. We have taken advantage of the collection of olfactory neuron-specific genes that have been isolated in recent years to isolate the transcriptional initiation regions for these genes and examine in molecular detail the DNA sequences that could play a role in their regulated expression. In particular, we have attempted to define common DNA sequences among the control regions for the olfactory genes and, subsequently, identify proteins which bind to these putative *cis*-acting regulatory sites.

We focused our efforts on identification of these sites in five olfactory-specific genes by identifying, cloning and sequencing the promoter regions of each of these genes from rat genomic DNA. Our initial studies identified a sequence in the promoter region for the *OcNC* gene that was bound by a protein present only in nuclear extracts from olfactory tissue. The sequence of this site displayed marked similarities to a site identified in the olfactory marker protein (*OMP*) gene by Margolis and co-workers (Stein-Izsak et al 1991, Kudrycki et al 1993). They have named the protein factor that binds to the site in the promoter of *OMP*, Olf-1. Our studies indicate that each of the olfactory-specific genes we have examined—those coding for G_{olf}, type III adenylate cyclase, OcNC and OMP, and the 50.06 gene—contains at least one binding site for the Olf-1 protein. The presence of an Olf-1-binding site in each of these genes leads us to propose that the olfactory-specific genes are controlled coordinately by the putative transcriptional activator, Olf-1, and that Olf-1 may be a critical transcription factor involved in activating the expression of the signal transductory components and other genes in the sensory neurons.

Identification of olfactory neuron-specific transcription factors

We have used a variety of biochemical and molecular genetic approaches to identify the genes encoding sequence-specific DNA-binding activities. Several have been cloned by direct protein purification and microsequencing. The relatively low abundance of sensory neurons in olfactory tissue preparations (10–25%), combined with the limited amounts of tissue that can be obtained, suggested that purification of the Olf-1-binding activity by standard protein purification methods would prove difficult. Alternatively, several DNA-binding proteins have been cloned by screening recombinant bacteriophage expression libraries with radiolabelled concatamerized sites. Our attempts to screen olfactory cDNA expression libraries with concatamerized, radiolabelled OcNC Olf-1-binding sites were not successful in identifying candidate clones encoding proteins with binding activity like that of Olf-1. The special difficulties associated with identifying cDNA clones encoding Olf-1 activity in a complex tissue with potentially only a fraction of the cells expressing the factor led us to devise a scheme which takes advantage of the powerful genetic selections available in the yeast *Saccharomyces cerevisiae*. Conceptually, this system consists of a yeast strain unable to grow in the absence of histidine, which carries a plasmid expression vector comprising a yeast minimal promoter adjacent to three copies of the Olf-1-binding site and upstream of the yeast *HIS3* gene. In the absence of a transcriptional activator protein capable of binding to the Olf-1 site, this yeast strain remains unable to grow in the absence of histidine. However, if one introduces a cDNA expression library derived from olfactory tissue into the yeast strain, those rare cDNA clones capable of increasing transcription of the *HIS3* gene and allowing the cells to grow in the absence of histidine can be selected for directly by plating out the cells on His⁻ media. This system is formally similar to the two-hybrid system of Fields & Song (1989), but there is a distinct difference in the kinds of protein interactions that are being examined (DNA–protein versus protein–protein).

When an olfactory cDNA library of 3.6 million clones was constructed in the yeast expression vector and transformed into a His3⁻ yeast strain carrying the *HIS3* reporter plasmid, a single HIS⁺ colony arose from the initial two million transformants. Analysis of the olfactory cDNA present in this transformant (Y11) revealed an open reading frame encoding a 570-amino acid protein which begins with a methionine at nucleotide 44 preceded by an in-frame stop codon. Secondary structure analysis of the Y11-encoded protein sequence predicts an α-helix separated by a turn or loop followed by another α-helix. Surprisingly, the two helices are nearly identical at the amino acid level and bear modest similarity to the second helix of the basic helix-loop-helix (bHLH) family of transcription factors, which includes MyoD. However, several aspects of the primary protein sequence suggest that it belongs to a distinct family from the previously characterized HLH proteins. In particular, the basic region, characteristic of those MyoD-related factors that activate transcription, is absent in Olf-1. Additionally, the Olf-1 protein appears to homodimerize to activate transcription.

The message encoded by the putative *Olf-1* cDNA is confined to olfactory tissue, as expected on the basis of the distribution of Olf-1-binding activity. When its mRNA is transcribed and translated in rabbit reticulocyte lysates, the expressed protein is able to bind to the Olf-1-binding sites in each of the olfactory-specific genes. To obtain a more definitive localization of the Olf-1 protein, we generated antibodies to the C-terminal portion of the predicted protein and used these for immunohistochemistry on sections of olfactory epithelium. Our results indicate that Olf-1 is localized to the nuclei of olfactory neurons and their presumed precursors. As revealed by counterstaining with DAPI (diaminophenylindole), a non-specific DNA-binding dye, nuclei of the sustentacular cells and the basal cells (presumed to be the most undifferentiated cells) do not contain appreciable amounts of Olf-1 immunoreactivity.

Implications

We have successfully exploited the yeast genetic system to clone an olfactory-specific transcription factor with known target genes. Although we have demonstrated the ability of Olf-1 to activate a synthetic promoter carrying multiple copies of the Olf-1 site, it will be interesting to learn whether Olf-1 expression is sufficient to induce high transcription levels in a reporter construct when it is driven by the intact *OcNC* promoter. When MyoD is expressed in fibroblast cells, it directs the expression of several muscle-specific genes and induces a myocyte-like morphology. If similar processes can be induced in appropriate cell lines by the Olf-1 factor, some processes associated with olfactory neuronal differentiation could be examined in culture.

Preliminary Southern blot analysis of rat genomic DNA suggests that there are multiple members of the Olf-1 family. The role of these other family members in olfactory neuronal differentiation or, potentially, in the differentiation of other neuronal tissues remains to be examined.

The odorant receptors described by Buck & Axel (1991) appear to be expressed specifically by the olfactory neurons. Unlike the other genes encoding the downstream components of the signal transduction cascade, an individual receptor gene appears to be expressed by only a small fraction of the mature neurons. Several investigators have suggested that these complex expression patterns are achieved by having a single expression locus in the genome and moving genetic information for receptors from resident silent copies to the expression site. In such a model, the presence of Olf-1-binding sites at the expression locus would impart special properties to that site and permit the specific expression of a receptor in only a fraction of the cells, while the remaining components of the cascade were expressed uniformly by the sensory neurons. While the mechanism by which specific regulation is achieved remains unknown, the *Olf-1* product may play a role as a factor directing expression of this receptor family in the sensory neurons.

Acknowledgements

This work was supported by the Howard Hughes Medical Institute. M.M.W. was supported by a Medical Scientist Training Program grant.

References

Bakalyar HA, Reed RR 1990 Identification of a specialized adenylyl cyclase that may mediate odorant detection. Science 250:1403–1406

Buck L, Axel R 1991 A novel multigene family may encode odorant receptors: a molecular basis for odor recognition. Cell 66:175–187

Davis RL, Weintraub H, Lassar AB 1987 Expression of a single transfected cDNA converts fibroblasts to myoblasts. Cell 51:987–1000

Dhallan RS, Yau KW, Schrader KA, Reed RR 1990 Primary structure and functional expression of a cyclic nucleotide-activated channel from olfactory neurons. Nature 347:184–187

Fields S, Song O-K 1989 A novel system to detect protein–protein interactions. Nature 340:245–256

Jones DT, Reed RR 1989 G_{olf}: an olfactory neuron specific-G protein involved in odorant signal transduction. Science 244:790–795

Kudrycki K, Stein-Izsak C, Behn C, Grillo M, Akeson R, Margolis FL 1993 Olf-1 binding site: characterization of an olfactory specific promoter motif. Mol Cell Biol 13:3002–3014

Pace U, Hanski E, Salomon Y, Lancet D 1985 Odorant-sensitive adenylate cyclase may mediate olfactory reception. Nature 316:255–258

Rogers KE, Dasgupta P, Gubler U, Grillo M, Khew-Goodall Y-S, Margolis FL 1987 Molecular cloning and sequencing of a cDNA for olfactory marker protein. Proc Natl Acad Sci USA 84:1704–1708

Stein-Izsak C, Grillo M, Behn C, Sakai M, Corbin J, Margolis FL 1991 Trans-acting elements and olfactory neuron gene transcription. Soc Neurosci Abstr 5:1286

DISCUSSION

Margolis: Does the presumptive *Olf-1* cDNA clone that you have isolated show the kinds of developmental changes or responses to bulbectomy that we would anticipate for an mRNA directing the synthesis of a transcription factor involved in olfactory neuron-specific gene supression?

Reed: Northern analysis indicates that its abundance goes down five days after bulbectomy, but I think that's what you would expect from the immunohistochemical analysis—that is, there are a lot of mature olfactory neurons that appear to be *Olf-1* positive. If you were to deplete that population you would expect to lose mRNA. The question is, do you lose as much mRNA as you lose for *OMP*?

Buck: Have you looked for E12 or other members of the helix-loop-helix (HLH) family of DNA-binding proteins in olfactory neurons or the olfactory epithelium?

Reed: We have not looked specifically for E12 in the neurons. It's a very complicated picture, but you could probably find E12 or just about every transcription factor present there.

Michael Wang analysed about 50 cDNAs that we pulled out from an olfactory epithelium cDNA library (unpublished results). There is one differential splice that includes eight amino acids similar to the HLH proteins. Other than that, we couldn't find any related family members by PCR or restriction digests. Genomic Southern blots have shown clearly that there are proteins closely related to the HLH family in the olfactory epithelium.

Buck: Is there any way that you could look for other members of this group to see whether there were homologues of the E12 type or ID type?

Reed: E12 is expressed in practically all cells and so to find E12 message in olfactory epithelium wouldn't be especially significant.

In muscle cells—and perhaps differentially in the various types of muscle cell—there are transcription factors related to MyoD, such as myogenin, *Myf-5* and *Myf-2*. One of the things we are interested in is whether what we expect to be related members of this HLH family are also expressed in olfactory tissue (our initial experiments there all appear to be negative), or in other neuronal systems, such as the retina and the auditory system. Maybe these related homologues are going to play some role in differentiation in other systems.

Margolis: In view of the report that there are olfactory receptor-like molecules in the testis (Parmentier et al 1992), have you looked there for the presence of Olf-1?

Reed: Yes; we haven't found any.

Margolskee: Have you looked in the vomeronasal organ?

Reed: No.

Margolskee: Does the vomeromodulin gene have an Olf-1-binding site?

Margolis: It's unlikely, because vomeromodulin is synthesized primarily by the lateral nasal gland and not olfactory neurons.

Schmale: If there really is odorant receptor expression in testis and you don't find Olf-1 there, does that mean that the olfactory receptor genes do not have to have Olf-1-responsive elements for transcription?

Reed: We have no evidence that olfactory receptor gene transcription is regulated by Olf-1. You could say that we have no evidence that any of these genes are regulated by Olf-1 *in vivo*, although I think it's a reasonable suggestion. The question of what directs the expression of members of this olfactory receptor family in testis is very complicated. We did a simple experiment trying to answer this: we isolated a cDNA for a given olfactory receptor, and looked at whether it was expressed in testis. It was not.

Margolskee: When you express Olf-1 in COS cells, do you activate the endogenous olfactory-specific genes?

Reed: Michael Wang has prepared the RNA to perform that experiment. All the cells that we have used, COS-1 or HEK-293, are of human or monkey origin. As we want to use PCR to do this experiment, we would like to use oligonucleotides that span an intron so we can discriminate between genomic contamination and Olf-1-induced expression of RNA. We have these for the channel and now for G_{olf} also, so we are set up to look at it.

Bargmann: Do you have a sense of how many different clusters of olfactory receptors there are, for instance, how many different chromosomes contain several different olfactory receptor clones?

Reed: We have looked at three different human genomic clones carrying olfactory receptors; they map to three different chromosomes. We know that on chromosome 19, there are in the order of 50–100 receptor genes. As far as we can tell, all 50–100 receptors are located on the central third of the short arm of chromosome 19, in a region which is probably of the order of 5–10 Mb long.

Bargmann: Are there enough data yet to tell whether that sort of a cluster might correspond to the expression domains?

Reed: Linda Buck needs to do the chromosome mapping of the different members of the cluster! We have the map for one mouse receptor; it turns out to be very closely linked to the mouse erythropoietin receptor. This is interesting because that region is at the distal end of the 19p region that we know from all the human receptors we've used. So it looks like that whole region may be syntenic between mouse and human.

Buck: You said that you had evidence for olfactory receptor gene location on two other chromosomes: can you exclude receptor expression on the rest of the chromosomes?

Reed: We took three genomic phages and we did fluorescence *in situ* hybridization. One mapped to chromosome 19, one mapped to chromosome 17 and one mapped to chromosome 11.

Buck: So the receptor genes could be present on other chromosomes as well?
Reed: Yes.
Lancet: Chromosome 17 is clearly a place where there is a cluster of olfactory receptor genes in humans.
Buck: Do you know how many?
Lancet: A few dozen, I would say.
Preliminary experiments show that a general human genomic olfactory receptor probe labels chromosomes 19 and 17, but others are not excluded. This is a difficult experiment because the probe is very short, there is a lot of noise and you need amplification.
Margolis: What is the evidence that these genomically mapped olfactory receptor genes are actually expressed?
Reed: Some of the members of gene clusters have multiple frame shifts and so are predicted to be non-functional genes. The receptor genes that we've looked at in the 19 cosmid cluster, which are different genes, look like they probably have 5' untranslated introns, but there is no direct evidence that they are functional genes. It's the problem of going from mouse to human; it's impossible to figure out which of the human genes correspond to a gene in, say, hamster. Basically, you have to re-invent the entire system when you go from species to species.
Buck: What proportion of the human olfactory genes might be pseudogenes, then?
Reed: We've sequenced eight, and probably two to three are pseudogenes, so about 30% might be the figure.
Lancet: We have sequenced the central regions of 12 receptors (between transmembrane segments 2 and 7); two of these have deletions that would cause frame shifts, indicating that they are pseudogenes. There may be other sequence locations or non-coding regions for some others. As for expression, we have looked at human olfactory mRNA and we can show that at least one member of the olfactory receptor 17th chromosome cluster is expressed (unpublished results).

Reference

Parmentier M, Libert F, Schurmans S et al 1992 Expression of members of the putative olfactory receptor gene family in mammalian germ cells. Nature 355:453–455

A new tool for investigating G protein-coupled receptors

Michael R. Lerner, Marc N. Potenza, Gerard F. Graminski, Tim McClintock, Channa K. Jayawickreme and Suresh Karne

Howard Hughes Medical Institute Research Laboratories, Yale University School of Medicine, New Haven, CT 06536-0812, USA

Abstract. Vertebrate olfactory receptors are members of the seven-transmembrane-domain G protein-coupled receptor family. They utilize intracellular signal transduction pathways which are activated by stimulation of odorant receptors and use the second messengers cAMP and/or inositol 1,4,5-trisphosphate and diacylglycerol. Studies of how odorants bind to and activate the receptors can be considered part of the more general problem of how chemicals interact with G protein-coupled receptors. This review describes the development of a new technique for assessing functional interactions between chemicals and these receptors in only minutes. Predicted uses of the system include structure–function analyses of both G protein-coupled receptors and their ligands, studies of receptor coupling to G proteins and cloning of cDNAs for these receptors.

1993 The molecular basis of smell and taste transduction. Wiley, Chichester (Ciba Foundation Symposium 179) p 76–87

G protein-coupled receptors are detectors for several diverse biologically relevant signals. The agents which these seven-transmembrane-domain receptors recognize and respond to include environmental stimuli such as light and odorants, neuromodulators such as dopamine and serotonin, hormones controlling events ranging from follicle stimulation to circadian rhythm, catecholamines and prostaglandins. Not surprisingly, the basic mechanisms by which these stimulants activate or block the receptors are of considerable scientific and medical significance. We have developed a new technique for studying these receptors, the basis of which comes from an unusual source—the ability of animals to change colour.

While the chameleon—which can alter its appearance from green to brown with variations in between— is probably the best known, the list of colour-changing animals includes reptiles, amphibians, fish and crustaceans (Noël 1985). Depending on the animal, not only is almost any skin colour attainable, but shades of colour and complex patterns can also be generated (Sumner 1911, Mast 1914). A striking example of the latter can be seen in the winter flounder, which camouflages itself by actively mimicking the terrain on which it lies. The cells responsible for these variations in appearance are called chromatophores, of which there are several varieties, including melanophores, erythrophores, xanthophores, leukophores and iridophores. Melanophores contain the brown to black pigment melanin, erythrophores and xanthophores contain red and yellow pigments, respectively, and leukophores and iridophores have ordered reflecting plates composed of crystalline guanine. Typically, these different cell types work together within organized structures called chromatophore units.

A change in appearance of an animal is effected by the controlled movement of pigment within chromatophores or rearrangement of guanine crystals within iridophores. For example, an anuran changes its hue from green to brown by dispersing the melanin-containing organelles—melanosomes—throughout the cytoplasm of its cutaneous melanophores (Bagnara & Hadley 1973). Conversely, a change from brown to green is achieved by the aggregation of pigment into small spheres within the melanophores. Depending on the animal, there are at least three means of activating pigment movement within chromatophores— neuronal excitation, hormonal influences and environmental agents. Examples of these stimuli are adrenergic neurons which synapse directly onto chromatophores, the pituitary peptide melanocyte-stimulating hormone (MSH) and light, respectively (Hogben & Slome 1931, van der Lek et al 1958, Abe et al 1969). In every case, the chromatophores receive the signals to re-deploy their pigment by way of cell surface receptors that signal via G proteins (Birnbaumer et al 1990, Gilman 1987, Stryer 1986).

The ease with which changes in the disposition of pigment within melanophores mediated by G protein-coupled receptors can be visualized prompted us to ask whether these cells could form the basis of a new method for detecting functional interactions between chemicals and recombinant forms of these receptors. To answer this question, we developed an immortalized line of frog melanophores. This review describes how the pigment cells can be used to evaluate functional interactions between chemicals and recombinant G protein-coupled receptors whose stimulation leads to activation of either adenylate cyclase or phospholipase C.

Results and discussion

The ability of melanophores to move their melanosomes either centripetally or centrifugally depends on their intracellular concentrations of cAMP (Lerner et al 1988, Abe et al 1969, Krause & Dubocovich 1991). For example, melatonin induces a fall in cAMP concentration and causes frog melanophores to aggregate their pigment granules into compact spheres located near the centre of the cells. On the other hand, treatment with MSH activates adenylate cyclase to bring about pigment dispersion. The ability to alter pigment position within the cells depends on the cAMP-dependent phosphorylation of a 57 kDa

FIG. 1. Macroscopic appearance of confluent melanophores in the pigment-aggregated and pigment-dispersed states. Two 10 cm tissue culture dishes, each containing approximately equal numbers of melanophores, were fixed and photographed after 30 min of incubation with 1 nM melatonin (left) or 100 nM melanocyte-stimulating hormone (right).

melanosome-associated protein (Rozdzai & Haimo 1986, Lynch et al 1986). A high concentration of intracellular cAMP leads to protein phosphorylation and pigment dispersion; a low concentration leads to protein dephosphorylation and pigment aggregation. Figure 1 shows how pigment distribution within melanophores provides a visual assay of intracellular cAMP levels (Potenza & Lerner 1992). Each 10 cm dish is covered with the same number of *Xenopus laevis* pigment cells. The plate on the left was treated with melatonin for 30 min, while the one on the right received MSH. This visual effect is associated with a fivefold difference in intracellular cAMP concentrations.

Because a rise above the natural baseline intracellular cAMP levels within melanophores can be visualized as the darkening of a collection of cells, we tried to see whether cells expressing a recombinant receptor that coupled to G_s would disperse their pigment upon exposure to an agonist for that receptor. To test this, we created a stable pigment cell line that expresses the human β_2-adrenergic receptor (β_2-AR) (Potenza et al 1992). The cells were plated into 96-well microtitre plates and treated with melatonin to pre-aggregate their pigment. Next, a range of concentrations of the β_2-AR-selective agonists,

A tool for investigating receptors 79

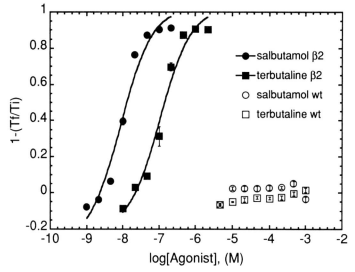

FIG. 2. Pigment translocation responses of melanophores exposed to β_2-selective adrenergic agonists. Wild-type (wt) cells and cells stably transfected with the β_2-adrenergic receptor growing in 96-well microtitre plates were incubated with melatonin for 30 min, then with a range of concentrations of a selective β_2 agonist, either salbutamol or terbutaline. The amount of light transmitted through each well was measured just after addition of the agonist (T_i) and 60 min later (T_f). Values represent the averages of triplicate samples; error bars depict the corresponding standard errors of the mean.

salbutamol and terbutaline, was added to the wells (Fig. 2). The β_2-AR-expressing cells responded to both chemicals by darkening in a dose-dependent manner, as quantitated by the equation $1 - T_f/T_i$ (where T_i is the amount of light transmitted by wells just after the addition of agonists and T_f is the amount of light transmitted at 30 min). The dose–response curves provide accurate EC_{50} values for the ability of agonists to activate the β_2-AR (Table 1). Table 1 also includes EC_{50} values for the β-AR-selective drugs fenoterol and metaproterenol. The rank order agonist properties for the four drugs are the same as when they are applied to the β_2-AR in other systems. This demonstrates the fidelity of the melanophore-based assay for assessing ligand interactions with a recombinant G protein-coupled receptor that functions via G_s.

While the contribution of cAMP to the regulation of melanosome positioning within melanophores is well established, we wondered whether any other second messengers might play a role. The first suggestion that this was the case came when we applied the phorbol ester 12-*O*-tetradecanoylphorbol-13-acetate (TPA) to the cells (Graminski et al 1993). Within a short time, the chemical induced a robust pigment dispersion. Though TPA can have a number of effects on cells, its ability to activate protein kinase C by mimicking the second messenger diacylglycerol (DAG) is particularly striking. This raised the possibility

TABLE 1 EC_{50} values for the action of β adrenergic receptor agonists on melanophores stably transfected with the β_2 adrenergic receptor

Ligand	Class	EC_{50} (nM)
Fenoterol	$\beta_2 \gg \beta_1$	7.59
Salbutamol	β_2	10.7
Metaproterenol	$\beta_2 \gg \beta_1$	38.0
Terbutaline	β_2	126.0

EC_{50} values were determined from dose–response curves using the pigment dispersion assay. Values represent the concentrations causing half-maximal increases in pigment dispersion.

that stimulation of a G protein-coupled receptor that normally activated phospholipase C could effect melanophore darkening by inducing the production of DAG (along with inositol 1,4,5-trisphosphate [$InsP_3$]). To test this hypothesis, we transfected melanophores transiently with a plasmid coding for a murine bombesin receptor. Three days after the introduction of the vector, the cells were tested with melatonin to aggregate their pigment, at which point they were exposed to bombesin and related peptides for 30 min. Figure 3 demonstrates that the bombesin receptor-expressing cells dispersed their pigment in a dose-related manner in response to bombesin, litorin and neuromedin B. Table 2 presents EC_{50} values for these three peptides, as well as for gastrin-releasing peptide (18–27) and [Leu^{14}, ψ13–14]bombesin, as calculated from the dose–response curves. As was the case with the G_s-linked human β_2-AR, the rank order EC_{50} values for bombesin and the related peptides appropriately reflected their known properties when applied to cells expressing the bombesin receptor. High-performance liquid chromatography analysis detected agonist-induced generation of $InsP_3$ without a rise in cAMP (measured by radioimmunoassay), thus confirming that the bombesin receptor-mediated pigment dispersion results from the activity of phospholipase C and not adenylate cyclase.

The melanophore-based pigment dispersion assay appears to be generally applicable for investigating functional interactions between chemicals and any G protein-coupled receptor which, in its native environment, utilizes adenylate cyclase or phospholipase C in its signal transduction cascade. So far, eight such receptors have been expressed in the melanophores and in every case application of an agonist has led to appropriate effector coupling and dose-dependent pigment dispersion. The melanophores can also be used to investigate ligand interactions with G_i-linked receptors, as confirmed by experiments with human α_2 adrenergic, chick muscarinic 4 and human dopamine 2 and 3 receptors (M. R. Lerner, M. N. Potenza, G. F. Graminski, unpublished results). Because the system performs strongly in 96-well plates or other macroscopic

A tool for investigating receptors

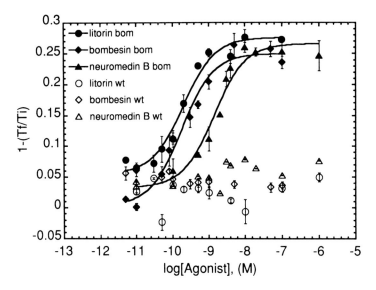

FIG. 3. Pigment translocation responses of melanophores exposed to bombesin or the related peptides, litorin and neuromedin B. Wild-type (wt) cells and cells transiently transfected with the bombesin receptor (bom) were grown in 96-well microtitre plates for three days after electroporation. Cells were pretreated with melatonin (120 min) then incubated with bombesin, litorin or neuromedin B. The amount of light transmitted through each well was measured just after addition of the agonist (T_i) and 30 min later (T_f). Values represent the averages of four samples; error bars depict the corresponding standard errors of the mean.

TABLE 2 EC_{50} values for the action of bombesin receptor agonists upon melanophores transiently transfected with the bombesin receptor

Ligand	EC_{50} (nM)
Bombesin	0.18
Litorin	0.20
Gastrin-releasing peptide (18–27)	0.23
Neuromedin B	1.4
[Leu$^{14}\psi$13–14] Bombesin	13.0

EC_{50} values were determined from dose–response curves using the pigment dispersion assay. Values represent the concentrations causing half-maximal increases in pigment dispersion.

configurations, it is ideally suited for large-scale investigations of interactions between chemicals and G protein-coupled receptors.

Once the utility of the melanophore-based pigment dispersion assay for rapidly and simultaneously assessing functional interactions between numerous chemicals and receptors using a standard format was established, we wanted to know whether the system could also be used for individual cells (McClintock et al 1993). If pigment movement in response to activation of a G protein-coupled receptor was a stable and reproducible event for a single melanophore, not just as an average for thousands of cells, an assay for cloning new receptors, as well as a powerful new tool for investigating the molecular basis of receptor activation, could be envisioned. In Fig. 4, the method that we used to test this idea is presented. Frame A shows a digitized video image of cells that were transiently expressing the murine bombesin receptor and had been treated with melatonin to aggregate their pigment. Frame B shows the same field of cells following incubation with bombesin for 30 min. Frame C presents the subtracted image A – B and frame D shows this same picture with a false red colour that has been superimposed on the original image seen in A. The pigment dispersion responses of many bombesin receptor-expressing cells can be monitored simultaneously. In practice, we have found that the regulation of pigment position within melanophores is so tight that pigment movement in up to 10 000 cells can be tracked within a single microscopic field. With the addition of a scanning stage, activation of G protein-coupled receptors can be assessed in over 200 000 single cells at the same time. From a practical standpoint, this means that cells expressing many sequence variants of a receptor, such as those generated by site-directed mutagenesis, can be compared rapidly for their ability to respond to different ligands.

The ability to follow pigment movement in hundreds of thousands of cells in parallel means that the system can also be used to isolate cDNA clones for G protein-coupled receptors. Figure 5 shows the effect of applying a β_2-AR-specific agonist to melanophores that have been transfected with plasmid DNA coding for the human β_2-AR which has been diluted on a colony to colony basis against plasmids containing a human liver cDNA library. Frame A shows the response of cells transfected with plasmids from a 1:100 dilution of the β_2-AR into liver cDNA, while frames B and C show the responses at dilutions of 1:1000 and 1:7000, respectively. Frame D shows a control field of cells. In practice, plasmids from at least 10 000 colonies can be screened for a clone of interest using a single field of transfected melanophores. In addition, if both an agonist and an antagonist are available for a receptor of interest, specific signals can be visualized for both activation and inactivation of the receptor and represented as two distinct false colours (data not shown). The intersection of these images generates a third colour and allows a clone coding for a G protein-coupled receptor to be detected against a background of over 100 000 others.

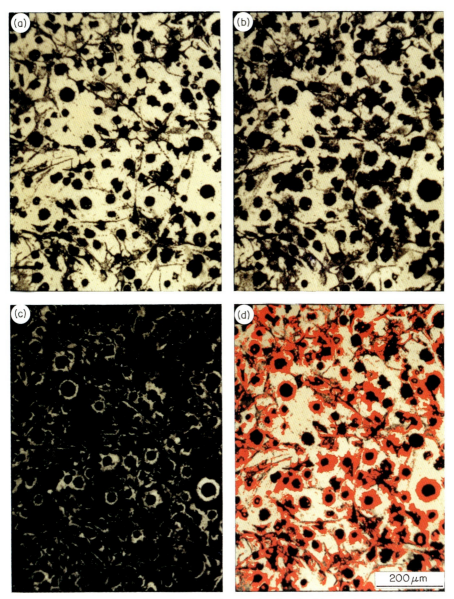

FIG. 4. Analysis of transfected melanophores stimulated with agonist. Melanophores were transfected with bombesin receptor plasmid DNA diluted 1:10 with DNA from plasmids containing no cDNA inserts. (A) A video image taken after pretreatment with melatonin and immediately after addition of 100 nM bombesin. (B) The same field of cells 30 min later. (C) Subtracted image, A − B. Grey scale values grade from no change (black pixels) to saturating changes (white pixels). (D) Red bitmap of pixels which became more than 10% darker overlaid on the image field shown in (A). Complete image fields are displayed.

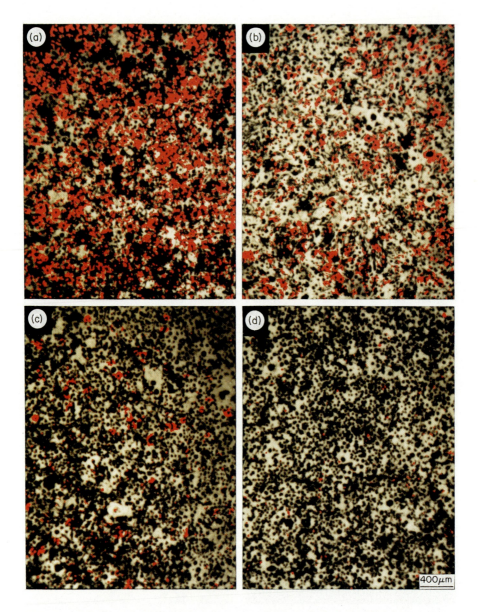

FIG. 5. Melanosome dispersion in response to activation of the β_2-adrenergic receptor in recombinant melanophores. Red pixels represent the change in melanosome dispersion caused by 100 nM salbutamol. (A) Response to salbutamol in melanophores transfected with plasmid DNA from a dilution of one bacterial colony containing β_2 receptor plasmid into 100 colonies containing a human liver cDNA library. (B) Response at a dilution of 1:1000. (C) Response at a dilution of 1:1700. (D) Lack of response to salbutamol in melanophores transfected with plasmid DNA prepared from the liver cDNA library. 25% of each image field is displayed.

In summary, we have developed a new means of rapidly detecting functional interactions between chemicals and G protein-coupled receptors. Predicted uses of the system include structure–function analyses of both G protein-coupled receptors and their ligands, studies of receptor coupling to G proteins and cloning of cDNAs for these receptors.

Acknowledgements

We thank Mrs Alison Roby-Shemkovitz and Mrs Lina Golovyan for excellent technical assistance. This work was supported by the Burroughs Wellcome Fund, G. D. Searle & Company, the Office of Naval Research and the Howard Hughes Medical Institute.

References

Abe K, Butcher RW, Nicholson WE, Baird CE, Liddle RA, Liddle GW 1969 Adenosine 3',5'-monophosphate (cyclic AMP) as the mediator of the actions of melanocyte stimulating hormone (MSH) and norepinephrine on the frog skin. Endocrinology 84:362–368

Bagnara JT, Hadley ME 1973 The nature of pigmentation. Prentice Hall, New York, p 4–45

Birnbaumer L, Abramowitz J, Brown AM 1990 Receptor-effector coupling by G proteins. Biochim Biophys Acta 1031:163–224

Gilman AG 1987 G proteins: transducers of receptor-generated signals. Annu Rev Biochem 56:615–649

Graminski GF, Jayawickreme CK, Shemkovitz-Roby A, Lerner MR 1993 Pigment dispersion in frog melanophores can be induced by a phorbol ester or stimulation of a recombinant receptor that activates phospholipase C. J Biol Chem 266:5957–5964

Hogben L, Slome D 1931 The pigmentary effector system. VI. The dual character of endocrine co-ordination in amphibian colour change. Proc R Soc Lond Ser B Biol Sci 108:10–53

Krause DN, Dubocovich ML 1991 Melatonin receptors. Annu Rev Pharmacol Toxicol 31:549–568

Lerner MR, Reagan J, Gyorgyi T, Roby A 1988 Olfaction by melanophores: what does it mean? Proc Natl Acad Sci USA 85:261–264

Lynch TJ, Wu BY, Taylor JD, Tchen TT 1986 Regulation of pigment organelle translocation. J Biol Chem 261:4212–4216

Mast SO 1914 Changes in shade, color and pattern in fishes, and their bearing on the problems of adaptation and behavior, with special reference to the flounders Paralichthys and Ancylopsetta. Bull US Bur Fish 34:173–238

McClintock TS, Graminski GF, Potenza MN, Jayawickreme CK, Shemkovitz-Roby A, Lerner MR 1993 Functional expression of recombinant G-protein coupled receptors monitored by video imaging of pigment movement in melanophores. Anal Biochem 209:298–305

Noël PY 1985 Chromatophores direct photoreactivity. In: Bagnara J, Klaus SN, Paul E, Schartl M (eds) Pigment cell 1985. Biological, molecular and clinical aspects of pigmentation. University of Tokyo Press, Tokyo (Proc XIIth Int Pigm Conf, Giessen, Germany) p 219–227

Potenza MN, Lerner MR 1992 A rapid quantitative bioassay for evaluating the effects of ligands upon receptors that modulate cAMP levels in a melanophore cell line. Pig Cell Res 5:372–378

Potenza MN, Graminski GF, Lerner MR 1992 A method for evaluating the effects of ligands upon G_s protein coupled receptors using a recombinant melanophore-based bioassay. Anal Biochem 206:315–322

Rozdzai MN, Haimo LT 1986 Bidirectional pigment granule movements of melanophores are regulated by protein phosphorylation and dephosphorylation. Cell 47:1061–1070

Stryer L 1986 Cyclic GMP cascade of vision. Annu Rev Neurosci 9:87–119

Sumner FB 1911 The adjustment of flatfishes to various backgrounds. J Exp Zool 10:409–504

van der Lek, De Heer J, Burgers CJ, Van Oordt GJ 1958 The direct reaction of the tailfin melanophores of *Xenopus* tadpoles to light. Acta Physiol Pharmacol Neerl 7:409–504

DISCUSSION

Breer: Is there a particular reason why you used photosensitive melanophores, rather than the chromatophores or the iridophores for this assay?

Lerner: We used the melanophores because they were the only source of cells from which we were able to create an immortalized line. The iridophores in frogs don't move their guanine crystalline plates, although in some fish they do and it's pretty spectacular. In the frog, the only cell that can translocate its pigment is the melanophore.

Bargmann: How long does the stimulus have to be present for the melanophores to change colour? Do you get desensitization of the response to any of the transmitters?

Lerner: It depends on ligand concentration and time. It takes 30 min or so to get a full response, but we can see a response within 5 min. However, we can't see a response in milliseconds, which could be a problem for using this assay to study olfactory receptors.

In terms of desensitization, no receptor that we have put into the cells for which we know the ligand has ever shown any desensitization. A number of endogenous receptors do show desensitization, such as the β adrenergic, the endothelin C and the serotonin receptors. I suspect that it's only a matter of time before we put in an exogenous receptor and one of the enzymes in the cell causes desensitization.

Hatt: Desensitization is determined by the relationship between the peak and steady-state responses after the rapid application of agonist. Only rarely will you get a fully desensitizing response without any equilibrium activity, as in the case of some synaptic glutamate receptors (Dudel et al 1990). You normally see a residual activity. I think, therefore, that dose–response curves of a desensitizing system only show that the drug is effective and the effect is specific.

Lerner: It's only the EC_{50} values that you can determine using these assays. You can't use them to determine binding constants. Also, they are transient assays—it would be pointless trying to measure equilibrium constants with them.

Breer: The receptors you express activate the endogenous G_s, G_i and G_q proteins—is there any G_o in the melanophores?

Lerner: There is a G_o, although it's not easy to determine what is coupling to what. The only $G\alpha$ subunit that we haven't found so far is transducin.

Lindemann: Is the melatonin response coupled to phosphodiesterase (PDE)?

Lerner: Yes; indirectly. It inhibits adenylate cyclase and, like PDE, lowers cAMP, presumably through G_i.

Carlson: Have you used antibodies against any of the receptors that are expressed, in order to block or possibly stimulate a response?

Lerner: No.

Buck: You did some experiments using a series of ethanols as odorants. Did you try any other odorants in this assay system?

Lerner: Yes, we also used more standard odorants such as citral or β-ionone.

Buck: Do the cells respond to them?

Lerner: The concentration of odorants that Doron Lancet and Sol Snyder had used in their studies—about 100 μM—was sufficient to cause pigment dispersion in melanophores, but we didn't follow up on that work.

Reed: Have you expressed olfactory receptors in these cells?

Lerner: We made one attempt at expressing olfactory receptors: we used PCR to obtain a few clones, but these were not checked for fidelity. After transfecting the cells, we screened against approximately 200 odorants; we did not see any effects. There are all sorts of reasons these experiments could have failed.

Lancet: What happened to the original basal responses of these cells, which might occur even without olfactory receptors?

Lerner: We screened at 1 μM concentration of odorants. We would not have been able to screen at 100 μM, because the cells would have given non-specific responses to the odorants.

Firestein: So besides the olfactory receptors, have any other attempts to express receptors failed?

Lerner: None have failed where we have known the ligand.

Reed: Have you been able to screen complex libraries to pull out clones for receptors?

Lerner: We did try that when we were looking for a glucagon receptor in the liver; after screening about a million clones we didn't see anything, so we decided not to continue trying. Liver cells may not be a good choice: because they're making so many things, the percentage of any one message would be small.

Reed: In a complex library, can you pull out a clone for the β2-adrenergic receptor from a tissue that is known to express it?

Lerner: If we dilute a known plasmid encoding a β2-adrenergic receptor 10 000-fold into a complex library of expression clones, such as an olfactory bulb library or a liver library, we can certainly see the β2 receptor against that background.

Reed: But wouldn't you expect the expression frequency of the β2-adrenergic receptor to be at least one in a few tens or hundreds of thousands of clones?

Lerner: You might.

Reed: Do you see anything without the addition of the exogenous β2 plasmid?

Lerner: No; if you just have an olfactory bulb cDNA library with 10 000 clones, you won't detect the expression of β2-adrenergic receptor.

Margolis: Your estimate then is that about one in 10 000 would be the threshold of the assay.

Lerner: It's the limit if you only have an agonist. If you have an antagonist for the receptor as well, then it's one in 100 000.

Getchell: Is it the lack of an antagonist that's the rate-limiting step in this case?

Lerner: It depends on the sensitivity you want to screen with.

Reed: If you screened a million clones, would you just get, say, nine false positives?

Lerner: No. You are limited to screening 10 000 cells at a time, because of the nature of the assay. However, with a scanning stage, one can effectively screen about 200 000 cells at a time.

Reed: So the limit is the size of field you are looking at?

Lerner: No; the limit is the false positive rate per 10 000 cells.

Reed: Which is one per 10 000?

Lerner: One or two. You could screen a bigger field than 10 000 cells, but it wouldn't actually help you, because that's the limit at which you start to see the false positives.

Carlson: Have you tried using cell sorting in order to purify responding cells?

Lerner: No.

Margolskee: Are these stable transfectants?

Lerner: The only stable line that we've created is for the β2-adrenergic receptor. We tried creating a stable line that expresses the dopamine 2 receptor and failed to do so; we are working on that now.

Margolskee: So screening of expression libraries is based upon transient transfection.

Lerner: Yes; the whole assay is.

Margolskee: What efficiency of transfection do you get?

Lerner: We use electroporation, so, as you might expect, transfection efficiency varies from person to person and from day to day: the range is between 30 and 100%.

Margolis: If I understand correctly, in comparison with attempts in my lab to use functional expression in *Xenopus* oocytes to clone olfactory receptors, if this assay worked, it would certainly be more efficient. However, it still sounds like a lot of work if you have to screen pools of 10 000 cells.

Lerner: If this assay would work for an olfactory receptor, you should be able to use it, but you're right, it would still be a lot of work.

Reed: The problem you have with having to use 10 000 clones would be greatly aided by a receptor library. Have you thought about making such a library?

Lerner: We have; the issue has been one of manpower and where to focus. We have concentrated our efforts on developing the technology. That's why we haven't done all that many biochemical experiments. The only real science has been the discovery of the endothelin C-specific receptor of these cells, the demonstration that the second messenger pathway based on diacylglycerol can activate pigment dispersion and functional characterization of the human dopamine 3 receptor.

Reference

Dudel J, Franke C, Hatt H 1990 Rapid activation, desensitization and resensitization of synaptic channels of crayfish muscle after glutamate pulses. J Biophys 57:533–545

General discussion I

Caprio: Are all olfactory receptors G protein-linked or are there direct ligand-gated channels?

Margolis: Labarca et al (1988) published a paper suggesting that there were direct ligand-gated channels.

Teeter: Vodyanoy & Murphy (1983) also described K^+ channels that were regulated directly by odours.

DeSimone: These were channels resulting from fusing vesicles from homogenized olfactory epithelium with bilayer lipid membranes. The mean open state time was nearly 30 s (Vodyanoy & Murphy 1983). This is longer than normally observed *in situ* with patch-clamp recording methods. Whether these channels truly represent native odorant-gated channels is still an open question.

Teeter: Additionally, M. Zviman at Michigan State University has shown activation of what appear to be K^+ channels by diethyl sulphide and two analogues using reconstituted frog olfactory cilia membranes (Zviman 1993).

Margolis: Again, this was with reconstituted lipid bilayers. I understand that there has been some criticism of working with these bilayers in terms of the nature of the background and problems with non-specific responses: are these major concerns?

Teeter: Not if you do controls. I don't think they're major problems if you apply all of your stimuli at the appropriate concentration to the bilayers before membrane vesicles are added to make sure that the lipid bilayer is not responding.

van Houten: Concerning the specificity of individual receptor cells, Linda, you indicated that you would expect a given olfactory receptor cell to express one receptor protein. Would you therefore expect only one signal transduction pathway to be expressed in those cells?

Buck: I like the idea of one gene expressed per cell, but there could be more than one.

Reed: I guess Judith van Houten is asking whether an individual cell has one receptor that couples, for example, through the G_{olf} cyclase pathway and a second receptor that couples through, say, the $InsP_3$ pathway.

Ache: Our data in the lobster clearly indicate that one olfactory receptor cell is capable of expressing at least two types of receptor, one coupled through the adenylate cyclase pathway and one through the $InsP_3$ pathway. In the lobster (as is now known to occur in several vertebrate species) different odours can excite and inhibit the same cell. Excitation is mediated by the $InsP_3$ pathway, while inhibition is mediated by the cyclase pathway (Michel & Ache 1992, Fadool & Ache 1992). Whether there is more than one receptor for each of the pathways in any one cell, we do not know. However, an odour

General discussion I

that inhibits one cell can excite another and vice versa, indicating that there are minimally two different receptors for any one odour in the population.

van Houten: Linda, do you envisage that all the members of the same receptor subfamily couple through the same signal transduction pathway? An ancillary question is: within the regions that you see that reflect subfamily expression, is there a substructure? Do individual members of the same family have their own distribution patterns within a region?

Buck: It's likely that all the members of a subfamily would couple to the same G protein because their sequences are so similar.

Firestein: We know that the third cytoplasmic loop in these receptors is essential for coupling to G proteins. Are there differences in this cytoplasmic loop that might be correlated with known receptors coupling to other systems?

Buck: No, there's actually a lot of sequence variability in this region of the odorant receptors.

Reed: But in the original Buck & Axel paper (1991), you implied that the structures of the 3rd cytoplasmic loop fall into two classes.

Buck: That's what we thought, but since then we have looked at many different receptors and the more we have looked at, the more diversity of structure we have seen in that region.

Firestein: Is that true for seven-transmembrane-domain receptors in general? For example, can you look at $\beta1$ versus $\beta2$ adrenergic receptors and look at the third cytoplasmic loop and say: here's the difference between G_s and G_i?

Reed: You can correlate structure with function, for instance, between the various muscarinic receptors, as these are highly homologous to start with. But it seems that you can't compare the putative coupling regions of muscarinic and dopamine receptors, or β adrenergic and any of the peptide hormone receptors, and predict that they are going to activate adenylate cyclase.

Margolskee: Randy, have you gone back and looked for G_q or G_q-like G proteins that show increased expression in olfactory epithelium?

Reed: Jennie Davis, in my laboratory, spent about a year and a half on that project, looking initially for $G\alpha$ subunits and later γ subunits, in analogy with the visual system. She pulled out and sequenced about 180 $G\alpha$ subunits: including G_{11}, G_q, G_2, every known $G\alpha$ subunit, but no novel olfactory-specific ones. We pulled out 1–2 new $G\gamma$ subunits, but their expression is not confined to olfactory neurons.

Margolskee: But the G protein doesn't have to be novel or olfactory specific; it just has to have elevated expression in olfactory neurons.

Reed: G_o clearly fits that bill and was one of the first five that we found. It is enriched in olfactory neurons and is present in the olfactory cilia. It is also pertussis toxin sensitive, which is consistent with Heinz Breer's experiments on rat cilia (this volume: Breer 1993).

From immunocytochemical localization studies using anti-G_{olf} or anti-type III adenylate cyclase antibodies and from looking at cyclic nucleotide-activated channels in isolated rat olfactory neurons, it is our impression that every olfactory neuron expresses the cAMP pathway.

There is a recent report that one of the phospholipase C β subunits can be activated by G protein $\beta\gamma$ subunits (Park et al 1993). This raises the possibility that there's a bifurcating pathway in which $G_{olf}\alpha$ activates adenylate cyclase while the released $\beta\gamma$, perhaps in conjunction with G_o, activates phospholipase C.

Lancet: This would go against the finding by Heinz Breer that the activation of the two different pathways is mutually exclusive.

Breer: All the odorants we have analysed so far at a concentration of 1 μM or less elicited either a cAMP or an $InsP_3$ response. None of them activated both systems.

Lancet: I am aware of that subtlety here. But if what Randy Reed is saying is right, any odour that activates adenylate cyclase will lead to the dissociation of the $\beta\gamma$ subunit, so you should see activation of the other pathway as well.

Breer: The reports demonstrating that certain types of phospholipase C are activated by $\beta\gamma$ subunits (Camps et al 1992, Berstein et al 1992, Park et al 1993) indicate that only specific $\beta\gamma$ subunits of certain G proteins activate the enzyme. On the activation of, for instance, a β-adrenergic receptor, it has never been seen that the α subunit of G_s activates the adenylate cyclase and at the same time the subunits stimulate phospholipase C. Moreover, there is no indication that a $\beta\gamma$-activated system operates in olfactory cells.

Ache: Our laboratory, in collaboration with Heinz Breer and using the same rapid quench-flow methodology he used in his original studies, has recently shown that the same odour can stimulate the production of both cAMP and $InsP_3$ in the outer dendrites of lobster olfactory receptor cells. The two odours that we have tested in detail were selective, but not exclusive, for one pathway or the other. The selectivity of the two odours nicely parallels the frequency with which they are found to be excitatory or inhibitory in our physiological experiments. The selectivity persists in a concentration-dependent manner down to nanomolar concentrations (unpublished results).

Ronnett: We work with two systems, one of which is a primary culture system of olfactory neurons in which we look at not subsecond, but at 1–60 s activation of adenylate cyclase. With a series of odorants, we found that every odorant we tested in this whole-cell system stimulated cAMP production as well as $InsP_3$ production, but with different potencies (Ronnett et al 1993). Some odorants are slightly better cAMP activators (10–100-fold difference in potencies), while others are better at activating $InsP_3$, but every odorant affected both systems.

Ache: That's consistent with what we found in the lobster.

Breer: Mike Lerner said that if he uses high enough concentrations, even melanophores respond to every odour!

Ronnett: We have used nanomolar concentrations and found activation of these systems. We've just recently started to do these experiments on a subsecond scale; we've not looked at the InsP$_3$ responses yet, but all the odorants we have tested stimulate cAMP production (Ronnett et al 1993).

Margolis: There appears to be a real discrepancy between the observations in different labs. Is this really a problem or is it a function of differences in stimuli, species and methodology?

Ronnett: Some of the odorants that we have used are the same as those that Dr Breer has used, some are different. For example, we have both looked at isovaleric acid. In our whole-cell system (Ronnet et al 1993), isovaleric acid stimulates both InsP$_3$ and cAMP production. We have also used the rapid stop-flow system and similarly have found that all odorants tested stimulate both cAMP and InsP$_3$ production. In the cilia, 10 nM isovaleric acid stimulates cAMP production. We usually test odorants at eight or nine concentrations, because the dose–response curves are complicated and usually biphasic. Thus, testing at just one concentration might yield misleading results.

Hatt: In our experiments on salamander olfactory neurons, we never saw InsP$_3$ responses, but we didn't look very carefully because our main interest was in the cAMP-activated channel.

Firestein: John, when you said you saw an InsP$_3$ response in catfish neurons, were you looking at current or voltage clamp?

Teeter: Both. Under current clamp, about 90% of the catfish neurons we examined displayed a transient 20–40 mV depolarization when 10 μM InsP$_3$ was included in the whole-cell pipette. Under voltage clamp, 20–70 pA currents were observed at −60 mV. Similar depolarization and currents were observed in rat olfactory receptor cells, but they tended to be sustained rather than transient responses.

Firestein: We've only looked at voltage clamp: if the current is only a few pA, which would be enough to change Ca^{2+} concentrations significantly, we couldn't see it. But we would see a voltage deflection. These cells have a high resistance, so a 2 or 3 pA current would give a significant voltage change, whereas one wouldn't see the current. This could be part of the reason that we don't see it and you do.

Ache: We find InsP$_3$ receptors—actually two different types of channels—in the plasma membrane of cultured lobster olfactory cells (Fadool & Ache 1992). These channels appear to mediate the odour-evoked inward current in this instance, which can be quite large in the cultured cell—up to several tens of pA—and readily detectable. As you say, though, even a few pA could functionally depolarize a tight cell. More relevant to the general issue here is that during this past week, in collaboration with Hans Hatt, we were able to find InsP$_3$ receptors of very similar slope conductance and pharmacology in the outer dendrite of cells *in vivo*. This suggests that the InsP$_3$ pathway we characterized in the cultured cell is indeed functional *in vivo*, together with the cyclic nucleotide pathway that we earlier established *in vivo* (Michel & Ache 1992).

Firestein: Nobody has found the receptor in invertebrates yet. We could question whether they have this receptor at all, or whether there is a family of receptors that is as yet undiscovered.

Buck: I think there are some G protein-coupled receptors that will couple to several different kinds of G protein.

Reed: For example, α2-adrenergic receptor activates adenylate cyclase, phospholipase C and modulates ion channels.

Ache: In the case of olfaction, however, the odour specificity of the two pathways in a given cell would appear to be different.

DeSimone: There can be more than one type of receptor for ligands on a cell. For example, pancreatic acinar cells have receptors for secretin, cholecystokinin, acetylcholine and other ligands that cause secretion (Chey 1980).

Margolis: Neurons express receptors for lots of different kinds of ligands and that's all these olfactory receptor cells are—specialized neurons. The outside world is effectively the presynaptic side of these neurons. Therefore it shouldn't be surprising that they may each express a number of different receptors. I find the possibility that each cell only expresses one or two kinds of receptor far more intriguing, because that would make them different from the majority of neurons.

Lancet: Many neurons express only a limited number of receptors of a given kind; maybe one peptide receptor and one neurotransmitter receptor. It's true that in general cells express a lot of receptors, but when you go into a given class, there is a lot of exclusion going on. Immune cells, of course, are the most prominent example of this.

Kinnamon: I'm not sure why we expect an olfactory neuron to express just one type of receptor, because if you look at all the classic intracellular physiological studies, neurons appear to be broadly tuned to odours (Gesteland et al 1965).

Buck: This raises another question about odorant concentration. Presumably, these receptors could be highly specific at low concentrations, but much less specific at higher concentrations. Experiments using high concentrations of odorants may not reflect the physiological situation at all.

Siddiqi: We have mapped the single unit responses of the olfactory hairs on the antenna of *Drosophila* to six test odours (Siddiqi 1987). One can classify the neurons into types that respond to one, two, three or more chemicals in fixed combinations. Not all possible combinations of responses are found. Moreover, the hairs with different types of neurons are not distributed randomly. They occupy predictable positions in continuous zones or islands on the antennal surface. This suggests that there is a well-defined mechanism for specifying the receptors on the neurons as well as the topographic distribution of the chemoreceptor types.

Caprio: There's good evidence from electrophysiological cross-adaptation (Caprio & Byrd 1984) and binding (Bruch & Rulli 1988) studies in the channel catfish (*Ictalurus punctatus*) that acidic, basic and neutral amino acids bind to different receptor sites. Whole-cell patch-clamp recordings from six isolated olfactory receptor neurons in the channel catfish (T. Ivanove & J. Caprio, unpublished results) indicated that these neurons respond to all three classes of amino acids, suggesting the activation of different amino acid receptor types in each of these cells. The most effective odorant, based on the magnitude of the inward current, was unique for each of the six receptor neurons tested.

Firestein: Although it's a little dangerous to compare olfaction with the photoreceptor system, it's interesting that in photoreception, the three pigments have a relatively wide tuning spectrum and these spectra overlap (Baylor et al 1987). Using overlapping receptors is in some ways a better system for defining a specific colour than a system that has a narrow peak.

Lancet: The real problem is that much of what has been said in the last five minutes addresses the question of responses to odorants in given cells; what we really want to get at is the question of receptor molecule expression. As long as we don't really know the spectrum of each given receptor molecule with respect to various odorants, we may assume that some of these receptors are pretty broadly tuned, as Stuart Firestein has said. Eventually this will be resolved when the actual receptors are identified by immunolocalization, *in situ* hybridization and single-cell PCR.

Ache: From a molecular point of view, is there any reason you could not have broadly tuned receptors? Could the multigene family encode broadly tuned receptors with partially overlapping response spectra?

Buck: I don't think there's any evidence one way or the other. There are a lot of receptors, so there's the possibility for high specificity, but it doesn't have to be that way. Indeed, we can smell more odorants than we have receptors, so it is unlikely that each receptor recognizes only a single odorant.

Margolis: All ligand–protein interactions are represented by a spectrum; nothing is absolute. This is true for antibodies and it's true for the cytochrome P_{450} complex, which can deal with essentially every odour molecule and comprises 50–100 enzymes, each of which is broadly tuned, with overlapping specificities (Margolis & Getchell 1991).

Bartoshuk: The human experience with odours is described quite differently than some of you working on receptors might think. It's very difficult for a human to learn to name a new odour. Master perfumers practice for months to do this. The learning of and the memory for odours follows a model of what we call 'holistic learning'. I can give you an analogy from vision: if you think about a big blobby shape that doesn't have features that are particularly noticeable, we learn and forget that type of shape perception in much the way we learn and forget odours. It might be worthwhile for some of you to look into the cognitive psychology of this, because odour learning in humans does not look like an analytical system with many individual receptors that add up in an analytical fashion. Taste is an analytic sense, but for the most part olfaction is not.

Reed: It's important to caution against the impression that we actually know something about this entire receptor family. Even Linda Buck's experiments, which are probably the biggest tour through this receptor family, represent only maybe 20% of all the receptors. We've made an antibody against the C-terminus of one of the receptor subfamilies we have looked at. We would expect it to be quite specific—this antibody only represents the cloned receptor to which it was raised, but it sees about 10–20% of all the receptor neurons in the epithelium. The notion that there may be broadly expressed receptors and highly restricted receptors is a real possibility. We just haven't stumbled upon the general receptors in Linda Buck's tour or in the few we've done (R. Reed, unpublished results). It's possible that we haven't put all the technologies together yet and that there is going to be some fraction of the receptors, say 5%, that is broadly expressed in a lot of the cells. This could be missed in the kinds of experiments we've done.

Margolis: We are also assuming that these molecules are all odour receptors and that they serve no other function in the cells. I don't know whether there's any evidence for or against the possibility that they may also be involved, not as odour receptors, but in targeting back into the olfactory bulb.

Buck: Although Randy Reed's group and our group have seen evidence for odorant receptors on olfactory axons, it's hard to imagine that these receptors could be used for targeting synapse formation in the bulb.

Carlson: How many of them have been looked at for expression early in development?

Buck: Nef et al (1992) looked at one.

Breer: We have analysed the expression of several receptor subtypes at various stages of development in rat. Three or four receptor types were found to be already expressed at embryonic day 14 (E14). So the onset of odorant receptor expression in rat is even earlier than E14.

Buck: We've seen them expressed in the mouse olfactory epithelium as early as E12.

Carlson: Parmentier et al (1992) found expression of olfactory receptors in testis. What fraction of the receptors that have been looked for in testis are actually expressed there?

Reed: Our lab has looked at four receptors. We made specific oligonucleotides to 5' and 3' untranslated regions and carried out a PCR experiment with RNA templates from a spectrum of tissues. When the products were hybridized with a radiolabelled receptor probe, we saw intense hybridization in olfactory tissue, but no evidence of expression in testis or any other tissue. Parmentier et al (1992) identified olfactory receptors in testis by Northern blotting: this is not a very sensitive technique and the probes they used were not specific for one member of the family.

Buck: It's hard to interpret their experiments because they didn't say how much RNA they used, whether it was polyA$^+$ RNA or total RNA, or what the conditions of hybridization were.

Lancet: The question of tissue specificity of a particular clone or a group of olfactory receptors is compounded by DNA contamination and no matter how much you try to control for it, it is still a problem.

Buck: In PCR analyses, DNA contamination is a serious problem even if polyA$^+$ RNA is used. The contamination is usually, but not always, eliminated by pretreatment of the RNA with DNase. It is absolutely critical that the appropriate control be performed to exclude DNA contamination as the source of amplified PCR product. It has been shown that the PCR product is *only* obtained when reverse transcriptase is used. Unfortunately, although Parmentier et al (1992) state that they pretreated their RNA with DNase, they don't mention the control experiment, so it's not clear whether the molecules they amplified derive from RNA or the contaminating DNA.

References

Baylor DA, Nunn BJ, Schapf JL 1987 Spectral sensitivity of cones of the monkey *Macca fascicularis*. J Physiol 390:145-160

Berstein G, Blank JL, Jhon D-Y, Exton JH, Rhee SG, Ross EM 1992 Phospholipase C-β1 is a GTPase-activating protein for $G_{q/11}$, its physiologic regulator. Cell 70:411-418

Breer H 1993 Second messenger signalling in olfaction. In: The molecular basis of smell and taste transduction. Wiley, Chichester (Ciba Found Symp 179) p 97-114

Bruch RC, Rulli RD 1988 Ligand binding specificity of a neutral L-amino acid olfactory receptor. Comp Biochem Physiol B Comp Biochem 91:535-540

Buck L, Axel R 1991 A novel multigene family may encode odorant receptors: a molecular basis for odor recognition. Cell 66:175-187

Camps M, Hou C, Sidiropoulos D, Stock JB, Gierschik P 1992 Stimulation of phospholipase C by guanine nucleotide binding protein $\beta\gamma$ subunits. Eur J Biochem 206:821-831

Caprio J, Byrd RP 1984 Electrophysiological evidence for acidic, basic, and neutral amino acid olfactory receptor sites in the catfish. J Gen Physiol 84:403-422

Chey WY 1980 Gastrointestinal hormones and pancreatic, biliary and intestinal secretions. In: Glass J (ed) Gastrointestinal hormones. Raven Press, NY, p 565-586

Fadool DA, Ache BW 1992 Plasma membrane inositol 1,4,5-trisphophate activated channels mediate signal transduction in lobster olfactory receptor neurons. Neuron 9:907-918

Gesteland RC, Lettvin JY, Pitts WH 1965 Chemical transmission in the nose of the frog. J Physiol 181:525-559

Labarca P, Simon SA, Anholt RB 1988 Activation by odorants of a multistate cation channel from olfactory cilia. Proc Natl Acad Sci USA 85:944-947

Margolis FL, Getchell TV 1991 Receptors: current status and future directions. In: Muller PM, Lamparsky D (eds) Perfumes, art, science and technology. Elsevier Science Publishers, Amsterdam, p 481-498

Michel WC, Ache BW 1992 Cyclic nucleotides mediate an odor evoked potassium conductance in lobster olfactory receptor cells. J Neuro Sci 12:3979-3984

Nef P, Hermans-Borgmeyer I, Artieres-Pin H, Beasley L, Dionne V, Heinemann SF 1992 Spatial pattern of receptor expression in the olfactory epithelium. Proc Natl Acad Sci USA 89:8948-8952

Park D, Jhon D-Y, Lee C-W, Lee K-Y, Rhee SG 1993 Activation of phospholipase C isozymes by G protein $\beta\gamma$ subunits. J Biol Chem 268:4573–4576

Parmentier M, Libert S, Schiffmann S et al 1992 Expression of members of the putative olfactory receptor gene family in mammalian germ cells. Nature 355:453–455

Ronnett GV, Cho H, Hester LD, Wood SF, Synder SH 1993 Odorants differentially enhance phosphoinositide turnover and adenylyl cyclase in olfactory receptor neuronal cultures. J Neurosci 13:1751–1758

Siddiqi O 1987 Neurogenetics of olfaction in *Drosophila melanogaster*. Trends Genet 3:137–142

Vodyanoy V, Murphy RB 1983 Single-channel fluctuations in bimolecular lipid membranes induced by rat olfactory epithelial homogenates. Science 220:717–719

Zviman M 1993 Responses of reconstituted olfactory receptors to diethyl sulfide derivatives. PhD dissertation, Michigan State University, University Microfilms, Ann Arbor, MI

Second messenger signalling in olfaction

Heinz Breer

University Stuttgart-Hohenheim, Institute of Zoophysiology, 7000 Stuttgart 70, Germany

Abstract. Odorous molecules are recognized by specific receptor proteins located in the ciliary membrane of olfactory receptor neurons. These receptors have been identified using molecular cloning—they are members of the seven-transmembrane-domain G protein-coupled receptor superfamily. Specific receptor subtypes are expressed in subsets of olfactory neurons spatially segregated within certain areas of the olfactory epithelium. Interaction of odorants with receptors initiates the primary reaction of olfactory signalling. Intracellular reaction cascades are activated via specific G proteins, leading to a rapid and transient rise in second messenger levels; odorous compounds elicit mutually exclusive cAMP or inositol 1,4,5-trisphosphate responses. Odorant-induced second messenger signalling is terminated via kinase-mediated negative feedback loops uncoupling the reaction cascades by phosphorylation of receptor proteins. Strong odour stimuli elicit a delayed response of another messenger system, the nitric oxide/cGMP cascade. cGMP may control some adaptive reactions in olfactory receptor neurons.

1993 The molecular basis of smell and taste transduction. Wiley, Chichester (Ciba Foundation Symposium 179) p 97–114

Organisms continuously monitor their chemical environment, employing highly specialized chemosensory systems. Small volatile molecules—odorants—are detected by olfactory systems composed of bipolar chemosensory neurons that encode the strength, duration and quality of odorous stimuli into distinct patterns of afferent neuronal signals. Over the last decade, considerable evidence has accumulated suggesting that olfactory perception is the result of a complex cascade of biochemical and electrophysiological processes (Reed 1992, Breer & Boekhoff 1992, Firestein 1992). Second messenger processes are supposed to mediate the transduction of chemical signals into electrical responses in olfactory receptor neurons. The starting point for a thorough exploration of the molecular machinery underlying the transduction of olfactory signals was the observation that olfactory cilia contain high levels of adenylate cyclase that is activated upon exposure to odorous compounds (Pace et al 1985). This finding was soon followed by the discovery of a cAMP-gated cation channel in frog olfactory cilia (Nakamura & Gold 1987). However, it was observed that only certain odorants stimulate adenylate cyclase; others have no effect (Sklar et al 1986). In addition, two important aspects had mostly been overlooked: the high doses of odorants required for stimulation and the temporal aspects of the reaction.

Rapid kinetics of second messenger signals

We have studied the time course of odorant-induced second messenger responses using a rapid kinetic methodology which allows us to monitor odorant-induced formation of second messengers in olfactory preparations on a subsecond time scale (Breer et al 1990). Mixing a suspension of isolated rat olfactory cilia with submicromolar concentrations of odorants (e.g. citralva or isomenthone) elicited a rapid increase in the cAMP concentration. The second messenger response reached a peak within 50 ms and thereafter rapidly decayed to the basal level within a few hundred milliseconds (Fig. 1). This odorant-induced 'pulse' of cyclic nucleotide thus clearly precedes the electrical response of olfactory receptor cells. The delay may be due to the time taken for the build-up of cAMP. From a presumed intraciliary cAMP concentration of about 5 μM, the observed 5–10-fold rise brings the concentration well above the K_d value determined for the cAMP-gated ion channel in ciliary membranes (Firestein et al 1991). Thus, the rapid and transient increase in cAMP may well account for the odour-induced generator current. The specificity of the observed second messenger response is emphasized by the fact that it satisfies all the criteria for a G protein-mediated process. The 'on-set' kinetics of the second messenger 'pulse' is independent of the dose of odorant applied: however, strong stimuli (e.g. 10 μM isomenthone) lead to a different 'off-set' kinetic; the decaying rate levelled off after about 50–100 ms, resulting in an intermediate (tonic) cAMP concentration, which persisted for several seconds. This tonic response is probably due to the time required for inactivation of high concentrations of odorants; active odorous molecules may be around when desensitized receptors are reactivated.

Alternative second messenger pathway

We tested the second messenger response of rat cilia to a variety of different odorous compounds and found that only certain odorants induced a rapid rise in cAMP. Subsequently, we demonstrated that 'inactive' odours, such as pyrazine, affected an alternative second messenger pathway; they induced a rapid and transient change in inositol 1,4,5-trisphosphate (InsP$_3$) levels (Boekhoff et al 1990). These results confirm and extend the observation that odorants stimulate phosphoinositide turnover (Huque & Bruch 1986) (Fig. 2). The functional importance of odour-induced rapid InsP$_3$ responses

Second messenger signalling in olfaction 99

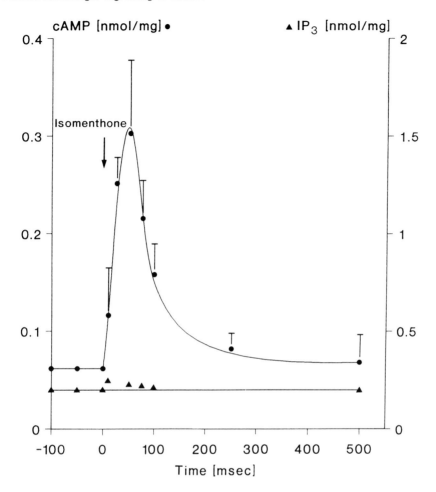

FIG. 1. Odorant-induced second messenger response. Challenging rat olfactory cilia with a low concentration (1 μM) of isomenthone results in a rapid increase in cAMP concentration which reaches a dose-dependent maximum after 50 ms; thereafter the cAMP concentration decays to approximately the basal level within 250 ms. This 'pulse' of cAMP is supposed to activate cyclic nucleotide-gated cation channels, thus mediating the chemo-electrical transduction process. IP$_3$, inositol 1,4,5-trisphosphate. Concentrations are nmol/mg of protein.

is emphasized by the recent identification of InsP$_3$-gated ion channels in the plasma membrane of olfactory cells (Restrepo et al 1990, Cunningham et al 1992, Fadool & Ache 1992).

The physiological implications for a dual system of olfactory signalling are still elusive. An interesting role for the two second messenger pathways was

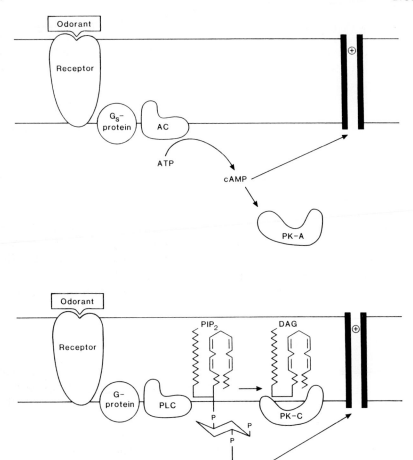

FIG. 2. A schematic representation of the alternative chemo-electrical transduction pathways in olfactory neurons. Interaction of odorants with specific receptor proteins triggers the activation of an intracellular reaction cascade via specific G proteins, generating second messengers (cAMP produced by the action of adenylate cyclase [AC]; inositol trisphosphate [IP_3] and diacylglycerol [DAG] produced from phosphatidylinositol 4,5-bisphosphate [PIP_2]). The resulting transient rise in second messenger levels has a dual function: firstly, activation of second messenger-gated cation channels, leading to depolarization of the membrane, which then elicits the generator potential; secondly, stimulation of protein kinases (PK-A, PK-C) which uncouple the reaction cascade via phosphorylation of activated receptors (signal termination). PLC, phospholipase C.

recently discovered in the antennal cells of the spiny lobster, where both the InsP$_3$ and the cAMP pathways operate (Fadool & Ache 1992). In the lobster, odorant-induced excitation of the chemosensory receptor cells is mediated by specific InsP$_3$-activating cation channels in the plasma membrane. In contrast, odorant-induced hyperpolarization is mediated via the cAMP cascade, apparently through the action of cAMP-gated chloride channels (Michel & Ache 1992).

Coexistence of both olfactory signalling pathways in the same cell would allow an efficient interaction between the two, providing the potential for considerable positive or negative 'cross-talk'. Thus, convergent integration of different odour stimuli may begin at the level of primary reactions in olfactory receptor cells.

It has been hypothesized that compounds of different flavour may activate different second messenger pathways and that members of a certain odour class may activate the same second messenger system. By assaying a variety of odorous compounds, we found that all odorants elicit mutually exclusive cAMP or InsP$_3$ pulses in olfactory cilia (Breer & Boekhoff 1991). So far, there is no apparent correlation between the flavour of a compound and the second messenger pathway activated (Table 1).

Termination of the olfactory signal

One of the characteristic features of the odorant-induced second messenger response is its transient nature. Pulsatile changes in second messenger concentrations are essential for sensory receptor neurons to respond precisely to iterative stimulation. Termination of the second messenger response is not due to odorant inactivation, which is accomplished via cytochrome P_{450}-mediated and/or glucuronosyltransferase reactions (Lazard et al 1991), but rather seems to be caused by a rapid desensitization of the system. Desensitization of many hormone and neurotransmitter systems is achieved by phosphorylation of the receptor proteins, which uncouples the reaction cascade by preventing further G protein activation (Lefkowitz et al 1990). When inhibitors of protein kinase A were applied to olfactory cilia, the rising phase of an odorant-induced cAMP signal was prolonged and the increased second messenger concentration persisted for minutes (Fig. 3). Apparently, the 'turn off' reaction is blocked by the kinase inhibitor. A similar effect was observed for the InsP$_3$ signal in the presence of inhibitors of protein kinase C (Boekhoff & Breer 1992). The odorant-induced second messenger response is terminated only by the kinase controlled by the second messenger generated in that particular cascade—the cAMP pathway by protein kinase A and the InsP$_3$ pathway by protein kinase C. These observations suggest that odorant-induced second messenger signalling is turned off via a negative feedback reaction which uncouples the reaction cascade through the phosphorylation of an element of the transduction apparatus.

TABLE 1 Odorant-induced increase in second messenger concentration

Odorant	cAMP[a]	Inositol 1,4,5-trisphosphate[b]
Fruity		
Citralva	100	4 ± 7
Citraldimethylacetal	63 ± 10	3 ± 5
Citronellal	61 ± 7	3 ± 5
Citronellylacetate	62 ± 10	1 ± 1
Lyral	0	100
Floral		
Hedione	108 ± 21	6 ± 7
Geraniol	62 ± 25	6 ± 7
Acetophenone	36 ± 8	4 ± 6
Phenylethylalcohol	16 ± 9	8 ± 7
Lilial	1 ± 2	106 ± 22
Herbaceous		
Eugenol	70 ± 18	4 ± 4
Isoeugenol	49 ± 13	2 ± 3
Ethylvanillin	0	63 ± 15
Putrid		
Furfurylmercaptan	37 ± 9	8 ± 7
Triethylamine	4 ± 2	141 ± 60
Phenylethylamine	6 ± 2	137 ± 78
Pyrrolidine	0	70 ± 25
Isovaleric acid	0	68 ± 24

Odorant concentration = 1 μM.
[a]Data for cAMP are expressed as a percentage of the concentration induced by citralva.
[b]Data for inositol 1,4,5-trisphosphate are expressed as a percentage of concentration induced by lyral.

Recent evidence indicates that another kinase, a β-adrenergic receptor kinase (βARK)-like enzyme is involved in turning off the olfactory cascade. A characteristic subtype—a βARK2-like kinase—is preferentially expressed in the olfactory neuroepithelium and this kinase is specifically involved in the rapid desensitization of olfactory signalling (Dawson et al 1993, Schleicher et al 1993). The βARK2-like kinase seems to represent another example of particular isoforms constituting the signal transduction apparatus in olfactory neurons, as previously shown for a G protein (G_{olf}) and adenylate cyclase (Reed 1992).

The proposal that the olfactory transduction machinery is uncoupled through the phosphorylation of important elements of the cascade implies that the reaction pathway may be reactivated by dephosphorylation of the modified elements. In this context, phosphatases should play an important role in reactivating and thus increasing the efficiency of the system. In the presence of specific phosphatase inhibitors (such as okadaic acid), the intensity of the response was significantly reduced. Thus inhibition of phosphatases results in a reduced reactivity of the signalling pathway owing to elements remaining in the phosphorylated, inactive state.

Odour-induced phosphorylation of ciliary proteins

In attempting to identify the constituents that are phosphorylated during desensitization of olfactory signalling, it is interesting to note that desensitization phenomena in photoreceptors and hormone-sensitive cells have also been attributed to phosphorylation of the receptor proteins, such as rhodopsin or β-adrenergic receptors. It is therefore conceivable that in the olfactory system, the receptor proteins for odorants are phosphorylated. Experiments with [^{32}P]γATP have shown that the incorporation of ^{32}P is significantly enhanced upon odour stimulation (Fig. 3). Furthermore, phosphorylation induced by odours activating the cAMP pathway is prevented by protein kinase A inhibitors, whereas protein kinase C inhibitors block ^{32}P incorporation induced by InsP$_3$-generating odours (Boekhoff et al 1992). The odour-induced surplus of incorporated ^{32}P was mainly found in polypeptide bands with an apparent molecular mass of about 50 kDa. The predicted molecular size of putative odorant receptors is about the same (Buck & Axel 1991, Raming et al 1993). Recent immunoprecipitation experiments with receptor-specific antibodies have confirmed that receptor proteins are phosphorylated upon odorant stimulation.

Putative role of cGMP in olfaction

While the rapid and transient generation of cAMP or InsP$_3$ pulses is considered to be a primary reaction in olfactory transduction, there is new evidence that suggests that another second messenger, cGMP, may play a role in olfactory signal processing. Stimulation of olfactory cilia with high odour doses leads to a delayed and sustained increase in cGMP concentrations. Attempts to explore the underlying mechanisms indicated that L-N^G-nitroarginine, a selective inhibitor of nitric oxide (NO) formation, completely prevents the cGMP response in rat olfactory cilia. Haemoglobin, which binds and inactivates NO, does as well. This suggests that the NO–cGMP system may be operating in chemosensory neurons (Breer et al 1992).

These observations have several implications for signal processing in olfactory systems (Breer & Shepherd 1993). NO generated in stimulated receptor cilia may activate guanylate cyclase in adjacent cells. The generated cGMP, in turn,

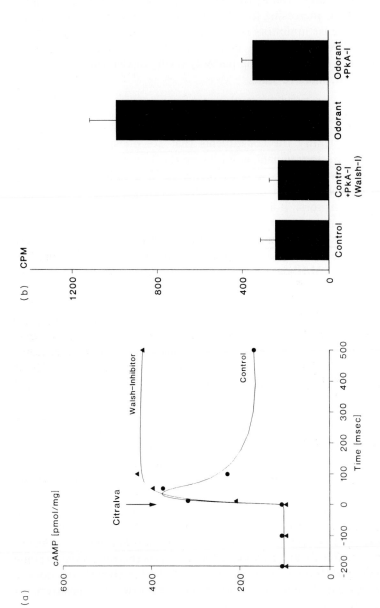

FIG. 3. Protein kinase inhibitors prevent desensitization, i.e. rapid termination of olfactory second messenger signalling as well as odorant-induced phosphorylation of proteins in rat olfactory cilia. (a) The rapid and transient cAMP signal elicited by 1 µM citralva is not terminated after 50 ms in the presence of the specific inhibitor of protein kinase A. (b) Odorant stimulation of rat olfactory cilia induces a significant increase in the [^{32}P] phosphate incorporation into ciliary proteins. In the presence of a protein kinase A inhibitor (3.8 µM Walsh inhibitor), ^{32}P incorporation is reduced to control level and the odorant-induced phosphorylation is blocked.

is able to activate cyclic nucleotide-gated cation channels. Thus, adjacent cells may be recruited to convey the information of a very strong stimulus to the brain.

In addition, NO may act in the generator cell itself; increases in cGMP are supposed to attenuate responsiveness and may be involved in sensory adaptation of olfactory neurons. It has been found that pretreatment of cilia preparations with permeable derivatives of cGMP significantly reduces second messenger responses upon odour stimulation. This observation agrees with results of electrophysiological studies in which pretreatment of isolated bullfrog olfactory epithelia with permeable cGMP analogues for several minutes significantly attenuated the net inward current induced by odour stimulation (Persaud et al 1988). Definitive answers concerning the crucial roles and the mode of actions of cGMP in olfactory signalling await further studies.

Receptors for odorants

Exploring the nature and diversity of receptors for odorants has long been considered as the key to understanding the molecular basis of olfaction. The recent discovery of a novel gene family supposed to encode odorant receptors (Buck & Axel 1991) has opened the way to explore this critical element of the transduction apparatus. In order to prove that encoded proteins bind specific odorants and interact with second messenger systems, we have expressed identified cDNAs in non-neuronal surrogate cells. Sf9 cells (ovary, *Spodoptera frugiperda*) were infected with baculovirus harbouring the receptor-encoding cDNA downstream from the strong polyhedrin promoter. After about 48 h the cells contained high levels of receptor-specific mRNA. Membrane preparations of these cells were assayed for odorant-induced second messenger formation. A large panel of odorous compounds, including representatives of different odour and chemical classes, was analysed. Although most of the odorants were inactive, some odorous aldehydes induced significant generation of $InsP_3$; no changes in cAMP concentrations were observed (Raming et al 1993). Thus, candidate odorant receptors can be expressed in surrogate cells and the second messenger response elicited by odorants can be used to monitor the affinity and specificity of the ligand–receptor interaction.

The identification of cDNA clones encoding odorant receptors has allowed us to explore the spatial segregation of olfactory neurons expressing particular receptor genes. Using *in situ* hybridization with antisense RNAs as specific probes, we analysed tissue sections from the nasal neuroepithelium for cells containing receptor-encoding transcripts. Under high stringency conditions, a punctate pattern of labelling was observed in certain regions of the sections; large areas were non-reactive. These observations indicate that cells expressing a particular receptor type are not randomly distributed throughout the neuroepithelium, but rather, are segregated in a defined but wide area. Some of the

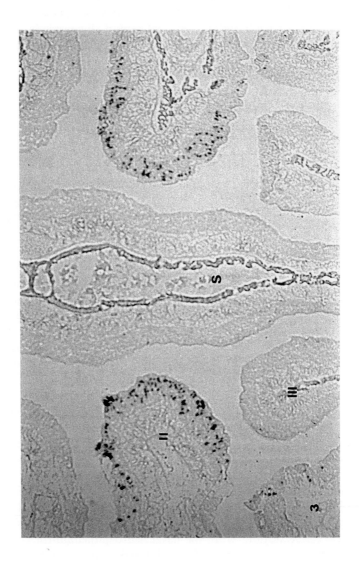

FIG. 4. Spatial distribution of olfactory receptor neurons expressing a particular odorant receptor subtype (OR37). Coronal sections of the rat nasal neuroepithelium were probed with digoxigenin-labelled antisense RNA riboprobe. In sections of the olfactory epithelium, OR37 transcripts can be visualized in a subset of chemosensory neurons segregated in restricted areas on endoturbinate II (II) and ectoturbinate 3 (3). The clusters of reactive cells are located symmetrically on both sides of the septum (S) in the nasal cavities. III, endoturbinate III.

subtypes were found within the same regions, others in complementary zones. In all cases, only a small proportion of the cells was reactive. Whether the non-reactive cells within the same area express related or very different receptor subtypes is of particular importance for understanding the functional organization of the chemosensory epithelium. One of the receptors was found to be expressed only in a spatially very restricted subset of cells (Strotmann et al 1992). The specific transcript was found exclusively in clusters of cells located on the tip of endoturbinate II and ectoturbinate 3, with the two reactive zones facing each other (Fig. 4). A detailed mapping of the nasal neuroepithelium for the spatial expression pattern of related and divergent odorant receptor subtypes may help to unravel the chemotopic organization of the olfactory system.

Conclusions

Recent advances in physiology and biochemistry support the concept that second messengers provide the critical link between the initial odour recognition and the elicitation of generator currents in olfactory receptor cells. Upon interaction of odorous molecules with specific receptors in the chemosensory membrane of olfactory neurons, key enzymes of second messenger pathways (adenylate cyclase or phosphoinositide-specific phospholipase C) are activated via specific G proteins, leading to a rapid and transient signal of either cAMP or $InsP_3$. Whether multiple second messenger signalling pathways operate in an individual receptor cell is still unclear, however, they have been described in a variety of cell types. The diversity of signalling mechanisms available suggests that complex cross-talk mechanisms may be involved in processing olfactory stimuli in receptor cells. The molecular basis of the regulation and possible mutual interaction of these pathways requires further investigation.

Termination of odour-induced responses apparently occurs via the uncoupling of the reaction cascades by kinases catalysing the phosphorylation of activated odorant receptors. The phosphorylation status of odorant receptors, controlled by kinase/phosphatase systems, determines the sensitivity of olfactory receptor cells.

The availability of DNA clones encoding specific elements of the olfactory transduction apparatus (including odorant receptors) that can be expressed in surrogate cells opens the possibility for reconstitution of the olfactory cascade. This is a prerequisite for exploring important aspects of olfactory signalling such as specificity of odorant–receptor and receptor–G protein interactions, and also questions concerning the initiation, regulation and termination of second messenger signals.

Acknowledgement

The work from this laboratory was supported by the Deutsche Forschungsgemeinschaft.

References

Boekhoff I, Breer H 1992 Termination of second messenger signaling in olfaction. Proc Natl Acad Sci USA 89:471–474

Boekhoff I, Tareilus E, Strotmann J, Breer H 1990 Rapid activation of alternative second messenger pathways in olfactory cilia from rats by different odorants. EMBO (Eur Mol Biol Organ) J 9:2453–2458

Boekhoff I, Schleicher S, Strotmann J, Breer H 1992 Odor-induced phosphorylation of olfactory cilia proteins. Proc Natl Acad Sci USA 89:11983–11987

Breer H, Boekhoff I 1991 Odorants of the same odor class activate different second messenger pathways. Chem Senses 16:19–29

Breer H, Boekhoff I 1992 Second messenger signalling in olfaction. Curr Opin Neurobiol 2:439–443

Breer H, Shepherd GM 1993 Implications of the NO/cGMP system for olfaction. Trends Neurosci 16:5–9

Breer H, Boekhoff I, Tareilus E 1990 Rapid kinetics of second messenger formation in olfactory transduction. Nature 345:65–68

Breer H, Klemm T, Boekhoff I 1992 Nitric oxide mediated formation of cyclic GMP in the olfactory system. NeuroReport 3:1030–1032

Buck L, Axel R 1991 A novel multigene family may encode odorant receptors: a molecular basis for odor recognition. Cell 66:175–187

Cunningham AM, Reed RR, Ryugo DK, Snyder SH 1992 An $InsP_3$ (inositol-1,4,5-trisphosphate) receptor is localized in the ciliary surface membrane in olfactory sensory neurons and may mediate odorant-induced signal transduction. Chem Senses 17:608 (abstr)

Dawson TM, Arriza JL, Jaworsky DE et al 1993 β-adrenergic receptor kinase-2 and β-arrestin-2 as mediators of odorant-induced desensitization. Science 259:825–829

Fadool DA, Ache BW 1992 Plasma membrane inositol 1,4,5-trisphosphate-activated channels mediate signal transduction in lobster olfactory neurons. Neuron 9:907–918

Firestein S 1992 Electrical signals in olfactory transduction. Curr Opin Neurobiol 2:444–448

Firestein S, Zufall F, Shepherd GM 1991 Single odor-sensitive channels in olfactory receptor neurons are also gated by cyclic nucleotides. J Neurosci 11:3565–3572

Huque T, Bruch RC 1986 Odorant- and guanine nucleotide-stimulated phosphoinositide turnover in olfactory cilia. Biochem Biophys Res Commun 137:36–42

Lazard D, Zupko K, Poria Y et al 1990 Odorant signal termination by olfactory UDP glucuronosyl transferase. Nature 349:790–793

Lefkowitz RJ, Hausdorf WP, Caron MG 1990 Role of phosphorylation in desensitization of the β-adrenoceptor. Trends Pharmacol Sci 11:190–194

Michel WC, Ache BW 1992 Cyclic nucleotides mediate an odor-evoked potassium conductance in lobster olfactory receptor cells. J Neurosci 12:3979–3984

Nakamura T, Gold GH 1987 A cyclic nucleotide-gated conductance in olfactory receptor cilia. Nature 325:442–444

Pace U, Hanski E, Salomon Y, Lancet D 1985 Odorant-sensitive adenylate cyclase may mediate olfactory reception. Nature 316:255–258

Persaud KC, Heck GL, DeSimone SK, Getchell TV, DeSimone J 1988 Ion transport across the frog olfactory mucosa: the action of cyclic nucleotides on the basal and odorant-stimulated states. Biochim Biophys Acta 944:49–62

Raming K, Krieger J, Strotmann J et al 1993 Cloning and expression of odorant receptors. Nature 361:353–356

Reed RR 1992 Signaling pathways in odorant detection. Neuron 8:205–209

Restrepo D, Miyamoto T, Bryant BP, Teeter JH 1990 Odor stimuli trigger influx of calcium into olfactory neurons of the channel catfish. Science 249:1166–1168

Schleicher S, Boekhoff I, Arriza J, Lefkowitz RJ, Breer H 1993 A β-adrenergic receptor kinase-like enzyme is involved in olfactory signal termination. Proc Natl Acad Sci USA 90:1420–1424

Strotmann J, Wanner I, Krieger J, Raming K, Breer H 1992 Expression of odorant receptors in spatially restricted subsets of chemosensory neurons. NeuroReport 3:1053–1056

Sklar PD, Anholt RH, Snyder SH 1986 The odorant-sensitive adenylate cyclase of olfactory receptor cells: different stimulation by distinct classes of odorants. J Biol Chem 261:15538–15543

DISCUSSION

Ache: I'd like to emphasize the fundamental importance of Professor Breer's finding that the extent of cAMP and InsP$_3$ production is dependent on the composition of the odour. This finding implies that the two second messenger-mediated inputs function in coding *per se* and not in cell processes such as adaptation that would presumably be independent of the particular odour being detected. We can get a hint as to how two input pathways might function in coding from the lobster results I mentioned earlier (p 88). In the lobster, the two second messengers drive separate but opposing ionic conductances, one which increases and one which decreases the probability that the cell will discharge. This scheme allows the animal to expand its potential for coding odour information across the population of receptor cells into the negative range (as opposed to a purely excitatory system). The mechanistic details are not as important as the fundamental strategy. One could achieve the same effect in other animals by different mechanisms. For example, increased internal Ca^{2+} concentration through activation of the InsP$_3$ pathway could suppress the sensitivity of the cyclic nucleotide-gated channel to its ligand, as recently demonstrated by Kramer & Siegelbaum (1992).

DeSimone: So you are suggesting that coding involves ensuring that certain cells do *not* become excited as well as ensuring that others do. This could occur if the cells that should remain quiet become hyperpolarized and remain so during the course of odorant application. Then the chance that those cells may be excited will be remote—less than the chance that they would fire spontaneously, as they do occasionally in the absence of stimuli. This would have the effect of sharpening the signal and increasing the information content of the response. Professor Breer's work and your own support such a scheme.

Ache: I envision not so much that a receptor cell drops out in response to inhibitory input (although I suppose that could happen given a disproportionately large signal), but rather that the inhibitory input reduces the magnitude of excitation of the cell in a finely graded manner. The idea of introducing inhibition into the system by modulating excitation is not problematical because natural odours are mixtures of many different chemicals that have a high probability of co-activating both the excitatory and inhibitory input pathways.

Breer: In his work on isolated olfactory neurons from mudpuppy, Vincent Dionne has demonstrated that the chemosensitive cells exhibit two primary response characteristics: excitation and inhibition (Dionne 1992).

Kurahashi: Stuart, have you seen this inhibitory response in your patch-clamp experiments?

Firestein: No, but I've only tested a limited number of odours. It's possible that with a different choice of odours there would be an inhibitory response.

Breer: Did you ever try this with strong $InsP_3$-producing odours such as triethylamine or pyrazine?

Firestein: No; that would be one way of looking at this. I could see a role for a hyperpolarizing response: we know, from recordings of the voltage-gated currents of these cells, that at the cell's normal resting potential, 80–90% of the Na^+ current is inactivated. Therefore, there's a tremendous reserve of Na^+ channels that stay inactivated most of the time. One reason for this could be that if you hyperpolarized the cells, additional Na^+ channels would be released from inactivation and be brought back into the pool of activatable Na^+ channels, which would then make that cell more likely to fire with only a very small stimulation. I can see a role like that for hyperpolarization where you over-sensitize the cell by releasing voltage-gated currents from inactivation.

Kinnamon: I'd like to comment on your observation that Na^+ channels are inactivated at negative potentials. When we went from whole-cell recording to perforated-patch recording, the inactivation voltage shifted from -70 mV to -40 mV. Steve Kleene saw exactly the same thing in olfactory neurons when he did the same switch. He thinks that there's a soluble second messenger that's being leached out on whole-cell recording that may be keeping the result artificially negative. Therefore, there might be inactivation voltages *in situ* of -40 mV.

Margolis: Heinz, you are working with suspensions of isolated olfactory cilia. Do you know the state of these cilia? Are their membranes intact or can molecules enter the cells fairly readily?

Breer: As in all experiments using subcellular fractions, we expect to get a mixture of resealed structures and membrane fragments. To increase the accessibility of exogenous compounds, we sometimes add small quantities of detergent.

Margolis: My concern is, if you can arrange conditions for molecules outside to enter, then there must be molecules on the inside that are going to exit.

Breer: Isolated olfactory cilia, like many other membrane preparations, seem to retain soluble cytosolic constituents. Even excised patches from rod outer segments were found to retain the functional phototransduction cascade (Ertel 1990). The molecular machinery for signal transduction remains in place, probably because of the lattice structures of the cytoskeleton.

Lancet: The way to look at isolated cilia experiments is that there are cytoplasmic components in the bulk solution that are concentrated enough to act. We are not talking about intact cells and I don't think we should look at it as a vectorial system with compartments. We have a mixture of membranes derived from cilia which have both sides exposed and components that may derive from the cells' cytoplasm around them in the solution. I see the question you are asking Frank, but I think, by and large, we shouldn't be worried.

Reed: It's always been my impression that most of the cilia preparations are broken into fragments of membranes that are probably completely exposed on both surfaces.

Ronnett: When the ciliary system was initially devised by Anholt et al (1986), extensive transmission electron microscope studies were carried out in Solomon Snyder's laboratory. They showed a number of well preserved ciliary structures, but also a number of large membrane fragments. This suggests that although the cilia may be maintained largely as intact ciliary structures, they are no doubt not entirely sealed. In our experience, if you take the initial homogenate and process it further to obtain purified cilia, you lose about 50% of phosphodiesterase (PDE) activity with each wash, so I think that the cilia are indeed open. Any comparison of biochemical experiments may be difficult because different laboratories may prepare the cilia in different ways.

Breer: I think this is a principal problem of working with subcellular fractions.

DeSimone: Another methodological question: the stopped-flow method measures cumulative cAMP for a given reaction time interval. A fall in cAMP measured with increasing time must mean that there are processes that decrease the cAMP concentration as well as those that increase it. Presumably a PDE is present, but I don't recall any mention of it.

Breer: It's there. In fact, there is a very high PDE activity in the olfactory cilia preparation. However, this enzyme activity is not under the control of odour stimulation. Thus, the build-up of the cAMP levels is due to odour-induced activity of adenylate cyclase. If the cascade of cAMP generation elicited by an odour stimulus is turned off, the cAMP concentration returns to the basal level due to the PDE activity.

DeSimone: So a state of constant cAMP concentration means that cAMP is actually still being produced, but this is balanced by PDE-mediated cAMP removal.

Firestein: In a couple of other systems, photoreceptors and M1 muscarinic receptors, there's another feedback loop onto the Gα subunit that increases the rate of its inherent GTPase activity (Arshavsky & Bownds 1992, Berstein et al 1992). This is another example of where a response is made transient with a feedback mechanism. Could this occur here as well?

Breer: It is quite possible that a feedback loop from the effector enzyme onto the Gα subunit increasing the GTPase activity is active in the olfactory system as well. For transducin, which is stimulated by PDE (Arshavsky & Bownds 1992), and for G_q, which is stimulated by phospholipase C isoenzyme β_1 (Berstein et al

1992), it has recently been shown that the effector proteins act as GTPase-activating proteins (GAPs) for these regulatory G proteins. Having effector proteins as GAPs may be important as a mechanism for controlling the lifetime of the activated G protein, thus keeping the second messenger response transient—a prerequisite for a sensory cell that may be repeatedly stimulated.

Firestein: It's interesting that there seem to be multiple, maybe redundant, systems for making this response transient. They work on slightly different, perhaps overlapping, time scales and there's presumably some careful modulation.

Hatt: We have to remember that these transient responses are measured biochemically and that the electrophysiological cell response can be decoupled completely from the biochemical activation, especially temporally. A well known example of this is the transmitter response of the cholinergic synapse on skeletal muscle fibres in vertebrates. Acetylcholine is released from the presynaptic side, but the increase in concentration in the synaptic cleft is extremely transient because acetylcholine is removed immediately (within 100 μs) by a high concentration of esterase. However, the response of acetylcholine-activated channels is in the order of milliseconds! It means that if an acetylcholine channel is opened, it must stay open for some milliseconds, independent of the presence of agonist. The time course of the synaptic current depends more on the biophysical properties of the channel protein than the time of transmitter release (Dudel et al 1992).

The cAMP-activated olfactory channel of humans, rats and salamanders behaves in the same way. A short (10 ms) application of cAMP is enough to open the channel and hold it open for about half a second (Zufall et al 1993). This seems to be a very effective functional mechanism: it gives the biochemical machinery in the cell time to recover after a stimulus whilst the electrical response of the cell persists.

Breer: I agree; the kinetics of the second messenger-gated channels may contribute greatly to the time course of generator current. However, the second messenger signal must be fast enough to initiate channel activity. Furthermore, the molecular signal must disappear before a subsequent stimulus can be processed.

Hatt: Could you comment on the role of cGMP in the olfactory system?

Breer: We suppose that cGMP may have a dual function in the olfactory system (Breer & Shepherd 1993). On strong odour stimuli, NO is formed, which leads to the formation of cGMP in adjacent target cilia, where cGMP may activate cyclic nucleotide-gated ion channels. In addition, NO produced in the generator cilia may also induce formation of cGMP in other compartments (dendrite, soma) of the receptor cell, where guanylate cyclase is not inhibited and where there are no or few cGMP-gated channels. At these sites, cGMP could initiate reaction cascades, e.g., via cGMP-activated kinases, which reduce the responsiveness of the receptor neurons: this may be a mechanism for adaptation of these cells.

Hwang: Have you tried using superoxide dismutase to see if it potentiates the effects of NO?

Breer: Yes; it does potentiate the odour-induced cGMP response.

Hwang: I would be interested to find out where the NO synthase may be localized in the olfactory system. Could it be located in the cilia or in other cellular elements constituting the olfactory epithelium?

Breer: We have not yet analysed the subcellular distribution of NO synthase activity in olfactory receptor cells. The biochemical analyses were performed with either isolated cilia or dissociated cells. The odorant-induced NO-mediated formation of cGMP was observed in both preparations. As yet, we can't say whether NO synthase is located primarily in the cilia.

Hatt: I think the steady-state concentration of cGMP in the cell must be much higher than that of $InsP_3$. Do you know the concentrations of cAMP and cGMP under resting conditions?

Breer: Our experiments have shown that the concentration of cGMP is significantly lower than that of cAMP. However, this is not a problem, since we know that the affinity of the cyclic nucleotide-gated channels for cGMP is higher than for cAMP. So, the endogenous concentration of cGMP may still be in the right range to activate the channels.

Firestein: Will the same PDE break down cGMP and cAMP, or are there different ones?

Reed: There are all flavours of PDE; several occur in cilia.

Ronnett: We looked at the different PDE activities present in rat cilia (Borisy et al 1992). There are several species of PDE present. The one novel finding was that there is a high affinity cAMP-specific PDE which is Ca^{2+}/calmodulin dependent. There was also a high-affinity species that did not appear to be Ca^{2+} dependent.

Caprio: We heard earlier (p 101) that odours are specific to certain second messenger pathways; a particular odour affects one pathway preferentially over another. Yesterday, I thought someone was saying that odours affect both pathways and that every odour does both. How do we correlate the two—is the effect dose dependent?

Ronnett: We've tested about 10 odours so far and our work shows that they all stimulate both cAMP and $InsP_3$ second messenger pathways with up to 10- to 100-fold different potencies (Ronnett et al 1993).

Getchell: Gabriele, there appears to be a discrepancy between the whole-cell response that you are recording and what Heinz Breer has shown biochemically, using similar odours or categories of odours to those that you have used, where there appears to be a preferential stimulation of the $InsP_3$ pathway.

Ronnett: A lot depends on what concentrations you use; if you do screening or you only look at one concentration, you may miss a small but physiologically relevant activation. Our experience is that in cells, the dose–response curves are not linear, they are at least biphasic. It's different for each odour.

Lancet: On the whole, we agree that there is some selectivity here and that the data sets that we see on both sides of the room are not really contradictory.

Kurahashi: I'm very curious about the onset kinetics of cAMP or other second messenger production in your rapid assay system, because almost all data showed very rapid production of cAMP. But there is a very large latency in the electrical responses of your isolated preparations, sometimes up to 500 ms. Can you explain this?

Breer: First of all, when you do electrophysiological measurements, you record from the soma of a single cell. The biochemical experiments are performed on cilia preparations from the whole epithelium of a rat.

Secondly, to elicit a measurable current in a cell, a certain critical level of cAMP has to be reached; this threshold will be approached much more rapidly if the stimulus is strong than with a weak stimulus. This could account for the differences in latencies.

References

Anholt RRH, Aebi U, Snyder SH 1986 A partially purified preparation of isolated chemosensory cilia from the olfactory epithelium of the bullfrog, *Rana catesbeiana*. J Neurosci 6:1962–1969

Arshavsky VY, Bownds MD 1992 Regulation and deactivation of photoreceptor G protein by its target enzyme and cGMP. Nature 357:416–417

Berstein G, Blank JL, Jhon D-Y, Exton JH, Rhee SG, Ross EM 1992 Phospholipase C-β1 is a GTPase-activating protein for $G_{q/11}$, its physiologic regulator. Cell 70:411–418

Borisy FF, Ronnett GV, Cunningham AM, Julifs D, Beavo J, Snyder SH 1992 Ca^{2+}/calmodulin-activated phosphodiesterase expressed in olfactory receptor neurons. J Neurosci 12:915–923

Breer H, Shepherd GM 1993 Implications of the NO/cGMP system for olfaction. Trends Neurosci 16:5–9

Dionne VE 1992 Chemosensory responses in isolated olfactory neurons from *Necturus maculosus*. J Gen Physiol 99:415–433

Dudel J, Franke C, Hatt H 1992 Rapid activation and desensitization of transmitter liganded receptor channels by pulses of agonists. In: Narahashi T (ed) Ion channels, vol 3. Plenum Press, New York, p 207–260

Ertel EA 1990 Excised patches of plasma membrane from vertebrate rod outer segments retain a functional phototransduction enzymatic cascade. Proc Natl Acad Sci USA 87:4226–4230

Kramer RH, Siegelbaum SA 1992 Intracellular Ca^{2+} regulates the sensitivity of cyclic nucleotide-gated channels in olfactory receptor neurons. Neuron 9:897–906

Ronnett GV, Cho H, Hester LD, Wood SF, Snyder SH 1993 Odorants differentially enhance phosphoinositide turnover and adenylyl cyclase in olfactory receptor neuronal cultures. J Neurosci 13:1751–1758

Zufall F, Hatt H, Firestein S 1993 Rapid application and removal of second messengers to cyclic nucleotide-gated channels from olfactory epithelium. Proc Natl Acad Sci USA 90, in press

Membrane currents and mechanisms of olfactory transduction

Stuart Firestein* and Frank Zufall

Section of Neurobiology, Yale University School of Medicine, 333 Cedar Street, PO Box 208041, New Haven, CT 06520-8041, USA

Abstract. The term olfactory transduction refers to the mechanisms that transform chemical information into electrical signals. With the patch-clamp technique it is possible to record those signals and to infer something about the mechanism that produced them. The direct activation of a cation-permeable channel by cAMP is the final step in producing the odour-induced ionic current. Because it occupies a critical position in the transduction process, measurements of the ion channel's activity provide useful insights into the molecular processes underlying olfactory transduction. In addition to its activation by cAMP and cGMP, the channel is modulated by both extracellular and intracellular Ca^{2+} ions and by extracellular Mg^{2+} ions, all at physiological concentrations. These effects are probably important in promoting signal reliability. An unusual feature of this channel is its termination kinetics—it can remain active for hundreds of milliseconds after the agonist has been removed. This is likely to add to the integrating properties of the olfactory sensory neuron.

1993 The molecular basis of smell and taste transduction. Wiley, Chichester (Ciba Foundation Symposium 179) p 115–130

Olfactory transduction is the process that transforms chemical information in the environment into electrical activity in the brain. The process involves the modulation of membrane ion channels, resulting in a change in membrane potential and leading, finally, to the generation of action potentials and their propagation to the central nervous system. Figure 1 outlines the steps from odour detection to action potential generation. The initial (and most critical) steps occur in the specialized cilia of the olfactory receptor neuron (Firestein et al 1990), where a cascade of biochemical reactions leads to the production of the second messenger, cAMP (Reed 1992). The final step in the transformation of chemical processes into electrical signals occurs when the cAMP binds to an ion channel, causing it to open and allow the flux of positive ions into the cell (Nakamura & Gold 1987). This ionic current depolarizes the membrane and drives the cell from its resting potential to the threshold for action potential generation.

Present address: Department of Biological Sciences, Columbia University, New York, NY 10027, USA.

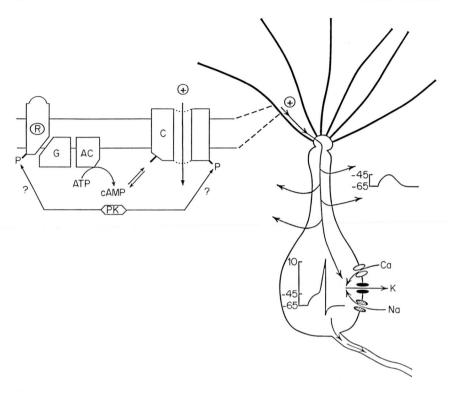

FIG. 1. Schematic diagram of the generation of electrical activity in the olfactory neuron. The ciliary membrane contains the various proteins of the second messenger system, including the receptor (R), G protein (G), adenylate cyclase (AC) and the cyclic nucleotide-gated ion channel (C). Positive ions enter the cell through the activated channel on the cilia, causing the membrane to depolarize from −60 mV to about −45 mV. This graded depolarization, sometimes called the generator potential, spreads passively through the cell to the dendrite and soma. In the somatic membrane there are a variety of ion channels sensitive to membrane voltage. When the membrane is depolarized to −45 mV, the Na$^+$ channels are activated, initiating the upstroke of the action potential. Subsequently, the activation of Ca^{2+} and K$^+$ channels repolarizes the cell.

This scheme emphasizes that the regulation of the channel activity by cAMP is central to the transduction process. Fortunately, there are techniques available for making quite sensitive measurements of the activity of single ion channels, or of the currents flowing through thousands of ion channels activated simultaneously. Thus considerable insight into the transduction machinery can be obtained by stimulating the cell with odours or other pharmacological agents and assaying ion channel activity. This chapter will present first a description of the ion channel and its functioning and then show how the summed activity of many channels results in the electrical response to odours.

Membrane currents in olfactory transduction 117

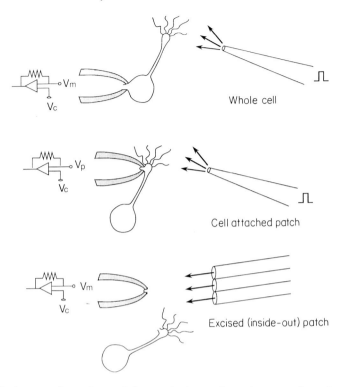

FIG. 2. Various configurations of the patch clamp. In each case a glass electrode is brought into contact with the plasma membrane and a very strong seal forms between the lipid and the glass. Pulling the pipette away suddenly from the cell rips off a small patch of membrane that may contain ion channels. Alternatively, the membrane in the electrode can be ruptured, providing access to the cell interior and the ability to voltage-clamp the cell while recording current flow anywhere in the cell.

Methods

The data described here were obtained by using one of various configurations (Fig. 2) of the patch clamp (Neher 1988). For recording the opening and closing of single channels, we excised patches of membrane from the dendrites of isolated olfactory receptors, such that the inside face of the membrane could be exposed to various bathing solutions. In some cases these solutions were changed very rapidly by means of a piezo-electric-driven device.

In experiments measuring the total current produced in the whole cell by exposure to odours, the 'whole-cell' voltage-clamp configuration was used. In this case, the cell remained intact, but attached to an electrode that permitted the measurement of the ion currents flowing into and out of the cell while a constant membrane potential was maintained. Odours were applied to the cell by pressure ejection or by immersing the cell in rapidly flowing streams of solution carrying different odours. All of these methods have been previously described in detail (Franke et al 1987, Menini et al 1988, Firestein et al 1991a, Zufall et al 1991a).

Results and discussion

The primary pathway for the odour-sensitive current is through cyclic nucleotide-gated channels

In patches of membrane taken from olfactory cilia, cAMP applied to the intracellular side of the membrane can directly open channels, permitting large cationic fluxes (Nakamura & Gold 1987). The same channels exist at a lower density on the dendrite of the receptor neuron and, in cell-attached patches, could be activated by exposing the cell to odours (Firestein et al 1991a). By excising these patches from the dendrite we have been able to pursue a biophysical characterization of the cyclic nucleotide-gated channel. In parallel, at the molecular level, several laboratories have cloned and characterized this channel (see Breer, this volume, Dhallan et al 1990, Ludwig et al 1990, Goulding et al 1992).

Although there is some variability from species to species and between different preparations, some consistent characteristics of channel activity have emerged from these various studies. The channel is sensitive to both cAMP and cGMP, with K_d's in the range of 2–40 µM. In all cases, the channel displays a slightly higher affinity for cGMP than cAMP, although cAMP appears to be the physiological ligand. The dose–response relationship indicates that activation is cooperative, with a Hill coefficient of 2–3, suggesting the required binding of at least three (and probably more) molecules of cAMP for complete activation. In general, the channel shows little voltage dependence in either its conductance or open probability. In the absence of divalent cations, the channel conductance is about 45 pS.

Divalent cations have several important effects under physiological conditions. It is known that Ca^{2+} ions permeate the channel (Kurahashi 1989, Frings & Lindemann 1991), but much more slowly than monovalent ions such as Na^+ or K^+. This gives rise to a rapid flickering effect in the opening of the channel and effectively results in a reduced conductance (Zufall & Firestein 1993). Indeed, at physiological concentrations of Ca^{2+}, the effective current may be reduced by as much as 99%. This block also imparts an apparent voltage dependence due to the influence of membrane potential on the strongly charged Ca^{2+} ions. Thus the block is strongest at negative membrane potentials near the resting potential, and is relieved as the membrane is made more positive and Ca^{2+} ions are repelled from the channel. Recently, we have determined that the binding site for the Ca^{2+} ions is about 10% through the voltage field, suggesting that it is within the channel pore itself. Similar blocking effects have been observed for Mg^{2+}, except that it appears that Mg^{2+} does not actually permeate the channel.

At first it might appear counterproductive to open a channel that is subsequently 99% blocked at the normal Ca^{2+} concentration and resting potential (-55 mV). However, two possible advantages may be gained. One is an increase in signalling reliability. At -55 mV, about 2.5 pA of current could pass through a 45 pS channel. With an input resistance of 2–5 GΩ, only 7.5 pA of current would be required to depolarize the olfactory neuron by the 15 mV necessary to reach spike threshold. This could be accomplished by the approximately synchronous opening of only 3–5 channels (Firestein & Werblin 1987, Lynch & Barry 1989). Now, the probability of a channel opening is, strictly speaking, a random occurrence that is simply biased one way or the other by the presence of the agonist (cAMP). We and others have estimated that the channel density on the cilia is of the order of 1000 μm^{-2}, meaning that a typical cell with 10 cilia might possess about 500 000 channels (Nakamura & Gold 1987, Zufall et al 1991a). Of these, it is not unreasonable to expect that on a random basis 3–5 channels might be open simultaneously, even in the absence of agonist, thereby initiating a spike in the absence of stimulus. A smaller conductance channel would require more channels to open synchronously to produce a depolarization sufficient to trigger an action potential. Thus, for a given amount of current, there is greater reliability associated with opening a large number of small channels, rather than a small number of large channels.

On the other hand, once the membrane begins depolarizing, it would be advantageous to increase the conductance rapidly; the voltage-dependent relief of the divalent block accomplishes just this goal. Once a sufficient number of small channels are activated and the membrane begins to depolarize, the block is relieved and the channels are able to pass current at the greater conductance level. This is a sort of regenerative process, not dissimilar to the well known voltage-dependent Mg^{2+} block seen at the NMDA (*N*-methyl-D-aspartate) receptor (Mayer et al 1984).

Ca^{2+} ions also have effects on the channel from the intracellular side. When the channel opens, both monovalent ions and Ca^{2+} ions permeate. The influx of Ca^{2+} ions may increase the intracellular Ca^{2+} concentration transiently, especially in the extremely tight confines of the very fine olfactory cilia. We have found that Ca^{2+} ions act on the intracellular side of the channel to stabilize a closed state and thereby reduce the probability of opening (Zufall et al 1991b). The effect of this on the overall response would be to reduce current flow and make the response more transient. This reduction in current occurs by a fundamentally different process from that of the open channel block described above. Here the conductance remains normal and the channel simply opens less frequently for a given concentration of cAMP when Ca^{2+} is present. Ca^{2+} is effective in the range of 1–3 μM, a physiologically relevant range, while Mg^{2+} is completely ineffective at concentrations up to 10 μM. This process may be important in adaptation, since under low Ca^{2+} conditions the channel shows virtually no desensitization, even during several-minute-long exposures to saturating concentrations of cAMP. A similar effect has recently been described in catfish olfactory neurons by Kramer & Siegelbaum (1992), who also showed that the current reduction due to Ca^{2+} could be competed away by increasing the cAMP concentration to sufficiently high levels. This suggests that the Ca^{2+} ions may be acting at the same site as cAMP. Alternatively, as suggested by Kramer & Siegelbaum, there may be an intermediate Ca^{2+}-activated enzyme that acts on the channel. The effects of divalent ions on the cyclic nucleotide-gated channel are summarized in Fig. 3.

Kinetics of the cyclic nucleotide-gated channel

In collaboration with Hans Hatt, we have studied the non-steady-state kinetics of the channel by using a piezo-electric-driven stimulus delivery system that

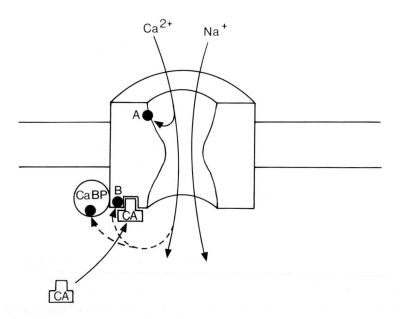

FIG. 3. Effects of divalent cations on the cyclic nucleotide channel. From the outside, Ca^{2+} and Mg^{2+} block the channel by binding to a site in the mouth of the pore. Ca^{2+} eventually permeates through the channel. At the inner face of the channel, Ca^{2+} acts to stabilize the channel in a closed configuration, even in the presence of agonist. CaBP, Ca^{2+}-binding protein; CA, cAMP.

can change the agonist concentration at the patch membrane surface within a few hundred microseconds. For these experiments, we measured single channels from salamander olfactory neurons, or macroscopic currents from patches containing the rat recombinant channel (kindly supplied by Dr Randall Reed). We found that the channel kinetics is surprisingly slow. After a jump in cAMP concentration, the current rises to its new level with a 10-90% time of about 5 ms at saturating concentrations of cAMP. Compared to the activation rise time for other channels, this is quite slow. For example, the nicotinic acetylcholine receptor (nACh) channel activates within $20\,\mu s$ of agonist application and voltage-gated channels within a few hundred microseconds of step changes in voltage (Hille 1992). Interestingly, the cyclic nucleotide-gated channel activation time is comparable to that for NMDA channels (Lester et al 1990).

Even more surprising was the deactivation kinetics at the end of a pulse. One would expect that after the rapid removal of the cAMP, the channel activity would decay exponentially with a time constant that reflected the closing rate constant for the channel. Instead, we found that channel activity persisted for as long as 200 ms after the removal of agonist and then decayed along a very slow exponential time course with a time constant of about 150 ms. This suggests a very slow off rate in the order of $7\,s^{-1}$. Thus the cyclic nucleotide-gated channel of olfactory receptors possesses rather sluggish kinetics, a feature that may play a critical role in shaping the odour response. This will be considered in further detail after a discussion of the odour-induced currents.

Characteristics of the odour-induced currents in intact receptor neurons

In vertebrate olfactory receptor neurons, the response to odour stimulation is typically an inward positive current of anywhere from a few pico-amps to a nano-amp, depending on the particular odour, its concentration and the duration of exposure. Measuring the cellular response to odour, the ionic current, is relatively straightforward with modern electrophysiological techniques. However, interpretation of the currents is confounded by our lack of control over the stimulus. Without knowledge of the magnitude (i.e. concentration) and time course of the input, sensible interpretation of the output is impossible.

We have sought to overcome this obstacle by several means. For rapid pulses of odour ($<100\,ms$), we have used a pressure ejection technique to deliver the stimulus. By loading the odour solution with a high concentration of KCl, we can use the cellular response to K^+ to monitor the time course and relative concentration of the odour solution at the cell membrane. The disadvantage of this method is that although stimulus can be delivered rapidly, it is removed only passively by diffusion. For longer steps of odour stimulation, we have, in collaboration with Anna Menini, used a technique in which the cell is immersed in a flowing solution containing the stimulus at a particular concentration. The cell can be moved into and out of the solution rapidly (solution change time < 30 ms) and the concentration remains stable for the length of the step (300 ms–3 s).

The following characteristics typify the odour-induced current in isolated cells stimulated by one of these techniques.

1) The odour-induced current is always an inward cationic current at negative potentials. The current–voltage relation for this current displays variable amounts of inward rectification in the hyperpolarized range, reverses near 0 mV and may be slightly outwardly rectifying at positive potentials (Firestein & Werblin 1989, Kurahashi & Shibuya 1989).

2) Currents are elicited with a broad specificity and relatively low affinity for odours. Typical EC_{50} values range from 3–60 μM and as many as 50% of the cells tested respond to more than one odour.

3) Olfactory receptor cells have rather narrow operating ranges. Dose–response curves are steep, covering typically only one log unit of concentration change for 10–90% of the response range.

4) There is a long concentration-dependent latency, ranging from 130–450 ms, between the arrival of the stimulus at the cilia membrane and the activation of the ionic current. Nonetheless, pulses of odour lasting less than the duration of the latency are still able to elicit responses. Thus once the odour–receptor binding occurs, the response process continues even in the absence of continued odour stimulus.

5) For times of up to about 1 s, the magnitude of the response current is dependent not only on stimulus concentration, but also on the duration of the stimulus. That is, the cell appears to integrate stimulus information over time, so that the response to a particular concentration of odour is larger for a stimulus duration of 1 s versus one of 500 ms. This is a key feature of the odour-elicited current and has important implications for the second messenger mechanism. These will be discussed in detail below.

6) The response to very long duration stimulation is transient. For maintained stimuli of 7–10 s, the odour-elicited current activates, reaches a plateau and then falls back towards baseline within about 4 s. Thus there are strong adaptation processes at the earliest stages of the odour response in the peripheral sensory neuron.

The shape of the odour response is determined by the kinetics of the second messenger cascade

All of these characteristics can, of course, be accounted for by the workings of a second messenger cascade intervening between odour–receptor binding and ion channel activation. There is now ample evidence for such a mechanism in olfactory neurons (this volume: Breer 1993). We can further extend this analysis by considering how each step in the cascade contributes to the physiology of the response. In particular, we might gain some insight by comparing the kinetics of the cAMP-sensitive channel to that of the overall odour-elicited current.

The fact that the odour response has a time dependence suggests that olfactory neurons do not merely sense the concentration of odour, but actually count molecules over some integrating time period. The question arises as to where the integration is taking place. One possibility is in the accumulating concentration of intracellular cAMP. In this case, the response of the cell is a function of the concentration of cAMP at any given time and as cAMP accumulates in the cell due to receptor activation, the response increases. This is, at least superficially, how most second messenger systems have been thought to operate. However, Breer and his colleagues (this volume: Breer 1993, Boekhoff et al 1990) have shown that cAMP production in response to odours is quite transient, reaching a peak in 50–75 ms and decaying within 300–500 ms. Furthermore, the kinetics of cAMP production do not show the same concentration dependence as the odour response.

An alternative locus of integration might be the cAMP-sensitive channel, which, with its sluggish kinetics, would be ideally suited to this role. In this model, the production of cAMP is pulsatile; the channels integrate this information by smearing out the time scale and summing the pulses over time. This scheme has several appealing features. One is that in the small confines of the olfactory cilia, a continuous accumulation of cAMP could lead rapidly to saturating concentrations and the end of responsivity. Another is that by producing sharp pulses of cAMP, the amplification aspect of the second messenger cascade can be retained without driving the system rapidly to saturation. That is, amplification is now translated into frequency (number of pulses of cAMP), rather than an accumulating quantity. This is a strategy that is common in the nervous system, for example, in the use of spike frequency rather than spike amplitude to code intensity.

This may be an appealing model, but is there any evidence to support it? Several years ago we conducted a series of double-pulse experiments that gave interesting but somewhat puzzling results (Firestein et al 1991b). In these experiments, two identical brief pulses of odour were delivered to a cell at varying intervals, ranging from 0.5 to 2 s apart. The idea was to deliver a second stimulus while the cell was still responding to the first. If, as we thought at the time, the odour-induced current was a function of intracellular cAMP concentration, then delivering a second stimulus pulse while the response to the first stimulus was ongoing would be the same as adding more cAMP into an already accumulated pool. We expected that the latency and onset kinetics of the second response would then be faster than the response to the first pulse because there was already a high level of intracellular cAMP. Instead, the kinetics remained identical for both pulses. The only parameter that was affected by the second pulse was the amplitude of the response, which more than doubled. In retrospect, these are precisely the results one would expect if the cAMP did not accumulate

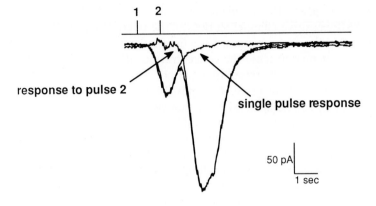

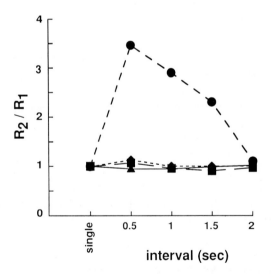

FIG. 4. Double-pulse experiments. The top panel shows the current elicited by two identical 50 ms pulses of odour delivered 1 s apart. The response to a single pulse was then subtracted by computer from the total response and the ratios of certain features were compared. In the lower panel, the ratio of latency, time to peak, half-rise time and peak amplitude of response 2 (R2) to response 1 (R1), are plotted against the interval between the two pulses. For intervals of 1 s or less, the response to pulse one was near its peak when the second pulse was delivered. Of the various features compared, only the amplitude changed significantly. (● = amplitude, ■ = latency, ▲ = half rise time, ♦ = time to peak.)

but rather was generated in a pulsatile fashion. Then the cAMP generated by the first odour pulse would have disappeared by the time the second pulse arrived. But the slow kinetics of the ion channel would sustain the current response even as the cAMP decayed away. Thus the kinetics of the second response would be unaltered, but the amplification would be preserved, just as is seen in Fig. 4.

Summary and conclusions

The response to odour stimulation results in the generation of an inward cationic current due to the activation of an ion channel by cAMP. The channel occupies a critical position in the transduction cascade and its kinetics serves to determine the shape of the odour response. Olfactory neurons faced with the task of recognizing a large number of disparate odour substances, but doing so without significant loss of sensitivity to low concentrations, use an amplification system that trades temporal resolution for increased sensitivity. This is accomplished by the integrating abilities of the cyclic nucleotide-gated channel, which derive from its sluggish kinetics.

Acknowledgements

The authors thank Drs Gordon Shepherd, Hans Hatt and Anna Menini, collaborators in much of the work reviewed here. Supported by the The Public Health Service, Office of Naval Research and NATO.

References

Boekhoff I, Tareilus E, Strotmann J, Breer H 1990 Rapid activation of alternative second messenger pathways in olfactory cilia from rats by different odorants. EMBO (Eur Mol Biol Organ) J 9:2453–2458

Breer H 1993 Second messenger signalling in olfaction. In: The molecular basis of smell and taste transduction. Wiley, Chichester (Ciba Found Symp 179) p 97–114

Dhallan RS, Yau K, Schrader KA, Reed RR 1990 Primary structure and functional expression of a cyclic nucleotide-activated channel from olfactory neurons. Nature 347:184–187

Firestein S, Werblin F 1987 Gated currents in isolated olfactory receptor neurons of the larval tiger salamander. Proc Natl Acad Sci USA 84:6292–6296

Firestein S, Werblin F 1989 Odor-induced membrane currents in vertebrate olfactory receptor neurons. Science 244:79–82

Firestein S, Shepherd GM, Werblin FS 1990 Time course of the membrane current underlying sensory transduction in salamander olfactory receptor neurones. J Physiol 430:135–158

Firestein S, Zufall F, Shepherd GM 1991a Single odor sensitive channels in olfactory receptor neurons are also gated by cyclic nucleotides. J Neurosci 11:3565–3572

Firestein S, Darrow B, Shepherd GM 1991b Activation of the sensory current in salamander olfactory receptor neurons depends on a G-protein mediated cAMP second messenger system. Neuron 6:825–835

Franke CH, Hatt H, Dudel J 1987 Liquid filament switch for ultra-fast exchanges of solutions at excised patches of synaptic membrane of crayfish muscle. Neurosci Lett 77:199–204

Frings S, Lindemann B 1991 Current recording from sensory cilia of olfactory receptor cells *in situ*. I. The neuronal response to cyclic nucleotides. J Gen Physiol 97:1–16

Goulding E, Ngai J, Kramer R et al 1992 Molecular cloning and single-channel properties of the cyclic nucleotide-gated channel from catfish olfactory neurons. Neuron 8:45–58

Hille B 1992 Ionic channels of excitable membranes, 2nd edn. Sinauer, Sunderland, MA

Kramer RH, Siegelbaum SA 1992 Intracellular Ca^{2+} regulates the sensitivity of cyclic nucleotide-gated channels in olfactory receptor neurons. Neuron 9:897–906

Kurahashi T 1989 Activation by odorants of cation-selective conductance in the olfactory receptor cell isolated from the newt. J Physiol 419:177–192

Kurahashi T, Shibuya T 1989 Membrane responses and permeability changes to odorants in the solitary olfactory receptor cells of the newt. Zool Sci 6:19–30

Lester RAJ, Clements JD, Westbrook GL, Jahr CE 1990 Channel kinetics determine the time course of NMDA receptor-mediated synaptic currents. Nature 346:565–567

Ludwig J, Margalit T, Eismann E, Lancet D, Kaupp UB 1990 Primary structure of cAMP-gated channel from bovine olfactory epithelium. FEBS (Fed Eur Biochem Soc) Lett 270:24–29

Lynch JW, Barry PH 1989 Action potentials initiated by single channels opening in a small neuron (rat olfactory receptor). Biophys J 55:755–768

Mayer M, Westbrook GL, Guthrie PB 1984 Voltage-dependent block by Mg^{2+} of NMDA responses in spinal cord neurones. Nature 309:261–263

Menini A, Rispoli G, Torre V 1988 The ionic selectivity of the light-sensitive current in isolated rods of the tiger salamander. J Physiol 402:279–300

Nakamura T, Gold GH 1987 A cyclic-nucleotide gated conductance in olfactory receptor cilia. Nature 325:442–444

Neher E 1988 The use of the patch clamp technique to study second messenger-mediated cellular events. Neuroscience 26:727–734

Reed RR 1992 Signalling pathways in odorant detection. Neuron 8:205–209

Zufall F, Firestein S, Shepherd GM 1991a Analysis of single cyclic nucleotide gated channels in olfactory receptor cells. J Neurosci 11:3573–3580

Zufall F, Shepherd GM, Firestein S 1991b Inhibition of the olfactory cyclic nucleotide gated ion channel by intracellular calcium. Proc R Soc Lond Ser B Biol Sci 246:225–230

Zufall F, Firestein S 1993 Divalent cations block the cyclic nucleotide gated channel of olfactory receptor neurons. J Neurophysiol 69:1758–1768

DISCUSSION

Lindemann: In salamander, you found that internal Ca^{2+}, at concentrations in the low micromolar range, had an inhibitory effect on cAMP-gated channels. This is not a flicker block, so there must be some biochemical change occurring that closes off the channels. We could not find this inhibitory effect in membrane patches excised from frog or rat olfactory cells. However, in rat, apart from the flicker block, Ca^{2+} has an inhibitory effect that can be overcome by increasing the cAMP concentration. Above $100\,\mu M$, Ca^{2+} will shift the dose–response curve of cAMP. This is a reversible effect: at very low Ca^{2+}

Membrane currents in olfactory transduction   127

concentrations the K_m is 3–4 μM, at 200 μM Ca^{2+} it is 30 μM and at 500 μM it is 90 μM. There is no change in the Hill coefficient. This may be a way of turning off the cAMP-gated channels, if the Ca^{2+} concentrations should ever reach such high levels inside the cilia (Lindemann et al 1992).

We then looked at the rate constants of cAMP activation using noise analysis. We could not resolve the very slow processes that you have described, but of those we could see in the 1–200 Hz band, both the on-rate and off-rate constants of cAMP activation were Ca^{2+} dependent (J. W. Lynch & B. Lindemann, unpublished results). Have you made similar observations?

Firestein: No.

Reed: In your dose–response curves for odours, you showed whole-cell currents that were in the order of 800 pA. How much current do you need to put into these cells to elicit action potentials?

Firestein: Nowhere near that much! There's some controversy over the exact amount. Under optimal conditions, something less than 10 pA would probably depolarize these cells, although I don't think they live in the most optimal conditions. In normal situations, where the resting potential is likely to be much less, i.e. more depolarized than −80 mV, and there is significant inactivation of some of the voltage-gated currents, it probably takes 30–40 pA to activate them. When you see these large currents, it's indicative of the existence of tremendous reserve and amplification mechanisms.

Reed: This suggests that in a physiological environment, the cell is actually detecting odorants well below these concentrations.

Firestein: Yes; that's probably true.

Reed: This questions the relevance of the 10 μM concentrations of odorants that you used. That cell would probably be firing action potentials in a good environment at an order of magnitude lower concentration.

Firestein: We would probably see responses to lower concentrations of odours if we were recording membrane potentials instead of current. A few picoamps of current might be enough to initiate spike generation and we would probably not detect such a small current. Then what we would be measuring would be the cable properties of the cell (how well current spreads from the site of influx to the site of impulse generation) and the threshold and activation parameters of the voltage-gated Na^+ channels in the soma. But is this what we would be interested in?

On the other hand, by measuring current, we are several steps closer to the actual transduction events and this is a much more direct measurement of the actual sensitivity of the receptors. The difficulty here is that we often confuse the concepts of threshold and EC_{50}, by equating both of them with some idea of sensitivity. I think there is no real threshold: there is a finite but small probability that even a single odour molecule could elicit a response. That is, if it happened to bump into the receptor molecule and stick for just a fraction of a second, but long enough to activate the cascade and lead to a few tens of channels opening. But what is the concentration of a single odour molecule? That is why the EC_{50} is a better measure. It really is a probability—the concentration at which, on average, half of the maximal response will be elicited.

Even measuring the current influx is several steps away from the primary event of receptor–ligand interaction, but at least it is still within what we might formally consider the transduction event, rather than in the signal (impulse) generation domain.

Hatt: One problem is that results from isolated, cell-free systems normally don't represent physiological conditions. F. Zufall, in my lab, showed that the conductance of the cAMP-activated olfactory channel is largely dependent on the extracellular Ca^{2+} concentration. Under experimental conditions of low Ca^{2+} concentrations ($<0.1\,\mu M$), the single-channel conductance is about 40 pS, but if the concentration is increased to 1 mM, the conductance decreases down to the fS range (Zufall & Firestein 1993). It means that an odorant-induced current of several hundred pA can be a consequence of the opening of relatively few big-conductance channels or of thousands of low-conductance channels. At present, we don't know precisely the free extracellular Ca^{2+} concentration under physiological conditions.

Lancet: On the other hand, one of the things we are interested in is the affinity of the receptors. The EC_{50} in these electrophysiological studies may occur at just 1:1000 saturation of the receptors, because of the second messenger amplification cascade. This has to be taken into account. It implies that the intrinsic association constant for odorant binding is much weaker than suggested by the EC_{50} (Lancet 1986).

Kurahashi: In order to understand the cell activity from electrophysiological studies, it is important to take into account the difference between current clamp and voltage clamp. Stuart Firestein's measurements were made under voltage clamp. However, real cells behave under current-clamp conditions. Since input resistances of the isolated receptor cells are very high, a small current is all that is needed to induce depolarization. However, the cell membrane itself shows strong outward rectification which is caused by the activation of voltage- and Ca^{2+}-activated K^+ channels. Thus, depolarization caused by odorant stimulation decreases membrane resistance. Furthermore, the odorant-activated conductance shows reversal potential at around 0 mV. Again, repolarization causes reduction of transduction current. Thus it is quite likely that, in terms of electrical activities, a saturating concentration of odorants causing several hundred pA of current under voltage-clamp conditions is still in the physiological range.

DeSimone: Couldn't the graded response that you see when you add the step odorant be fairly easily explained by the presence of an unstirred layer of 1 μm or so? The membrane would see such a region, under diffusion control, as variable, which would give rise to the type of response curves that you show.

Firestein: Two observations rule that explanation out. The first is the KCl trace, which shows that the concentration of KCl rapidly reaches a stable level and then falls off just as quickly. The second is that the cell's onset kinetics for each response are strongly concentration dependent. Each of these observations indicates that the cell saw the same concentration. The only thing that changed was the amplitude at the peak of the response, not the time of the peak.

Lancet: I'm not sure how integration over the time range of 10 ms–1 s could be relevant to olfaction. Odour pulses are about 1 s in duration: the odour comes into the nose, lasts for 0.5–1 s and goes away with a time constant of several hundred milliseconds. The channel may be sluggish: it has no reason to be fast because the stimulus is slow.

Reed: I disagree. Heinz Breer has already shown that in a cilium the entire second messenger signal is gone within 50–100 ms (this volume: Breer 1993).

Lancet: If the odorant is gone.

Reed: No; even if the odorant is still there!

Lancet: This is what you see in a stopped-flow experiment, but I am not sure that *in vivo* cAMP goes down to baseline concentration when the odorant is gone. Still, I agree that if cAMP dissociates slowly from the channel, the inactivation of the physiological response will be slowed down and this may be regarded as signal integration.

Ronnett: In our whole-cell experiments, the activation of adenylate cyclase does not go down that fast. Rather, adenylate cyclase is activated for 10–30 s upon continuous exposure to odorant. We feel that this is more consistent with the electrophysiological results. In addition, we have performed rapid stopped-flow experiments on isolated rat chemosensory cilia and found that although there is an initial subsecond burst of cAMP production, there is also a delayed elevation of cAMP which lasts at least 15 s. This confirms the results from our whole-cell experiments.

Firestein: In any sensory system, there is evolutionary pressure towards reaching an optimum transducer. But this optimum does not always have to be greater sensitivity and faster speed. In the olfactory system there is this immense task of recognizing a tremendous diversity of odours and doing it with reasonable sensitivity. In the physiology of these cells you see a series of trade-offs aimed at reaching that optimum: absolute sensitivity is traded-off for broad tuning. Some of the sensitivity is regained by amplification in the second messenger pathway, but this is at the cost of dynamic operating range. This can be regained by higher-order processing of cells with overlapping ranges. Finally, the cell gives up temporal resolution (for which it has no need) to regain sensitivity by integrating over time —sensitivity that is lost in deference to having optimally broad-tuned receptors.

References

Breer H 1993 Second messenger signalling in olfaction. In: The molecular basis of smell and taste transduction. Wiley, Chichester (Ciba Found Symp 179) p 97–114

Lancet D 1986 Vertebrate olfactory reception. Annu Rev Neurosci 9:329–355

Lindemann B, Frings S, Lynch JW 1992 The cAMP-gated channel from olfactory sensory neurons of frog and rat: blockage by divalent ions, modelling of ionic diffusion with the Eyring approach and modulation of sensitivity to cAMP. Eur J Neurosci Suppl 5:2023 (abstr)

Zufall F, Firestein S 1993 Divalent cations block the cyclic nucleotide gated channel of olfactory receptor neurons. J Neurophysiol 69:1758–1768

Olfactory receptors: transduction, diversity, human psychophysics and genome analysis

Doron Lancet, Nissim Ben-Arie, Shuki Cohen, Uri Gat, Ruth Gross-Isseroff, Shirley Horn-Saban, Miriam Khen, Hans Lehrach*, Michael Natochin, Michael North*, Eyal Seidemann and Naomi Walker

*Department of Membrane Research and Biophysics, The Weizmann Institute of Science, Rehovot 76100, Israel and *Department of Genome Analysis, Imperial Cancer Research Fund, Lincoln's Inn Fields, London WC2A 3PX, UK*

> *Abstract.* The emerging understanding of the molecular basis of olfactory mechanisms allows one to answer some long-standing questions regarding the complex recognition machinery involved. The ability of the olfactory system to detect chemicals at sub-nanomolar concentrations is explained by a plethora of amplification devices, including the coupling of receptors to second messenger generation through GTP-binding proteins. Specificity and selectivity may be understood in terms of a diverse repertoire of olfactory receptors of the seven-transmembrane-domain receptor superfamily, which are probably disposed on olfactory sensory neurons according to a clonal exclusion rule. Signal termination may be related to sets of biotransformation enzymes that process odorant molecules, as well as to receptor desensitization. Many of the underlying molecular components show specific expression in olfactory epithelium, with a well-orchestrated developmental sequence of emergence, possibly related to sensory neuronal function and connectivity requirements. A general model for molecular recognition in biological receptor repertoires allows a prediction of the number of olfactory receptors necessary to achieve efficient detection and sheds light on the analogy between the immune and olfactory systems. The molecular cloning and mapping of a human genomic olfactory receptor cluster on chromosome 17 provides insight into olfactory receptor diversity, polymorphism and evolution. Combined with future genotype–phenotype correlation, with particular reference to specific anosmia, as well as with computer-based molecular modelling, these studies may provide insight into the odorant specificity of olfactory receptors.
>
> *1993 The molecular basis of smell and taste transduction. Wiley, Chichester (Ciba Foundation Symposium 179) p 131–146*

The aim of the present chapter is to delineate the molecular understanding of olfaction that is emerging from work in our laboratory and others. In the last decade, olfactory science has evolved from a field with almost a complete lack of molecular models to one with a comprehensive molecular definition. This was brought about by the realization that ill-defined specificity and broad odorant spectra may still be compatible with stereospecific protein receptors; a notion that actually dates back to the pioneering work of Amoore (Amoore 1974, Whissell-Buechy & Amoore 1973), Beets (1971) and others. Convincing evidence for the existence of specific molecular reception pathways arose with the elucidation of G protein-coupled second messenger transduction cascades in olfactory cilia (Breer et al 1990, Pace et al 1985, Sklar et al 1986). Subsequently, what was mainly a hypothetical construct—a family of hundreds of G protein-coupled receptors (Lancet 1986, Lancet & Pace 1987, Reed 1990)—became a reality, with the breakthrough of the cloning of olfactory receptors (ORs) in several vertebrate species (Buck & Axel 1991, Parmentier et al 1991, Raming et al 1993). We describe here how such a molecular understanding may be combined with human psychophysical and genomic analysis, so as to obtain a better picture of human olfactory function.

Some open questions

The olfactory system is unusual in its ability to respond to millions of different chemicals (Lancet 1986, 1992, Shepherd 1991), in contrast to most other reception devices, which are much narrower in scope. Olfactory sensory neurons undergo membrane depolarization when exposed to almost any volatile organic chemical (Getchell 1986). Despite this, the pathway as a whole is at least as sensitive and selective as any other neuronal reception device. A long-standing open question has been how discrimination is maintained despite the very broad tuning of the individual sensory cells seen in single neuronal electrophysiological recordings (Sicard & Holley 1984). A favoured answer has been that the brain performs pattern analysis, whereby the exact combination of activities across the different sensory cells provides a definition of odour quality and concentration (Kauer 1991, Lancet 1986, Schild 1988, Shepherd 1985). None the less, the ill defined cellular response spectra suggested to some researchers that specific receptors could not be at work. Rather, non-specific odorant-induced mechanisms were proposed, such as membrane puncturing, lipid modulation or surface potential effects (Davies 1971, Nomura & Kurihara 1987). Alternatively, it has been argued that each sensory cell might bear numerous protein receptor types, hence the broad response spectra. Current opinions and evidence tend more towards the notion that each sensory cell might have one or very few receptor types (see below).

Another point of debate has been how many kinds of receptors are necessary for a functional olfactory system. An argument has been put forward that very few receptor types would suffice to code all odorants. A contrasting point of view, based on a probabilistic molecular recognition model (Lancet 1983, Lancet et al 1993a), claimed that 100–10 000 receptors would be necessary (Lancet 1986).

These questions remained unsettled for a long time, because continuous efforts failed to discover olfactory receptor proteins and their genes.

From transduction proteins to a receptor superfamily

The notion that olfaction is highly unusual and requires *ad hoc* paradigms gave way to the realization that odorant recognition utilizes widely conserved molecular devices. Odorant responses in olfactory cilia were shown to be mediated by heterotrimeric GTP-binding proteins (G proteins) and to involve the generation of second messengers such as cAMP, inositol trisphosphate and possibly nitric oxide (reviewed in Breer & Shepherd 1993, Lancet & Pace 1987, Reed 1990, Snyder et al 1988). Cation channels that respond to such second messengers through a direct-gating mechanism were found to underlie membrane depolarization in the sensory neurons (Dhallan et al 1990, Firestein et al 1991, Ludwig et al 1990, Nakamura & Gold 1987, Restrepo et al 1990). Most of these molecular components were found to be expressed rather specifically in olfactory sensory neurons. The preceding results led to the following deductions: (a) that olfactory receptor proteins must exist; (b) that one could narrow the search to the seven-transmembrane-domain receptor class, known to be universally coupled to G proteins; (c) that there should be a large family of receptors; and (d) that the receptors might be specifically expressed in olfactory epithelial tissue. A strategy based on these concepts finally resulted in the cloning of olfactory receptor genes (Buck & Axel 1991, this volume: Buck 1993).

Affinity of olfactory receptors: implications for reception and termination

Despite the rather low detection thresholds (nanomolar and lower) (Devos et al 1990), olfactory receptors are likely to have relatively low average affinities towards odorants—in the micromolar range (Lancet 1986). This is consistent with the notion that for most odorants (pheromones excluded), receptor–ligand interactions arise by chance, as novel xenobiotic compounds appear in the environment (Lancet et al 1993a). Amplification, for example by axonal convergence and second messenger cascades, then affords the detection of sub-nanomolar concentrations of air-borne odorants (Lancet 1986). As for many neurotransmitter receptors and contrary to the case of protein hormones and growth factors, signal termination may then occur via a dissociation-based mechanism: the ligand may leave the receptor binding site with a time constant of less than a second. In such a fast-equilibrium (or steady-state) system, signal modulation and termination may be mediated by perireceptor components (Carr et al 1990, Getchell et al 1984), such as carrier proteins (Bignetti et al 1985, Pevsner et al 1988, Raming et al 1989) and ligand-processing enzymes including biotransformation enzymes such as cytochrome P_{450} and UDP glucuronosyltransferase, which may mediate odorant signal termination (Carr et al 1990, Ding et al 1991, Lazard et al 1991, Nef et al 1989).

A low-affinity, probabilistic, multiple-receptor system also implies that each receptor will appear to have a rather broad ligand spectrum. Receptor promiscuity may actually be related quantitatively to the affinity and to the receptor repertoire size, through a formal receptor affinity distribution (RAD) model we recently developed (Lancet et al 1993a). That olfactory sensory neurons respond broadly to odorants is thus consistent with our expectations for the molecular properties of individual receptor proteins. It then follows that the broadly tuned sensory cells may well carry only one or very few receptor types—a phenomenon termed 'clonal exclusion' (Lancet 1986, Lancet et al 1993b). This possible scenario of one cell–one receptor type has implications for the way by which sensory neurons acquire their specificity and undergo unique axonal targeting (Lancet 1991). Specifically, it would be important to understand how the definition of a sensory neuron's identity through the particular OR gene(s) it expresses determines its synaptic target (Lancet 1992).

The genetic basis of human olfactory thresholds

Evidence for a genetic basis of olfactory threshold variations has been reported (Amoore & Steinle 1991, Gross-Isseroff et al 1992, Whissell-Buechy & Amoore 1973, Wysocki & Beauchamp 1984, 1991). The most likely molecular explanation is that threshold variations have to do with the different individuals having different OR genes. These differences could be due to point mutations, gene deletion or recombination. A more quantitative statement is the 'threshold hypothesis' (Lancet 1992), which suggests that an individual's olfactory threshold for a given odorant is determined by the receptor with the highest affinity towards this odorant. Specific anosmia (a decrement in the sensitivity to a particular odorant) may be explained by the absence of the gene for the highest affinity receptor, with the receptor gene with the next highest affinity 'taking over'. The latter then becomes threshold-determining. Specific hyperosmia might likewise be explained as the occurrence of a receptor with atypically high affinity, whose gene is present only in a few individuals. This simple model predicts that specific anosmia will behave as a recessive trait, as indeed has been reported (Whissell-Buechy & Amoore 1973). Likewise, specific hyperosmia is expected to behave as a dominant trait.

Olfactory receptor gene clusters

cDNA cloning of OR genes and preliminary comparisons with genomic clones indicated that OR coding regions are intronless and 900–950 bases long (Buck & Axel 1991, Levy et al 1991). This suggested that it would be relatively straightforward to carry out genomic cloning and mapping of such genes. As part

Olfactory receptor diversity and genome analysis 135

of an attempt to understand the genetics and molecular genetics of human olfaction, we have identified a cluster of about 20 OR genes within a contiguous stretch of ≈400 kb of DNA at the telomeric end of the p arm of human chromosome 17 (Ben-Arie et al 1992, N. Ben-Arie, D. Lancet, M. Khen, M. North & H. Lehrach, unpublished results) (Fig. 1). The cloning of this OR gene cluster was afforded through the screening of a single chromosome cosmid library (Nizetic et al 1991) with a general polymerase chain reaction (PCR)-generated OR probe, resulting in the isolation of 70 OR-positive cosmids. The cloned OR genes on chromosome 17 appear to be clustered with a density of about five genes per 100 kb, and several OR-coding regions are often found on one cosmid clone. Other OR gene clusters exist on chromosome 19 (Levy et al 1991) and possibly on several other chromosomes. Altogether, there may be several hundred OR genes in the human 'olfactory subgenome', perhaps occupying up to 10 Mb of DNA. Some of the open questions regarding such genes are listed in Fig. 2. A major challenge for the future will be to decipher the genomic structure and mechanisms that underlie OR diversity and control of gene expression within the OR gene clusters.

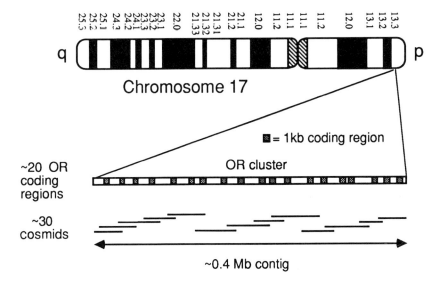

FIG. 1. A scheme depicting the cluster of olfactory receptor (OR) genes on human chromosome 17, at chromosomal band p13.3. 70 cosmids were isolated from a single chromosome cosmid library on the basis of their interaction with a PCR-generated probe generated by primers from regions conserved among many ORs. The localization is based on *in situ* hybridization and on the vicinity to published DNA markers. The entire DNA contiguous segment (contig) has been mapped and its OR genes have been partially sequenced (N. Ben-Arie, D. Lancet, M. Khen, M. North, H. Lehrach, unpublished).

Open questions	Predictions
1) Number of OR genes in an individual ?	N_{ind} =100-200
2) Number of genes in the population?	N_{pop}=500-1000
3) Allelic relations?	As in histocompatibility antigens
4) Gene pairs functionally identical? OR1 OR2 OR3 OR4 OR1* OR2* OR3* OR4*	Probably not OR1 OR2 OR3 OR4 OR5 OR6 OR7 OR8
5) Gene duplication?	Prevalent

FIG. 2. Open questions related to the number, arrangement and mutual relationships of olfactory receptor (OR) genes, with some hypothetical answers. It is proposed that OR polymorphism in the human populations is extensive, hence the number of OR genes in the population (N_{pop}) is considerably larger than in each human individual (N_{ind}). If the guess for N_{ind} is correct, then the human chromosome 17 OR cluster has 10–20% of the entire OR repertoire. It is proposed that allelic relations are as in histocompatibility antigens, whereby gene pairs at a given locus often constitute two alleles with multiple differences and possibly with a different ligand specificity. Gene duplication may be a major evolutionary mechanism for the generation of diversity in the OR repertoire and could also contribute to current inter-individual variation, as is the case for photoreceptor opsins (Nathans et al 1986).

Olfactory receptor polymorphism

Specific anosmia and hyperosmia are highly prevalent: practically any individual tested will display extreme olfactory threshold values for one in a few dozen odorants tested (Amoore & Steinle 1991). An explanation of such widespread individual differences in olfactory thresholds is that each human individual has only part of the entire OR gene repertoire. In other words, the number of alleles in the population may exceed greatly the number of gene loci in an individual (Fig. 2). This scenario holds for major histocompatibility complex genes (histocompatibility leukocyte antigen [HLA] in humans) (Klein et al 1983) and manifests itself in a high level of heterozygosity: at each locus there is a high probability of having two different alleles. The evolutionary regime that brings this about is known as balancing selection or overdominant selection (Altukhov 1991). The alleles generated by balancing selection are often different at multiple positions, and display a high ratio of non-synonymous to synonymous substitutions.

Individual variation in the ability to smell specific odorants is analogous to colour vision abnormalities. Most of the individual variations in colour sensitivity are related to the green and red pigments, because of the occurrence of these two cone opsin genes in a small cluster that usually contains genes for one red opsin and one to four copies of the green opsin (Nathans et al 1986). Variations arise due to unequal crossing over, resulting in an increase or decrease of the number of green opsin genes, in the loss of the green and/or red opsin, or in a red + green opsin recombination. It is quite likely that similar phenomena occur within OR gene clusters. The number of different variants that may arise in the latter case will be much larger, owing to the very high gene count. As a result, it is possible that no two human individuals (except monozygotic twins) are identical in their OR repertoire.

Olfactory receptor evolution

An important field of research addresses the ways by which the OR repertoire evolved. ORs appear to be more closely related to photoreceptor opsins than to other seven-transmembrane-domain receptors. This is mainly seen in the minimal intracellular and extracellular loop structures of both sensory receptor classes. For example, the third cytoplasmic loop (connecting transmembrane domains 5 and 6) is about 20 amino acids long in both ORs and photoreceptor opsins, but is usually larger in other superfamily members (up to 250 residues long in some of the muscarinic acetylcholine receptors). Likewise, the N- and C-termini of all sensory receptors are about 25 residues long, while in many other superfamily members one or both termini are much longer. This minimal structure raises the possibility that the sensory receptors are the most similar to the ancestral proteins in this receptor superfamily.

Macrosmatic mammals (e.g. dog and rat) may have many more OR genes than microsmatic ones (e.g. humans) and organisms with an even less broadly disposed olfactory system (e.g. fish and many invertebrates) could have even fewer ORs. A possible scenario is that early in evolution the number of OR genes of a species was small (10–30), as appears to be the case in fish (Ngai et al 1993a,b), *Caenorhabditis elegans* (this volume: Sengupta et al 1993) and *Drosophila* (this volume: Carlson 1993). In some mammals at later evolutionary stages, the number appears to have grown to several hundred (Buck & Axel 1991). Interestingly, this increase in OR diversity seems to have occurred solely at the germline level, without resort to ontogenetic DNA rearrangement, as is the case in immune receptors (Berek & Milstein 1988, Kronenberg et al 1986). Human evolution may have witnessed a recent decrease in OR gene number, as vision replaced olfaction as the dominant sense.

The original gradual increase in OR gene count may have occurred by gene duplication and diversification, as proposed for photoreceptor opsins (Nathans et al 1986). Two alternative scenarios may be envisioned: (1) OR genes were first individually scattered on different chromosomes, subsequently undergoing gene duplication; and (2) first, a small single OR gene cluster formed, then

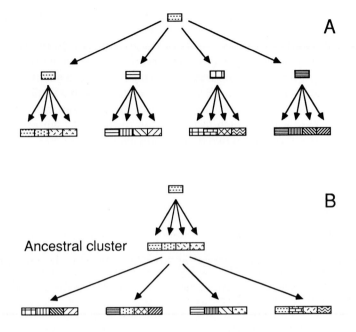

FIG. 3. Two alternative scenarios for the evolution of the olfactory receptor gene repertoire. A. An original single olfactory receptor (OR) gene is duplicated in distant genomic regions, perhaps also on different chromosomes. Each of these distant single genes then duplicates to form a cluster. The homology within each cluster is larger than that among members of different clusters. B. The original OR gene first forms a local cluster by duplication, termed the ancestral cluster. This cluster is then duplicated as a whole in different genomic regions, including on different chromosomes. In this case ORs within a cluster will not be particularly similar to each other and in some cases it will perhaps be possible to define homology groups *across* clusters, i.e. to find ORs in two different clusters that are more similar to each other than to other members of their respective clusters.

the cluster as a whole duplicated several times (Fig. 3). Homology relationships among the OR genes on human chromosome 17 and their comparison to OR genes in other species suggest that the second possibility is the correct one. This is because the OR genes in the chromosome 17 cluster show mutual differences nearly as large as those observed for randomly cloned OR genes (N. Ben-Arie, D. Lancet, M. Khen, M. North & H. Lehrach, unpublished results).

Conclusion

The cloning of the human chromosome 17 OR gene cluster may provide a good model system for studying the evolution of OR genes in humans and in other species. Our understanding of it will help in assessing the degree of genetic polymorphism among olfactory receptor genes. In the long run, correlating polymorphic genotypes with threshold variation phenotypes could help us to identify olfactory receptors with known odorant specificity. Molecular modelling based on known OR sequences, assisted by the predicted structures of other seven-transmembrane-domain receptors, could help us elucidate the structure of the OR odorant-binding site and its ligand pharmacology. This could eventually provide a molecular genetic basis for threshold polymorphisms, including specific anosmia and hyperosmia.

References

Altukhov YP 1991 The role of balancing selection and overdominance in maintaining allozyme polymorphism. Genetica 85:79–90

Amoore JE 1974 Evidence for the chemical olfactory code in man. Ann NY Acad Sci 237:137–143

Amoore JE, Steinle S 1991 A graphic history of specific anosmia. In: Wysocki CJ, Kare MR (eds) Chemical senses. Marcel Dekker, New York, vol 3:331–352

Beets J 1971 Olfactory response and molecular structure. In: Beidler LM (ed) Handbook of sensory physiology. Springer-Verlag, Berlin, vol 4:257–321

Ben-Arie, N, North M, Khen M, Margalit T, Lehrach H, Lancet D 1992 Mapping the olfactory receptor 'sub-genome': implications to human sensory polymorphisms. In: Myers R, Porteous D, Roberts R (eds) Genome mapping and sequencing. Cold Spring Harbor Laboratory Press, Cold Spring Harbor, NY, p 280

Berek C, Milstein C 1988 The dynamic nature of the antibody repertoire. Immunol Rev 105:5–26

Bignetti E, Cavaggioni A, Pelosi P, Persaud KC, Sorbi RT, Tirindelli R 1985 Purification and characterisation of an odorant-binding protein from cow nasal tissue. Eur J Biochem 149:227–231

Breer H, Boekhoff I, Tareilus E 1990 Rapid kinetics of second messenger formation in olfactory transduction. Nature 345:65–68

Breer H, Shepherd GM 1993 Implications of the NO/cGMP system for olfaction. Trends Neurosci 16:5–9

Buck L 1993 Receptor diversity and spatial patterning in the mammalian olfactory system. In: The molecular basis of smell and taste transduction. Wiley, Chichester (Ciba Found Symp 179) p 51–67

Buck L, Axel R 1991 A novel multigene family may encode odorant receptors: a molecular basis for odor recognition. Cell 65:175–187

Carlson J 1993 Molecular genetics of *Drosophila* olfaction. In: The molecular basis of smell and taste transduction. Wiley, Chichester (Ciba Found Symp 179) p 150–166

Carr WES, Gleeson RA, Trapido-Rosenthal HG 1990 The role of perireceptor events in chemosensory processes. Trends Neurosci 13:212–215

Davies JT 1971 Olfactory theories. In: Beidler LM (ed) Handbook of sensory physiology. Springer-Verlag, Berlin, vol 4:322–350

Devos M, Patte F, Rouault J, Laffort P 1990 Standardized human olfactory thresholds. IRL Press, Oxford

Dhallan RS, Yau KW, Schrader KA, Reed RR 1990 Primary structure and functional expression of a cyclic nucleotide-activated channel from olfactory neurons. Nature 347:184–187

Ding XX, Porter TD, Peng HM, Coon MJ 1991 cDNA and derived amino acid sequence of rabbit nasal cytochrome P450NMb (P450IIG1), a unique isozyme possibly involved in olfaction. Arch Biochem Biophys 285:120–125

Firestein S, Zufall F, Shepherd GM 1991 Single odor-sensitive channels in olfactory receptor neurons are also gated by cyclic nucleotides. J Neurosci 11:3565–3572

Getchell TV 1986 Functional properties of vertebrate olfactory receptor neurons. Physiol Rev 66:772–818

Getchell TV, Margolis FL, Getchell ML 1984 Perireceptor and receptor events in vertebrate olfaction. Prog Neurobiol 23:317–345

Gross-Isseroff R, Ophir D, Bartana A, Voet H, Lancet D 1992 Evidence for genetic determination in human twins of olfactory thresholds for a standard odorant. Neurosci Lett 141:115–118

Kauer JS 1991 Contributions of topography and partial processing to odor coding in the vertebrate olfactory pathway. Trends Neurosci 14:79–85

Klein J, Figueroa F, Nagy ZA 1983 Genetics of the major histocompatibility complex: the final act. Annu Rev Immunol 1:119–142

Kronenberg M, Siu G, Hood LE, Shastri N 1986 The molecular genetics of the T-cell antigen receptor and T-cell antigen recognition. Annu Rev Immunol 4:529–591

Lancet D 1983 What determines the size of the immune receptor repertoire? A probabilistic analysis of molecular interactions. In: First Congress of Israeli Societies of Life Sciences. Israel Biochemical Society, Jerusalem, p 169 (abstr)

Lancet D 1986 Vertebrate olfactory reception. Annu Rev Neurosci 9:329–355

Lancet D 1991 Olfaction. The strong scent of success. Nature 351:275–276

Lancet D 1992 Olfactory reception: from transduction to human genetics. In: Corey DP, Roper SD (eds) Sensory transduction. Rockefeller University, New York, p 73–91

Lancet D, Pace U 1987 The molecular basis of odor recognition. Trends Biochem Sci 12:63–66

Lancet D, Sadovsky E, Seidemann E 1993a Probability model for molecular recognition in repertoires of biological receptors: significance to the olfactory system. Proc Natl Acad Sci USA 90:3715–3719

Lancet D, Gross-Isseroff R, Seidemann E, Ben-Arie N 1993b Olfaction: from signal transduction and termination to human genome mapping. Chem Senses 18:217–225

Lazard D, Zupko K, Poria Y et al 1991 Odorant signal termination by olfactory UDP glucuronosyltransferase. Nature 349:790–793

Levy NS, Bakalyar HA, Reed RR 1991 Signal transduction in olfactory neurons. J Steroid Biochem Mol Biol 39:633–637

Ludwig J, Margalit T, Eismann E, Lancet D, Kaupp B 1990 Primary structure of cAMP-gated channel from bovine olfactory epithelium. FEBS (Fed Eur Biochem Soc) Lett 270:24–29

Nakamura T, Gold GH 1987 A cyclic nucleotide-gated conductance in olfactory receptor cilia. Nature 325:442–444

Nathans J, Piantanida TP, Eddy RL, Shows TB, Hogness DS 1986 Molecular genetics of inherited variation in human color vision. Science 232:203–210

Nef P, Heldman J, Lazard D, Margalit T, Jaye M, Hanukoglu I, Lancet D 1989 Olfactory-specific cytochrome P-450. cDNA cloning of a novel neuroepithelial enzyme possibly involved in chemoreception. J Biol Chem 264:6780–6785

Ngai J, Chess A, Dowling MM, Necles N, Macagno ER, Axel R 1993a Coding of olfactory information: topography of odorant receptor expression in the catfish olfactory epithelium. Cell 72:667–680

Ngai J, Dowling MM, Buck L, Axel R, Chess A 1993b The family of genes encoding odorant receptors in the channel catfish. Cell 72:657–666

Nizetic D, Zehetner G, Monaco P, Gellen L, Young BD, Lehrach H 1991 Construction, arraying and high-density screening of large insert libraries of human chromosome X and 21: their potential use as reference libraries. Proc Natl Acad Sci USA 88:3233–3237

Nomura T, Kurihara K 1987 Liposomes as a model for olfactory cells: changes in membrane potential in response to various odorants. Biochemistry 26:6135–6140

Pace U, Hanski E, Salomon Y, Lancet D 1985 Odorant-sensitive adenylate cyclase may mediate olfactory reception. Nature 316:255–258

Parmentier M, Libert F, Schurmans S et al 1991 Expression of members of the putative olfactory receptor gene family in mammalian germ cells. Nature 355:453–455

Pevsner J, Reed RR, Feinstein PG, Snyder SH 1988 Molecular cloning of odorant-binding protein: member of a ligand carrier family. Science 241:336–339

Raming K, Krieger J, Breer H 1989 Molecular cloning of an insect pheromone-binding protein. FEBS (Fed Eur Biochem Soc) Lett 256:215–218

Raming K, Krieger J, Strotmann J et al 1993 Cloning and expression of odorant receptors. Nature 361:353–356

Reed RR 1990 How does the nose know? Cell 60:1–2

Restrepo D, Miyamoto T, Bryant BP, Teeter JH 1990 Odor stimuli trigger influx of calcium into olfactory neurons of the channel catfish. Science 249:1166–1168

Schild D 1988 Principles of odor coding and a neural network for odor discrimination. Biophys J 54:1001–1011

Sengupta P, Colbert HA, Kimmel BE, Dwyer N, Bargmann CI 1993 The cellular and genetic basis of olfactory responses in *Caenorhabditis elegans*. In: The molecular basis of smell and taste transduction. Wiley, Chichester (Ciba Found Symp 179) p 235–250

Shepherd GM 1991 Sensory transduction: entering the mainstream of membrane signaling. Cell 67:845–851

Sicard G, Holley A 1984 Receptor cell responses to odorants: similarities and differences among odorants. Brain Res 292:283–296

Sklar PB, Anholt RR, Snyder SH 1986 The odorant-sensitive adenylate cyclase of olfactory receptor cells. Differential stimulation by distinct classes of odorants. J Biol Chem 261:15538–15543

Snyder SH, Sklar PB, Pevsner J 1988 Molecular mechanisms of olfaction. J Biol Chem 263:13971–13974

Whissell-Buechy D, Amoore JE 1973 Odour-blindness to musk: simple recessive inheritance. Nature 242:271–273

Wysocki CJ, Beauchamp GK 1984 Ability to smell androstenone is genetically determined. Proc Natl Acad Sci USA 81:4899–4902

Wysocki CJ, Beauchamp GK 1991 Individual differences in human olfaction. In: Wysocki CJ, Kare MR (eds) Chemical senses. Marcel Dekker, New York, vol 3:353–373

DISCUSSION

Siddiqi: We estimated the number of independent receptors for several chemicals in *Drosophila* by cross-adaptation or 'jamming' experiments. The largest number of receptors seems to be for acetate esters, probably five or six. There may be three or four receptors for alcohols (Siddiqi 1983, Börst 1983). We have also estimated the affinities of these receptors from single-unit response curves. The highest affinity is for ethyl acetate (half maximal dilution of

10^{-9} M). Amyl acetate and alcohol are in the middle of the range (half maximal dilution of 10^{-6} M). The lowest affinities (around 10^{-3} M) are for chemicals like cyclohexanone and aldehydes, for which, presumably, the fly has little use. These results seem to contradict your model.

Lancet: I wasn't careful enough to indicate that the model (Lancet et al 1993) applies only in those cases concerning odorants for which the receptors are not evolutionarily honed. Higher up in the evolutionary ladder (vertebrates, for instance), organisms begin to face behavioural tasks that require recognition of odorants that are previously unknown to them. We think that the olfactory system started with bacteria, which have a few receptors, progressed in organisms such as *Drosophila* and fish (which have 10–50) and then evolved into a system with 1000 or so receptors. Our model predicts that you get to a point where for every odorant encountered, at least one of the receptors has a significant affinity, at which point you don't need any more evolution for specific odorants. Traces of those evolutionary events may still prevail in us as well, so there are some odorants for which we have better receptors than others. The model is a generalized notion that can then be modified to accommodate some receptors that are evolutionarily related to particular ligands, as well as more generalized receptors. In *Drosophila*, I would assume that this equation tends towards the end of the spectrum where there are few receptors and they are geared evolutionarily for what the fly likes to do most—to look at flowers or fruit. So, in this case, the model may not apply in a simple way.

Margolis: Another potential complication in correlating the psychophysics with the molecular biology is the recent report from Wysocki et al (1989) that the observed psychophysics changes with exposure. Thus, some individuals initially anosmic for androstenone can subsequently detect it, once they have been exposed to this stimulus. This may mean that you are dealing with a changing database. How is that incorporated into your methodology?

Lancet: If it is proved that what you have said is always true for all odorants, it will be a problem. I believe, at present, that report remains an isolated case. However, when I say there is a genetic basis for olfactory thresholds, I am implying that there might be other determinants as well. The data are never perfect and may also include an environmental effect. The question is, can we purify the genetic component?

Bargmann: Doron, you say there may be as few as one receptor per cell. Stuart Firestein's results (this volume: Firestein & Zufall 1993) show that in the salamander one out of four receptor cells responds to cineol. Your model says that 1000 receptors can recognize almost any odorant. Does it tell you how many of these receptors recognize a particular odorant?

Lancet: Our model can actually give us a titration curve. If you are faced with 1000 receptors and one ligand, you titrate the ligand, starting with a low concentration that addresses high affinity receptors. As you increase the concentration of the odour, more receptors are recruited. The model tells us how many receptors will be recruited, on average, at each concentration. From the model, we can be comfortable with the idea that olfactory receptors are promiscuous, without feeling that we have made a compromise or invented something unusual. It predicts that all receptor repertoires are the same in this respect.

Bargmann: So you are not troubled by the idea that the model suggests that an odorant will be recognized by up to 25% of receptors.

Lancet: I'm not; that's what it suggests for low-affinity receptors.

Kinnamon: Stuart, what proportion of your cells show a high affinity to a particular ligand?

Firestein: When the same cell responds to different odours, it does so with differing affinities. For example, some odours exist as partial agonists. You could use these in competition experiments to determine whether you are looking at one receptor recognizing both odours, or different receptors, where you should see either additive affects or some competition (this volume: Firestein & Zufall 1993).

Kinnamon: What percentage of the cells have high micromolar affinity?

Firestein: There are very few cells that respond with such a high affinity. Normally, we don't see very high affinity in these cells.

We are only sampling three odorants and these are odours that, while relevant to the salamander, are not highly relevant in the sense of being pheromonal. They are what I presume to be nesting odours—they're fruity, floral, herbal, dirt-type odours. These odours probably occur at relatively high concentrations in the world of the salamander.

Lindemann: Doron, you said that the overall threshold for an odorant is determined by the receptor with the highest affinity. However, isn't it true that other factors co-determine the threshold, such as the number of receptors and other integrative processes, including the amplification of the second messenger cascade? If two cells have receptors of the same kind and one has more of them, the threshold of this cell might be lower.

Lancet: I agree.

Firestein: Any amplification system, for that matter, would serve to increase the sensitivity. I think that the concept of 'threshold' is probably a poor one; what really counts is whether you can get the signal out of the noise. That is, all of these receptors can respond to a single molecule—the question is how many molecules have to be present for an interaction to occur between the receptor and a single molecule. It is more of a question of probability than of absolute threshold.

Lancet: It is important to note that I'm not even beginning to imply that the threshold concentration has anything to do with the actual value of the binding constant. There are many factors in between. I'm saying that if an olfactory receptor has a high affinity for one odorant and a low affinity for another, there would be a difference in threshold that would have some correlation with the affinities.

Buck: We should remember that the immune system and the olfactory system are completely different in that the olfactory system is plastic and it may be shaped by experience.

Lancet: The immune system is plastic.

Buck: It's not plastic in the way the nervous system is. In the immune system there are always virgin B cells sitting and waiting in case an antigen comes along, whereas the nervous system is shaped by experience.

Lancet: It's exactly what Frank Margolis was alluding to: that experience or exposure might modify the olfactory responses. We are dealing with a complex system and are trying to dissect some of the parameters—to indicate the threshold might be a good way to address some of these at the molecular level. Clearly, there are complications with this approach; plastic changes might be one and another is that the affinity alone doesn't determine the threshold.

Siddiqi: The parameter that will be most closely related biologically to the number of receptors for an odorant is the range of concentrations over which the animal must be able to measure the existence of that odorant.

Firestein: That could be made up by different cells with different but overlapping affinity ranges and this information could be integrated at a higher processing level.

Lush: Is there any way in which you could test your concept that cytochrome P_{450} plays an essential part in the deactivation of odorant receptors?

Lancet: There are many experiments that can be done. We have better data for UDP glucuronosyltransferase (UGT) than for cytochrome P_{450}, showing that it inactivates odorants. We have tried doing experiments *in vivo* using inhibitors of P_{450}, but so far these haven't been successful.

Lush: Do you find conjugated forms of odorants in the mucus?

Lancet: With UGT, we can show odorant converted to its glucuronate in an olfactory epithelial explant.

Lush: But is it then secreted onto the cell surface?

Lancet: Yes; we have results that show this as well.

DeSimone: In the process of enzymic conversion, some of the early products themselves may be odiferous and may contribute an overtone to the stimulus. This could complicate the issue of what the stimulus is.

Lancet: Yes; there could be generation of, say, hydroxylated odorants *in situ*. In some cases this may be what we actually smell rather than the original compound or mixture.

Ache: But are the kinetics of those reactions such that you might expect them to happen within the time course of perception, which can be perhaps as short as 100 ms?

Lancet: That's a good point. They're fast enough to serve as termination devices, but they might be too slow to contribute to odour perception.

Getchell: Doron, I think your model is excellent in that it combines molecular genetics with human psychophysical experience. One of the hypotheses of your model is that the measure of affinity of the ligand or the odorant for the receptor is the threshold of the psychophysical experience. What is the evidence for the stability of psychophysical thresholds for odour perception in a population? In some of the earlier psychophysical data there appears to be a 'learning' process in odour perception (Doty 1991). How common is psychophysical plasticity in the population for taste and smell?

Bartoshuk: There is a problem with measuring thresholds in olfaction. Olfaction is odd in that the psychophysical functions are very flat. Flat psychophysical functions have an interesting property, because a slight change in the lower portion of the psychophysical function can shift the threshold way up and down and produce tremendous variability. As a result, the perceptual world and perceived intensity may be very stable, even if the threshold shows large variability. Your model might actually measure a psychophysical function, when a different attribute of the function would be a better way of quantifying what you are trying to measure.

Buck: What other parameter of perception should be measured?

Bartoshuk: The suprathreshold intensity.

Lancet: We've long been intrigued by the flatness of the olfactory psychophysical function and we think we may have a molecular explanation for it— the heterogeneity of receptors. This goes back again to our model, but can also be explained from first principles without the model (Lancet 1986). If you plot closely disposed standard saturation curves for several ligands and you measure the sum, the log/log plot (which is the same as the Hill plot) is rather flat (slope $\ll 1$). The electroolfactogram plot also has the same low slope when drawn as a log/log plot.

Buck: Are you implying that several different receptors are used for every response?

Lancet: That is true for suprathreshold responses. I would agree with Linda Bartoshuk about using suprathreshold measures, had it not been that it might then involve more than one receptor. I propose that we put up with the noise that we agree exists at the threshold level, just because it gives us a better chance of looking at individual receptors. The intercept of magnitude estimation curves is noisy, but when you measure threshold, you don't really use magnitude estimation. There are more objective means of obtaining thresholds and they are considerably less noisy.

Bartoshuk: But thresholds are *terribly* noisy in olfaction.

Buck: What do you mean by noise? Do you mean variability?

Bartoshuk: I mean when you do the same experiment on the same person over and over again and you keep getting different answers.

Firestein: Doron Lancet is right: that variance will decrease as you simplify the system and get down to a cellular or molecular level.

Pelosi: My own experience is that thresholds for an odorant are rather reproducible, when measured with the same method and averaged over a population of 30–50 subjects. They seldom differ by a factor larger than two, even if measured with different subjects (P. Pelosi, unpublished observations). However, individual olfactory thresholds can be much more variable and strongly depend on the physiological and emotional conditions of the subject.

Lancet: What you have to ask is this: looking at one of Amoore's population curves (Amoore & Steinle 1991), if I take an individual, what would the width of his curve be? As long as the width of an individual curve is appreciably lower than that for the entire population, you can begin to discern genetics. If it's equal, you can't.

Reed: In psychophysical experiments in vision, one of the ways of finding anomalous individuals is to look for people who, in independent experiments, make matches that display a very wide scatter. You can predict that someone who lacks the high affinity receptor will, if you test them 10 times, have a much greater scatter than normal.

Lancet: This is an interesting point that needs to be studied.

References

Amoore JE, Steinle S 1991 A graphic history of specific anosmia. In: Wysocki CJ, Kare MR (eds) Chemical senses. Marcel Dekker, NY, vol 3:331–352

Börst A 1983 Computation of olfactory signals in *Drosophila melanogaster*. J Comp Physiol 152:373–383

Doty RL 1991 Olfactory system. In: Getchell TV, Doty RL, Bartoshuk LM, Snow JB (eds) Smell and taste in health and disease. Raven Press, NY, p 175–203

Firestein S, Zufall F 1993 Membrane currents and mechanisms of olfactory transduction. In: The molecular basis of smell and taste transduction. Wiley, Chichester (Ciba Found Symp 179) p 115–130

Lancet D 1986 Vertebrate olfactory reception. Annu Rev Neurosci 9:329–355

Lancet D, Sadovsky E, Seidemann E 1993 Probability model for molecular recognition in repertoires of biological receptors: significance to the olfactory system. Proc Natl Acad Sci USA 90:3715–3719

Siddiqi O 1983 Olfactory neurogenetics of *Drosophila*. In: Chopra VL, Joshi BC, Sharma RP, Bansal HC (eds) Genetics: new frontiers. Oxford & IBH, New Dehli, p 242–261

Wysocki CJ, Dorries KM, Beauchamp GK 1989 Ability to perceive androstenone can be acquired by ostensibly anosmic people. Proc Natl Acad Sci USA 86:7976–7978

General discussion II

Margolis: I think it would be helpful for us to compare some of the technologies that have been used to investigate the transduction components in olfaction and taste, to get an idea of their relative strengths and weaknesses. This is because I suspect that some of the disagreements we have are due to the different methodologies we have used.

Ache: Heinz, do you think that the success you had in getting functional expression of the olfactory receptor may be because your assay got the components of the system together in closer physical intimacy than they would have been in a whole-cell assay?

Breer: It is difficult to say why the functional expression of a cloned receptor works in certain systems and not in others. We have tried several different cell lines and various receptor clones. You need quite a bit of luck to find a suitable expression system for the right receptor clone and its appropriate ligands.

Buck: I think the reason that Heinz Breer has been able to see a response in transfected cells, that none of the rest of us have, might have to do with the fact that he's homogenizing the cells and making membranes. There's an indication from Randy Reed's experiments that in some cell lines the receptor is not getting to the surface of the transfected cell, but staying stuck in the ER or the Golgi. It's possible that when Heinz homogenizes the cells and makes membranes, he is getting a mixing of the various components and lateral diffusion of proteins in the membrane, so that the receptors which were separated from the transduction components now have access to them.

Breer: The baculovirus/Sf9 cell system that we use for expression of odorant receptors (Raming et al 1993) is well known for its high efficiency in expressing recombinant genes (Miller 1988). It may simply be that this system produces just enough protein and targets it to the right sites so that its function can be monitored by the sensitive second messenger assay.

Reed: Traditionally, coupling to second messenger pathways is much more efficient in whole cells than it is in isolated membranes, so you might argue the other way. I think that the success that Heinz has had (Raming et al 1993) brings us to a point where we can try to begin to try and sort some of this out. One of the most interesting questions is going to be looking at closely related receptors (we have one pair that differs by six amino acids) and seeing how close the tuning curve is for the same ligands. This type of experiment may enable us to explain Stuart Firestein's results where he showed that receptors respond to a wide variety of odorants.

Lancet: Actually, Heinz Breer's results show a very broad spectrum of responses of receptors to odorants. One of the receptors tested responded to quite a few odorants and the other receptor is activated by hardly any odorants.

Firestein: Do you know for sure that the other receptor was expressed sufficiently?

Breer: Without any functional response, we have currently no alternative approach with which to assay odorant receptor expression. The availability of sequence-specific anti-peptide antibodies will allow us to monitor the synthesis of receptor proteins in heterologous systems using immunological approaches like enzyme-linked immunosorbent assays (ELISA) or Western blots.

Reed: We've taken a receptor closely related to Heinz Breer's OR5 receptor, expressed it in baculovirus, tested at least with lilial and lyral (the most responsive odorants for OR5) and we see almost no response. That could be because we can't get enough of our receptor on the plasma membrane, or those six amino acids are profoundly important in the ligand specificity. It seems unlikely that the cellular location of receptors with such extreme similarity is different.

Margolis: Is the ability to see a response a function of level of receptor expression or coupling efficiency or post-translational modification or insertion into the plasma membrane? If you do the transfection with one receptor clone that apparently works, in replicate batches, do you see the same level of response in all cases or is there a range? This might enable you to determine whether it is a question of passing a threshold of expression level.

Breer: When we did these experiments we got good dose–response curves and the batches of cells didn't show much variation.

Lindemann: Is the receptor phosphorylation dependent on continued binding of ligand?

Breer: The receptors obviously have to be liganded for the cascade to begin, but it's hard to tell whether the ligand has to remain in place for phosphorylation of the receptor protein to occur.

Ronnett: My laboratory has recently published the localization and functional studies implicating β-adrenergic receptor kinase 2 (βARK2) and β-arrestin 2 in olfactory desensitization (Dawson et al 1993). We had initially been collaborating with Dr Lefkowitz in order to look at the localization of the isoforms of βARK2 and β-arrestin 2 in the brain. We then decided to look in the olfactory neural epithelium and discovered that βARK2 and β-arrestin 2 are highly localized to the dendritic knobs and cilia of the olfactory receptor neurons. As the odorant receptors are thought to be members of the G protein-coupled receptor family, we thought it was reasonable that this kinase and arrestin may be involved in desensitization of the odorant response. Dr Lefkowitz provided us with neutralizing antibodies to both isoforms of βARK and arrestin. We found that the neutralizing antibodies to βARK1 and β-arrestin 1 attenuated desensitization of the odorant-induced cAMP response slightly, but that antibodies to the βARK2 and β-arrestin 2 isoforms not only attenuated the desensitization, but elevated the peak of the cAMP response approximately fourfold. Simultaneous addition of neutralizing antibodies to both βARK2 and β-arrestin 2 resulted in a synergistic effect, with the peak cAMP level rising approximately eightfold. These results suggest that these kinases and arrestin play a significant role in desensitization. Additionally, these results suggest that arrestins do not need a βARK-phosphorylated substrate to work, but can work independently. They also suggest that the absolute level of cAMP normally achieved in the cell does not represent the maximum. This suggests that the cyclase is quite active and that desensitization must dampen the signal.

References

Dawson TM, Arriza JL, Jaworsky DE et al 1993 β-adrenergic receptor kinase-2 and β-arrestin-2 as mediators of odorant-induced desensitization. Science 259:825–829
Miller LK 1988 Baculovirus as gene expression vectors. Annu Rev Microbiol 42:177–199
Raming K, Krieger J, Strotmann J et al 1993 Cloning and expression of odorant receptors. Nature 361:353–356

Molecular genetics of *Drosophila* olfaction

John Carlson

Department of Biology, Yale University, PO Box 6666, New Haven, CT 06511-8112, USA

Abstract. The functional organization of the *Drosophila* olfactory system has been studied using two complementary approaches. (1) A genetic screen has yielded olfactory mutants, of which one, *acj6*, shows greatly reduced electrical responses of its olfactory organs to all odorants tested except benzaldehyde (odour of almond). This mutant is also defective in olfactory behaviour at the larval stage. It is normal in tests of visual system physiology and behaviour. We aim to determine whether the *acj6* gene is expressed on a fraction of the antennal surface. (2) We have carried out an enhancer trap screen for olfactory genes which are expressed on regions of the antennal surface. From a screen of 6400 enhancer trap lines, 12 show expression of a *lacZ* reporter gene in both larval and adult olfactory organs, but relatively little expression elsewhere. Most lines identify regions of the adult antenna and some show labelling of very small numbers of cells. One line shows sexual dimorphism in its expression pattern and another pattern becomes increasingly restricted with adult age. Several lines also show staining of other chemosensory organs and one also shows staining in the male reproductive tract. We have generated deletions of DNA flanking these enhancer trap insertions to determine whether any of the lines define genes associated with responses to specific sets of odorants.

1993 The molecular basis of smell and taste transduction. Wiley, Chichester (Ciba Foundation Symposium 179) p 150–166

Drosophila melanogaster has a highly sensitive olfactory system (Siddiqi 1987, Carlson 1991). It responds to a wide variety of volatile chemicals and is capable of odour discrimination. *Drosophila* offers several advantages to the study of olfaction. First, its olfactory system is relatively simple, containing in the order of 10^3 receptor neurons, as opposed to $\approx 10^8$ in humans. Second, its olfactory system can be manipulated through a number of genetic and molecular approaches that would be difficult to use in most other organisms. Third, the function of the system can be analysed *in vivo* by measuring either electrophysiological or behavioural responses to odorants.

One of the most interesting questions concerning olfaction is how olfactory systems are organized so as to allow discrimination among different odorants

Molecular genetics of Drosophila olfaction 151

(for a review, see Shepherd 1985). This article describes work which may be useful in describing the functional organization of the *Drosophila* olfactory system and, in particular, how sensitivities to different odorants map onto the surface of the olfactory organs. It illustrates the utility of *Drosophila* as an experimental organism in which to examine olfaction, in that we have been able to take two experimental approaches which could not be used conveniently in vertebrate systems. The first approach is a genetic screen to isolate mutants defective in olfactory function; the second is an enhancer trap screen for olfactory genes. The first approach has yielded mutants that are defective in responses to subsets of odorants; genes defined by such mutations can be cloned and analysed to determine whether they are expressed in specific parts of the olfactory system. Conversely, genes that are expressed in different parts of the olfactory system can be mutated to determine whether they are required for responses to subsets of odorants.

Fruit flies sense odorants through the third segment of the antenna (Fig. 1a). This segment is covered with ≈ 500 sensory hairs, which fall into three morphological classes: the sensilla basiconica, sensilla coeloconica and sensilla trichodea (Venkatesh & Singh 1984, Stocker & Gendre 1988). Each class of hair

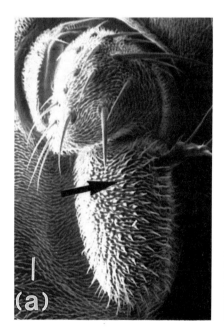

FIG. 1. (a) Antenna. Arrow indicates the third segment. Dorsal is at top; ventral is at bottom; lateral is at right; medial is at left. Scale bar = 25 μm. (b) Maxillary palp. The palp is ≈ 200 μm long. (a) is reproduced with permission from Ayer & Carlson (1992). ©1992, John Wiley & Sons, Inc.

has a characteristic distribution on the antennal surface and each hair is innervated by up to four neurons, which project directly to the antennal lobes of the brain. The sensilla basiconica and sensilla coeloconica contain pores or channels through which odorants are believed to pass; evidence conflicts as to whether the trichoid sensilla contain pores in *Drosophila*. There is physiological evidence that sensitivities to different odorants are not distributed uniformly on the antennal surface (Siddiqi 1983, 1987, Ayer & Carlson 1992).

Recently, a second organ of *Drosophila*, the maxillary palp, has been found to have olfactory function by physiological recording (Ayer & Carlson 1992). The maxillary palp (Fig. 1b), which is one of the mouthparts, is covered with basiconic and trichoid sensilla and sends projections to the antennal lobes (Singh & Nayak 1985, Stocker et al 1990).

Drosophila larvae also have a strong olfactory response (Monte et al 1989). This response is believed to be mediated through the antennal organ of the antenno-maxillary complex (AMC) (Singh & Singh 1984), located at the anterior tip of the larva.

Results and discussion

The acj6 mutant distinguishes two classes of odorant pathways

We found that fruit flies jump when exposed to high concentrations of any of several odorants (McKenna et al 1989). The response is mediated primarily through the antenna and is dose dependent. It is not merely a manifestation of trauma—flies recover quickly and can respond to a subsequent stimulus. We have used this behaviour as the basis of a genetic screen and have isolated a set of mutants defective in the jump response. Such mutants, called *acj* mutants (*acj* = *abnormal chemosensory jump*), could *a priori* be defective in any of a variety of steps, from odorant reception to motor response. To determine whether any *acj* mutants had defects in the peripheral steps of the olfactory pathway, we measured physiological responses of the olfactory organs to odorant stimulation. These measurements were made in extracellular recordings—electroantennograms (EAGs) and electropalpograms (EPGs)—which are believed to measure primarily the summed receptor potentials of olfactory neurons.

The *acj6* mutant shows a reduction in the amplitude of response to odorants in both the antenna and the maxillary palp, suggesting that the mutation blocks a step in odorant reception or signal transduction (Ayer & Carlson 1991, 1992). Interestingly, the defect shows odorant-specificity: this can be observed most clearly in recordings from the maxillary palp (Fig. 2). Whereas the amplitudes of response to acetone, propionic acid, butanol and ethyl acetate are drastically reduced in *acj6* mutants, response amplitude to benzaldehyde (odour of almond) is essentially normal in the maxillary palp over a broad range of stimulus concentrations.

Molecular genetics of *Drosophila* olfaction 153

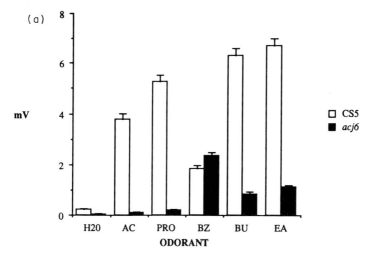

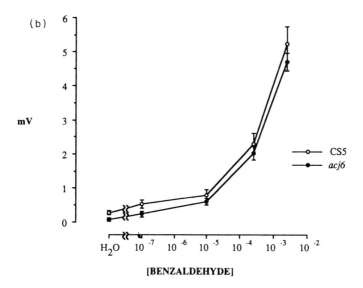

FIG. 2. Mutation in *acj6* produces an odorant-specific defect in olfactory physiology in the maxillary palps. (a) Amplitudes of wild-type (CS5) and mutant (*acj6*) responses in maxillary palps (from electropalpograms) to acetone (AC) (1:1000 dilution); propionic acid (PRO) (1:666 dilution); benzaldehyde (BZ) (1:4000 dilution); 1-butanol (B) (1:500 dilution) and ethyl acetate (EA) (1:2500 dilution). Dilutions were in water. $n = 25$ flies for each genotype. (b) Dose–response curves for dilutions of benzaldehyde in water. $n = 10-11$ flies for each genotype. Error bars represent SEM. CS5 is the parental wild-type for *acj6*. Reproduced with permission from Ayer & Carlson (1992). ©1992 John Wiley & Sons, Inc.

Consistent with the maxillary palp recordings, *acj6* reduces antennal response to ethyl acetate, but appears to cause little if any reduction for benzaldehyde (Fig. 3). These results are complicated by the fact that the wild-type antenna shows a substantial electrical response not only to stimulation with odorant, but also to stimulation with water, which is used as a solvent for odorants. *acj6* causes a decrease of ≈ 5 mV in the response to water stimulation. Figs 3a and 3b show that in *acj6* mutants the decrease in response to ethyl acetate (dissolved in water) ranged

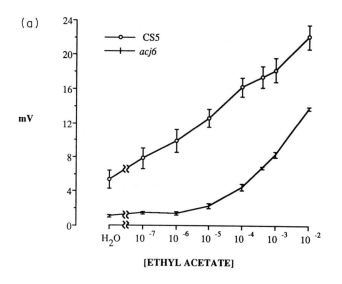

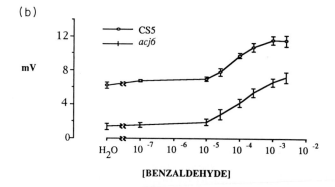

FIG. 3. Effect of *acj6* mutation on antennal response to ethyl acetate and benzaldehyde. (a) Response to dilutions of ethyl acetate in water. (b) Response to dilutions of benzaldehyde in water. (c) Ethyl acetate responses following subtraction of response amplitude for water stimulation. (d) Benzaldehyde responses following subtraction of

up to 12 mV, depending on ethyl acetate concentration—decreases which are much larger than those for water alone. By contrast, the decrease in response to benzaldehyde was smaller and dose independent: ≈ 5 mV at each concentration tested, the same decrease found for stimulation with water alone. The simplest interpretation of these results is that the *acj6* mutation reduces the response to water and to ethyl acetate vapour, but causes little if any reduction for benzaldehyde vapour.

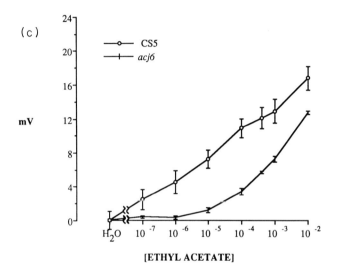

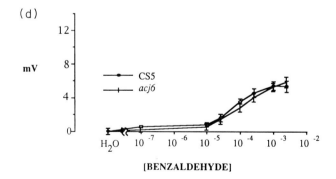

response amplitude for water stimulation. Each curve represents mean amplitudes for four flies for each genotype. CS5 is the parental wild-type for *ajc6*. Error bars represent SEM. All recordings made from the dorsomedial portion of the antenna (the medial portion of 'quadrant 1' of Ayer & Carlson 1992). Adapted from Ayer & Carlson (1992).

This interpretation can be drawn more directly from Figs 3c and d, in which the amplitudes of the water responses have been subtracted from each point of the dose–response curves shown in Figs 3a and b. *acj6* is seen to cause a large, dose-dependent decrease in response to ethyl acetate, but not to benzaldehyde. [In subtracting the water response, we are, however, making certain assumptions (Ayer & Carlson 1992).] These results are consistent with measurements made from a different location on the antenna (Ayer & Carlson 1991), which also provided evidence for a decrease in response to ethyl acetate vapour *per se*; in these experiments the decrease in response to benzaldehyde was again substantially smaller than that to ethyl acetate and could be accounted for predominantly, if not exclusively, in terms of reduced sensitivity to water stimulation.

How specific are the effects of the *acj6* mutation? The physiological response of the eye to flashes of light is normal, as determined by electroretinogram measurements (Ayer & Carlson 1991). The olfactory response of the larva is defective, as measured in a simple behavioural assay in which animals migrate across an agarose-containing Petri dish towards an olfactory attractant. By contrast, *acj6* larvae are normal in a test of visually driven behaviour. The *acj6* mutation also affects some non-olfactory functions. During EAG recording, the antenna appears to show a small response to the mechanosensory stimulation accompanying odorant delivery; *acj6* apparently reduces this response. This defect may explain, at least in part, our observation that *acj6* flies show reduced behavioural response to mechanical stimuli (e.g. agitation of culture vials) and reduced climbing activity. Flies also show reduced chemosensory jump response to benzaldehyde, despite the strong physiological response to this odorant. One interpretation of this finding is that *acj6* affects not only peripheral olfactory physiology for some odorants, but also an additional function that is required for jump response to all odorants.

The *acj6* mutation maps to position 49.4 on the X chromosome (Ayer & Carlson 1991). Efforts are currently underway to isolate additional alleles of *acj6*. It will be of interest to determine whether *acj6* is a null mutant and, if not, whether a null mutant (if viable) has normal electrical response to benzaldehyde.

The *acj6* defect is reciprocal to that of *ptg^{3D18}*, another recently characterized olfactory mutant (Helfand & Carlson 1989), in the following sense: whereas *acj6* severely affects the EPG in response to all odorants tested except benzaldehyde, *ptg^{3D18}* affects response to benzaldehyde, but not to other odorants, in two different behavioural assays. Other mutants showing benzaldehyde-specific defects were isolated some years ago (Rodrigues & Siddiqi 1978, Ayyub et al 1990), although none of these—nor *ptg^{3D18}*—has been demonstrated to have peripheral, rather than central, effects. Benzaldehyde is also distinctive in that it repels flies in a Y-maze at all tested concentrations, whereas propionic acid, butanol and ethyl acetate are attractive at all but the highest concentrations (Ayyub et al 1990). There is also evidence to distinguish benzaldehyde from certain other odorants from behavioural (Rodrigues 1980)

Molecular genetics of *Drosophila* olfaction

and EAG (Siddiqi 1983) studies in which odorants are tested alone and in pairs to determine whether response to one odorant is reduced by exposure to a second odorant.

A model suggested by these results is that benzaldehyde is sensed through an odorant pathway distinct from that of other odorants (Rodrigues 1980, Siddiqi 1983, Ayer & Carlson 1992). The *acj6* mutation, then, might affect a step in the peripheral part of the pathway (for example, a step in signal transduction) required for response to acetone, propionic acid, butanol and ethyl acetate, but not benzaldehyde. This step might also be required for mechanosensory signal transduction, but perhaps not for visual transduction, at either larval or adult stages.

Cloning of the *acj6* gene may be useful in addressing the functional organization of the olfactory system. It will be interesting to determine whether the *acj6* gene is expressed in a particular part of the olfactory system: if *acj6* in fact defines a gene essential to some, but not all, odorant pathways, then the expression pattern of the gene might help reveal whether different odorant pathways are spatially segregated in the olfactory system.

Use of an enhancer trap screen to identify genes expressed in distinct parts of the olfactory system

The genetic approach described above may help determine whether odorant-specific genes are expressed in specific regions of the olfactory system. This section addresses the same question, but from the opposite point of departure. The strategy described here was to identify genes expressed in specific areas of the olfactory system and then to ask whether such genes are required for response to specific sets of odorants.

Drosophila provides a convenient means of screening for genes expressed in restricted regions of the olfactory system: enhancer trapping (O'Kane & Gehring 1988). In this approach, a transposable element carrying a *lacZ* reporter gene is allowed to insert, essentially at random, into the genome. If the element inserts near an olfactory gene, the enhancer of the olfactory gene drives synthesis of the reporter gene in the olfactory system, which can be recognized by a simple staining procedure. Enhancer trapping has been used to study a variety of biological problems in *Drosophila* and there are now a substantial number of cases in which the pattern of reporter gene expression has been found to reflect the expression pattern of a flanking gene (Mlodzik & Hiromi 1992).

The function of the flanking genes can be investigated conveniently. A source of transposase is introduced into the enhancer trap line. The transposase causes the transposable enhancer trap element to excise, which in some cases occurs imprecisely, causing deletion of flanking DNA. The effects of deleting the flanking olfactory gene can then be assessed by measuring the response of the mutant to various odorants, using any of a variety of behavioural or physiological tests.

We have conducted a large-scale screen for enhancer trap lines which show expression in subsets of olfactory tissue (Riesgo-Escovar et al 1992). From a set of 6400 enhancer trap lines established in collaboration with several other laboratories (Spradling 1993), each containing an independent insertion of the PZ enhancer trap element (Klambt et al 1991), we identified 120 which showed staining associated with the AMC, but which showed relatively little staining elsewhere in late third-instar larvae. Figure 4a shows an example of one such line. Staining appears in cell nuclei because the *lacZ* reporter gene in the PZ enhancer trap is coupled to nuclear targeting sequences. When examined under Nomarski optics, some of the labelled cells appear to be neurons innervating the antennal organ.

As a secondary screen, we examined these 120 lines for staining at the adult stage and found 12 which showed staining associated with the third antennal segment, but relatively little staining elsewhere. The rationale underlying this two-step screening strategy is as follows: the larval and adult olfactory organs have a common function, but very different morphology and developmental origins. Thus, a gene which is expressed in both organs, but in few other tissues, seems likely to play a direct role in olfactory function. As a precedent to support this line of reasoning, the three opsin genes expressed in the larval photoreceptor organ are all expressed in the adult compound eye (Pollock & Benzer 1988).

Most of these 12 lines stain specific regions of the third antennal segment. Figure 4b,c shows examples of two lines with complementary staining patterns: line 3420 stains cells in the dorsomedial half of the third segment and line 2246 stains cells in the ventrolateral half. These staining patterns are similar to the distribution patterns of the sensilla basiconica and sensilla trichodea, which are concentrated in the dorsomedial and ventrolateral regions, respectively. These two lines seem unlikely to identify genes encoding structural components unique to either basiconic or trichoid sensilla, however, since both lines also show expression in the larval AMC.

Some lines show staining of small numbers of cells. Line 2609, for example, shows staining of only 36.5 ± 2.9 cells (SEM; $n=31$) (Fig. 5c). This number represents less than 1% of the total number of cells in the third antennal segment. The stained cells are distributed quite broadly across the surface of the third segment. By contrast, in line 6276, which also shows staining of a small number of nuclei, the stained cells are clustered in the dorsomedial region of the antenna. We do not know whether this clustering reflects a common function for the labelled cells, such as response to a common odorant.

Sexual dimorphism in expression is shown by another line, 4567 (Fig. 5a,b). Males show significantly more staining in the sacculus (a sensory invagination on the surface of the third segment) than females. These results suggest that this line may identify a gene whose product plays a role in an olfactory-driven sex-specific behaviour. No male-specific structures have been identified in the antenna or in the antennal lobes of *Drosophila*. Some differences have been documented

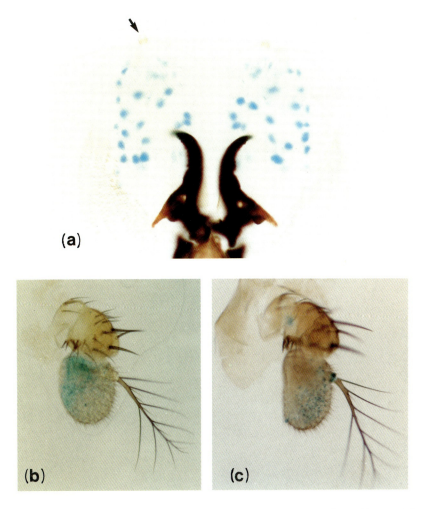

FIG. 4. (a) Anterior portion of larva of line 6288, stained for β-galactosidase activity, showing expression of *lacZ* reporter gene in cell nuclei associated with the AMC. The antennal organ is faintly visible (arrow). The dark cephalo-pharyngeal skeleton is also visible. (b) Antenna of line 3240, showing staining in the dorso-medial portion of the third segment. (c) Antenna of line 2246, showing staining in the ventro-lateral portion of the third segment. Some staining is also visible at the base of the arista (the feather-shaped structure on the right) and in the second segment. Panels (b) and (c) are reproduced with permission from Riesgo-Escovar et al (1992). ©1992, John Wiley and Sons, Inc.

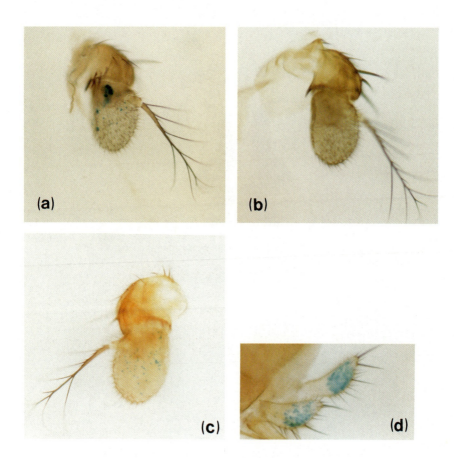

FIG. 5. (a) Male antenna of line 4567. (b) Female antenna of line 4567. (c) Antenna of line 2609. (d) Maxillary palps of line 7502. Panels (a), (b) and (c) reproduced with permission from Riesgo-Escovar et al (1992). ©1992, John Wiley and Sons, Inc.

between males and females in the numbers of basiconic and trichoid sensilla on the antennal surface (Stocker & Gendre 1988), but not in the number of sensilla in the sacculus.

A dynamic, age-dependent staining pattern was observed for line 6865. Shortly after eclosion, staining is observed over the entire surface of the third segment. Only a subset of cells is stained, but the subset is distributed across the entire surface. By contrast, animals that are a week old show staining in a restricted pattern: strong staining is observed in the sacculus and the surrounding dorso-medial region, but virtually no staining is observed in the ventrolateral region. This decrease in ventrolateral staining could reflect a decrease in gene expression, or possibly programmed cell death of expressing cells. We do not know what the significance of this change is, or whether it is related to the sexual maturation of the adult animal. Interestingly, the mushroom bodies of *Drosophila*, structures in the brain which play a role in olfactory learning, show increases in the number of axons during the first week of adult life (Technau 1984).

Although these 12 lines were identified from a collection of 6400 by virtue of the specificity of their expression in olfactory tissue, in no case is the specificity absolute: in all lines at least some staining can be seen outside the larval AMC and the third antennal segment. In some cases, staining is also seen in other chemosensory tissues. For example, some lines, including 3420, show staining in the maxillary palp, which, as discussed above, has olfactory function (Fig. 5d). Lines 3420 and 6865 show staining in the labellum, which contains taste hairs (Nayak & Singh 1983), as if these lines defined genes whose products function in both smell and taste. It is plausible that these two sensory modalities might share at least one common gene of a signal transduction pathway. We have previously found that *smellblind* mutants are defective in larval behaviour driven by both olfaction and contact chemoreception (Lilly & Carlson 1990) and our studies of the *ota1* (*olfactory trap abnormal 1*) mutant (Woodard et al 1989, 1992)—an allele of *rdgB* (*retinal degeneration B*)—have provided evidence for at least one common step between visual and olfactory physiology.

Some lines show staining in tissues which have not been associated with chemosensory function. Line 3420 shows staining in the male reproductive tract, which is of interest in light of the recent evidence that in mammals, a number of putative olfactory receptor genes are expressed in sperm (Parmentier et al 1992).

The enhancer trap insertions reside at a minimum of six distinct chromosomal locations, as determined by *in situ* hybridization to polytene chromosomes. Do the enhancer trap lines identify olfactory genes with expression patterns like those of *lacZ* receptor genes? We have cloned DNA flanking most of the 12 lines by plasmid rescue or other methods and have begun to investigate whether the flanking sequences hybridize to antennal RNA, either by *in situ* hybridization to RNA in tissue sections or by Northern analysis; this question can also be addressed by screening an antennal cDNA library.

Genetic analysis of these lines is currently underway. We have generated collections of sublines, each of which has suffered an excision of the transposable enhancer trap element. We have identified excision sublines which create flanking deletions and have begun testing some for defects in olfactory behaviour and physiology.

We are especially interested in determining whether olfactory genes expressed on different parts of the antennal surface are required for response to subsets of odorants.

Summary

The functional organization of the olfactory system has been studied by two approaches made possible by the powerful genetic and molecular tools available in *Drosophila*. The first approach has identified olfactory mutants, of which the *acj6* mutant shows a severe defect in its electrical response to a subset of odorants. The second has identified a set of enhancer trap lines, of which a number are likely to identify genes expressed in distinct parts of the olfactory system. Molecular analysis of the mutants and genetic analysis of enhancer trap lines may be informative in deriving a functional map of the *Drosophila* olfactory system. It will be interesting to see whether the two approaches eventually converge to provide similar pictures of functional organization and to provide insight as to how this organization arises during the course of development.

Acknowledgements

This work was supported by National Institutes of Health grants R01 GM36862 and P01 GM 39813, by a McKnight Scholars Award and by an Alfred P. Sloan Foundation Fellowship. I thank Elizabeth Vallen for helpful comments on the text.

References

Ayer RK, Carlson J 1991 acj6: a gene affecting olfactory physiology and behavior in *Drosophila*. Proc Natl Acad Sci USA 88:5467–5471
Ayer RK, Carlson J 1992 Olfactory physiology in the *Drosophila* antenna and maxillary palp: acj6 distinguishes two classes of odorant pathways. J Neurobiol 23:965–982
Ayyub C, Paranjape J, Rodrigues V, Siddiqi O 1990 Genetics of olfactory behavior in *Drosophila melanogaster*. J Neurogenet 6:243–262
Carlson J 1991 Olfaction in *Drosophila*: genetic and molecular analysis. Trends Neurosci 14:520–524
Helfand S, Carlson J 1989 Isolation and characterization of an olfactory mutant in *Drosophila* with a chemically specific defect. Proc Natl Acad Sci USA 86:2908–2912
Klambt C, Jacobs J, Goodman C 1991 The midline of the Drosophila central nervous system: a model for the genetic analysis of cell fate, cell migration, and growth cone guidance. Cell 64:801–815
Lilly M, Carlson J 1990 smellblind: a gene required for *Drosophila* olfaction. Genetics 124:293–302

McKenna M, Monte P, Helfand S, Woodard C, Carlson J 1989 A simple chemosensory response in *Drosophila* and the isolation of acj mutants in which it is affected. Proc Natl Acad Sci USA 86:8118-8122

Mlodzik M, Hiromi Y 1992 Enhancer trap method in *Drosophila*: its application to neurobiology. Methods Neurosci 9:397-414

Monte P, Woodard C, Ayer R, Lilly M, Sun H, Carlson J 1989 Characterization of the larval olfactory response in *Drosophila* and its genetic basis. Behav Genet 19:267-283

Nayak S, Singh RN 1983 Sensilla on the tarsal segments and mouthparts of adult *Drosophila melanogaster meigen* (Diptera: Drosophilidae). Int J Insect Morphol Embryol 12:273-291

O'Kane C, Gehring W 1988 Detection of *in situ* genomic regulatory elements in *Drosophila*. Proc Natl Acad Sci USA 84:9123-9127

Parmentier M, Libert F, Schumans S et al 1992 Expression of members of the putative olfactory receptor gene family in mammalian germ cells. Nature 355:453-455

Pollock J, Benzer S 1988 Transcript localization of four opsin genes in the three visual organs of *Drosophila*; RH2 is ocellus-specific. Nature 333:779-782

Riesgo-Escovar J, Woodard C, Gaines P, Carlson J 1992 Development and organization of the *Drosophila* olfactory system: an analysis using enhancer traps. J Neurobiol 23:947-964

Rodrigues V 1980 Olfactory behavior of *Drososphila melanogaster*. In: Siddiqi O, Babu P, Hall LM, Hall JC (eds) The development and neurobiology of Drosophila. Plenum, New York, p 361-371

Rodrigues V, Siddiqi O 1978 Genetic analysis of chemosensory pathway. Proc Indian Acad Sci Sect B 87:147-160

Shepherd G 1985 Are there labeled lines in the olfactory pathway? In: Pfaff D (ed) Taste, olfaction, and the central nervous system. Rockefeller University Press, New York, p 307-321

Siddiqi O 1983 Olfactory neurogenetics of *Drosophila*. In: Chopra VL, Joshi BC, Sharma RP, Bansal HC (eds) Genetics: new frontiers. Oxford & IBH, New Delhi, p 243-261

Siddiqi O 1987 Neurogenetics of olfaction in *Drosophila melanogaster*. Trends Genet 3:137-142

Singh RN, Nayak SV 1985 Fine structure and primary sensory projections of sensilla on the maxillary palp of *Drosophila melanogaster meigen* (Diptera: Drosophilidae). Int J Insect Morphol Embryol 14:291-306

Singh RN, Singh K 1984 Fine structure of the sensory organs of *Drosophila melanogaster meigen* larva (Diptera: Drosophilidae). Int J Insect Morphol Embryol 13:255-273

Spradling A 1993 Developmental genetics of oogenesis. In: Bate M, Martinez-Arias A (eds) Drosophilia development. Cold Spring Harbor Laboratory Press, Cold Spring Harbor, NY, p 1-69

Stocker RF, Gendre N 1988 Peripheral and central nervous effects of *lozenge³*: a *Drosophila* mutant lacking basiconic antennal sensilla. Dev Biol 127:12-24

Stocker RF, Lienhard MC, Borst A, Fischbach KF 1990 Neuronal architecture of the antennal lobe in *Drosophila melanogaster*. Cell Tissue Res 262:9-34

Technau G 1984 Fiber number in the mushroom bodies of adult *Drosophila melanogaster* depends on age, sex, and experience. J Neurogenet 1:113-126

Venkatesh S, Singh R 1984 Sensilla on the third antennal segment of *Drosophila melanogaster meigen*. Int J Insect Morphol Embryol 13:51-63

Woodard C, Huang T, Sun H, Helfand S, Carlson J 1989 Genetic analysis of olfactory behavior in *Drosophila*: a new screen yields the ota mutants. Genetics 123:315-326

Woodard C, Alcorta E, Carlson J 1992 The rdgB gene of *Drosophila*: a link between vision and olfaction. J Neurogenet 8:17-32

DISCUSSION

Lancet: I calculated that *Drosophila* might respond to 10^{-12}M amyl acetate. John Carlson and Stuart Firestein both suggested that *Drosophila* would only respond to concentrations of about 10^{-5}M or higher: is there a discrepancy?

Carlson: Our figures refer to dilution factors. It's very hard to compare results from different experiments, because the experimental paradigms are so varied. In some cases an airstream is bubbled through a solution and directed towards the animal; in other cases odorant is allowed to diffuse from a filter disk into a chamber. I don't have a good estimate of how many receptor molecules reach the receptor neurons in any case. Also, response thresholds may vary according to what is measured. Olfactory response can be measured as a single-unit recording, as an electroantennogram (EAG), or as a behavioural response.

Margolis: Are your *acj6* mutants at a locus that is responsible for primary events in chemotransduction?

Carlson: The *acj6* mutant has an electrophysiological defect in the antennae and could very well be defective in a primary event in chemotransduction.

Bargmann: John, there is an anosmic mutant, *lozenge*, in which one class of sensilla (the basiconica) in the antennae is knocked out (Stocker & Gendre 1988). What behavioural defects are associated with this mutation?

Carlson: Renée Vernard and Reinhard Stocker reported that deleting one particular class of sensilla reduces responses to a subset of odorants, which tells us something about which odorants are mediated through basiconica (Vernard & Stocker 1991). It would be very interesting if enhancer trap lines were to provide us with mutations specifically affecting the other classes of sensilla as well; the trichodea and coeloconica.

Siddiqi: In *lozenge* mutants, the EAG for most odorants is reduced by about 40%. The response to ethyl acetate is also reduced by about 30%, but the response to amyl acetate is barely affected (8–10% reduction in EAG). These results suggest that most of the responses present in the basiconica are also present in the coeloconica.

Breer: You mentioned screening an antennal cDNA library. Have you identified any genes yet?

Carlson: We have isolated two genes. They appear to encode proteins more closely related to the pheromone-binding proteins than to odorant-type binding proteins. Randy Reed mentioned a gene, *Olf-1*, the product of which had some characteristics of a transcription factor, including a nuclear localization signal with four out of five residues that are arginine. Our class C gene has an acid box and five arginines in a row after it, so it could conceivably also encode a transcription factor.

Hatt: In moths and cockroaches, there are many glomeruli for detecting fruit odours and only one so-called macroglomerulus for detecting pheromones. Does such a specialization also exist in *Drosophila*?

Carlson: There's no equivalent to the macroglomerulus in *Drosophila*. This may have something to do with the ecology of the species. The male moth has to detect a female that is a long way away, so it has an elaborate antennal system with extra sensilla which project to the macroglomerulus, a particular part of the antennal lobe in the brain. *Drosophila* is more of a social insect in the sense that the flies tend to congregate at food sources, so a specialized system for detecting long range pheromones may be less important.

Siddiqi: Our group has been working on the chemosensory neurogenetics of *Drosophila*, with objectives similar to those of Dr Carlson. Several facts reinforce Dr Carlson's conclusion that benzaldehyde differs from other odorants. It is a strong repellant to adult flies at all concentrations, whereas most other chemicals are attractants at lower concentrations but repel the imago at high concentrations. Dr Veronica Rodrigues carried out functional mapping of odour projections in the antennal lobe of *Drosophila* by studying the uptake of [^{3}H]deoxyglucose. The deoxyglucose maps for benzaldehyde are strikingly different from those for acetates and other chemicals in that benzaldehyde causes bilateral excitation, whereas the other chemicals cause ipsilateral excitation in distinct subsets of glomeruli (Rodrigues 1988). Correlated with this is the observation, first made by Axel Börst, that *Drosophila* exhibits a turning response (osmotropotaxis) to other chemicals but not to benzaldehyde (Börst & Heisenberg 1982).

One of the principal aims of chemosensory neurogeneticists is to identify genes encoding receptors and other molecules in the signalling pathway. This has been achieved in vertebrates (this volume: Buck 1993) but not yet in *Drosophila*.

My colleagues, Dr V. Rodrigues, Dr K. Vijayraghavan, Dr N. Shirsat and Dr G. Hasan have analysed a number of genes affecting olfaction and/or taste.

olf E was the first olfactory gene of *Drosophila* to be cloned and sequenced. The mutants are deficient in response to benzaldehyde. The *olf E* product is a protein of 1024 residues containing two short hydrophobic regions of 18 and 24 amino acids. The gene is expressed in the brain but not in the sensory neurons (Hasan 1989).

gust E has a specific effect on sensory physiology, blocking the response to Na$^+$ but not to other cations or sugars. The gene is expressed in the brain and probably not in the sensory neuron (S. Sathe & N. Shirsat, unpublished results).

bland affects the taste responses to sugars and salts. Its product is a secreted protein of 1046 residues with significant homology to serine proteases. The mutation affects axonal development (B. Murugasu, V. Rodrigues & W. Chia, unpublished results).

Mutations in the *east* gene lead to enhanced sensory thresholds for a variety of olfactory and taste stimuli. For all alleles, the defect is specific to the adult. The gene is expressed in neurons of the peripheral and central nervous system throughout development (Vijayraghavan et al 1992).

TABLE 1 (Siddiqi) **Autosomal mutations in *Drosophila* larvae which affect the electrophysiological response to three different acetate esters**

Autosomal mutation	Response to		
	Ethyl acetate	Isoamyl acetate	Butyl acetate
CS	Normal	Normal	Normal
A507	Reduced	Reduced	Normal
A345	Reduced	Reduced	Reduced
A16	Reduced	Normal	Reduced
A105	Reduced	Normal	Reduced
A402	Reduced	Normal	Normal
A534	Reduced	Normal	Normal
A413	Reduced	Normal	Normal
A701	Normal	Reduced	Reduced
A702	Normal	Reduced	Normal
A512	Normal	Reduced	Normal

Flies with mutations at the *scalloped* locus show defects in response to sugars and salts at the larval and adult stages. The Scalloped protein is a homologue of the human transcription factor TEF-1. This gene plays a role in the development of the peripheral nervous system and lethal alleles cause axonal defects (Campbell et al 1991, M. Inmadar, K. Vijayraghavan, V. Rodrigues, unpublished results).

malvolio was named after the Shakespearian character; flies with mutations in this gene lack the ability to taste different classes of sugars. Recently cloned and sequenced by Dr V. Rodrigues and Dr W. Chia, this is the only gene that might code for a transmembrane receptor. However, it has no homology to vertebrate receptor genes (P. Y. Chea, W. Chia & V. Rodrigues, unpublished results).

Why have we so far failed to obtain genes for receptors? One possibility is that these genes are indispensable. I think this is unlikely. A more likely explanation lies in the redundancy of receptors. Because of the multiplicity of receptors, I suspect that our selection for behavioural mutants is biased in favour of central against peripheral defects. The imago has a rich and complex system of chemoreceptor organs, the three types of olfactory hairs, with a variety of sensory neurons carrying overlapping sets of receptors. The larva, on the other hand, has a much simpler olfactory pathway. Our current approach is to look for larval olfactory mutants which exhibit altered electrophysiological responses.

We are concentrating on acetate esters which elicit strong responses in the larvae. Table 1 lists a set of autosomal mutations which produce differential responses to three different acetate esters. Two of these mutations, A105 and A402, give rise to changes in the electroantennogram of the adult. Several of these mutations also impair male courtship behaviour (R. Ranjan, A. Bala & Z. Rahman, unpublished results).

Lancet: If I understood what you said correctly, in *Drosophila* there are eight groups of sensory neurons and in each of the groups the cells behave almost identically. So, it is possible to identify classes of neurons, some of which have narrow and some which have broad specificity. It is not possible to do this in higher animals. How many cells were checked?

Siddiqi: Out of a total of 150 basiconica, the sample contains 101 neurons.

Lancet: So it's not unlikely that you have exhausted the different types of cells. If you go by the clonal exclusion hypothesis, that may mean that there are fewer than 10 distinct olfactory receptors in *Drosophila*.

Caprio: That's based solely on those stimuli you have tested; there's still the rest of the world out there.

Siddiqi: Probably not, because these stimuli were chosen as the strongest after screening about 100 chemicals.

Caprio: Your assay requires a behavioural response. You could still have physiological responses that didn't activate the behavioural system at those concentrations tested.

Siddiqi: We picked almost everything which evokes a reasonable electrophysiological response.

Buck: Did the strength of the physiological responses correspond to the strength of the behavioural responses?

Siddiqi: Roughly, yes. Ethyl acetate produces both the strongest behavioural and the strongest physiological response.

Getchell: John, using this rationale, could one conclude from your experiments that the *acj6* mutant leads to a deficit in a single receptor type for the ethyl acetate response?

Carlson: I don't think of it so much in terms of receptor types as pathways, but yes, *acj6* could affect some but not all receptor types.

Getchell: What were the components of the pathway, because you were very careful to say that you are not sure which peripheral components could be involved?

Carlson: We can't tell at this point, but if some odorants stimulate one transduction cascade and other odorants stimulate another cascade, then *acj6* could affect one transduction cascade and not the other.

Getchell: The technology is now available to test that hypothesis in *Drosophila*.

Hwang: There are a number of different mutations in the visual system that produce defects in the inositol trisphosphate pathway. Are there any olfactory mutants that also have defects in this pathway?

Carlson: One of the first mutants we found, in a completely different behavioural paradigm, is called *ota1*. We found that it was defective in a gene called *rdgB* (Woodard et al 1992), which turns out to have extensive sequence identity with the gene coding for phosphatidylinositol transfer protein (T. Vihtelic,

D. Hyde & J. O'Tousa, unpublished results). We found that certain *rdgB* mutants have a defect in the EAG as well as in visual transduction.

Lancet: Does the *rdgB* mutant have degeneration in the olfactory system, as seen in the visual system?

Carlson: We looked for degeneration in the antenna, but did not find any evidence for it.

References

Börst A, Heisenberg H 1982 Osmotropotaxis in *Drosophila melanogaster*. J Comp Physiol 147:479–484

Buck LB 1993 Receptor diversity and spatial patterning in the mammalian olfactory system. In: The molecular basis of smell and taste transduction. Wiley, Chichester (Ciba Found Symp 179) p 51–67

Campbell SD, Inamdar M, Rodrigues V, Vijayraghavan K, Palazzolo MJ, Chovnik A 1991 The *scalloped* gene encodes a novel, evolutionarily conserved transcription factor required for sensory organ differentiation in *Drosophila*. Genes & Dev 6:367–379

Hasan G 1989 Molecular cloning of an olfactory gene from *Drosophila melanogaster*. Proc Natl Acad Sci USA 86:2908–2912

Rodrigues V 1988 Spatial cloning of olfactory information in the antennal lobe of *Drosophila melanogaster*. Brain Res 453:299–307

Stocker R, Gendre N 1988 Peripheral and central nervous effects of *lozenge*[3]: a *Drosophila* mutant lacking basiconic antennal sensilla. Dev Biol 127:12–21

Vernard R, Stocker R 1991 Behavioral and electroantennogram analysis of olfactory stimulation in *lozenge*: a *Drosophila* mutant lacking antennal basiconic sensilla. J Insect Behav 4:683–705

Vijayraghavan K, Kaur J, Paranjape J, Rodrigues V 1992 The *east* gene of *Drosophila melanogaster* is expressed in the developing embryonic nervous system and is required for normal olfactory and gustatory responses of the adult. Dev Biol 154:23–36

Woodard C, Alcorta E, Carlson J 1992 The rdgB gene of *Drosophila*: a link between vision and olfaction. J Neurogenet 8:17–32

Perireceptor events in taste

H. Schmale, C. Ahlers, M. Bläker, K. Kock and A. I. Spielman*

*Institut für Zellbiochemie und klinische Neurobiologie, Universität Hamburg, Martinistrasse 52, D-20246 Hamburg, Germany and *Division of Basic Sciences, New York University College of Dentistry, 345E 24th Street, NY 10010, USA*

Abstract. The microenvironment at chemical receptor sites is important for ligand–receptor interaction as it can influence the entry, residence time or exit of odorant and sapid molecules. The perireceptor milieu at apical taste cell microvilli consists of taste pore mucus and secretions from salivary glands. The majority of taste buds are sheltered in epithelial folds of the foliate and circumvallate papillae where saliva is provided predominantly by the lingual von Ebner's glands (VEGs). To investigate possible saliva–tastant interactions, we have characterized a prominent 18 kDa secretory protein expressed in human, rat and pig VEGs. The human and rat VEG proteins share 60% sequence identity and, by virtue of their protein and gene structure, can be assigned to the lipocalin superfamily of lipophilic ligand carrier proteins. VEG proteins might function as transporters of hydrophobic molecules, for example bitter substances, like the nasal odorant-binding proteins that belong to the same protein family. Because binding experiments using various bitter substances have so far failed, and in light of the species-specific expression, other functions for VEG proteins must be considered. These include the protection of taste epithelia, pheromone transport and lipid binding.

1993 The molecular basis of smell and taste transduction. Wiley, Chichester (Ciba Foundation Symposium 179) p 167–185

Activation of taste receptor cells takes place at their apical microvilli, which are exposed to the oral cavity through taste pores and surrounded by an aqueous and mucous environment (Fig. 1). The perireceptor environment consists of taste pore mucus, presumably released by taste cells (Kinnamon & Cummings 1992, Witt & Reutter 1988), and of saliva, secreted in mammals by major and minor salivary glands. In addition to its essential role in mastication, digestion and the protection of oral tissues, saliva dissolves sapid chemicals and carries them to and from the taste receptor sites. There, taste receptor ligands must partition into the mucus covering the apical ends of taste cells to gain access to the membranes of the microvilli.

Saliva is a complex mixture of inorganic and organic substances secreted under parasympathetic and sympathetic nervous control. Salivary electrolytes, in addition to being taste stimuli, are thought to play several roles in taste perception (Spielman 1990, Rehnberg et al 1992), including acting as major

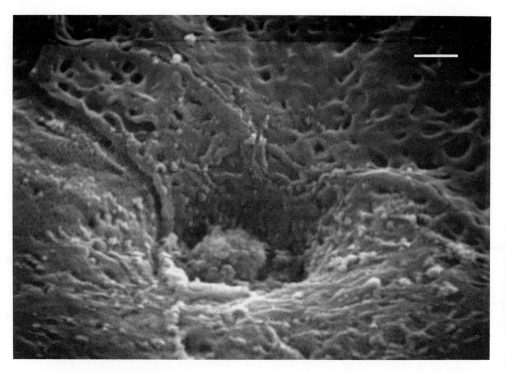

FIG. 1. High power scanning electron micrograph of the taste pore of a rat fungiform papilla. Scale bar represents 10 μm. (Reproduced, with permission, from Spielman et al 1991.)

carriers of current during depolarization of taste cells. Salivary ionic composition varies markedly in response to different gustatory stimuli and may therefore influence early electrical events in gustation (Dawes 1984). Whether any of the 60-plus salivary proteins present, for example, in human parotid saliva (Mogi et al 1986) have their primary function in taste perception is still an open question. Salivary proline-rich proteins (PRPs) have been proposed on the basis of genetic linkage studies to be involved in the ability of mice to taste bitter substances such as quinine, cycloheximide, strychnine and acetylated sugars (Azen et al 1986). PRPs bind dietary tannic acid, thereby reducing its bitter and astringent taste and thus influencing ingestive behaviour in mice (Glendinning 1992).

Over 90% of saliva is produced by the major parotid, submandibular and sublingual glands. However, the majority of taste buds sheltered in the epithelial folds of the circumvallate and foliate papillae in the posterior part of the tongue are in direct contact with serous secretions of the lingual salivary von Ebner's glands (VEGs) (Fig. 2). These complex VEG secretions, containing more than

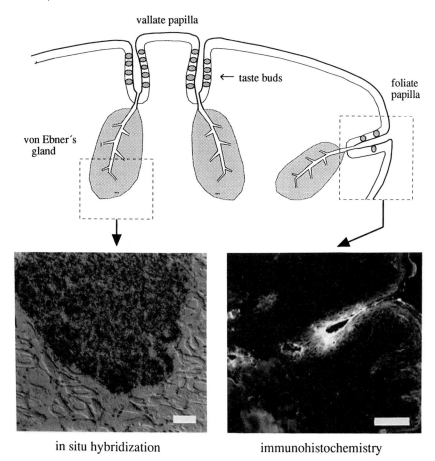

FIG. 2. Schematic representation of taste papillae with underlying lingual salivary glands, the von Ebner's glands (VEGs), on a coronal cross-section of rat tongue. Expression of VEG-P mRNA (see text) is shown by *in situ* hybridization using a ^{35}S-labelled RNA probe (left panel) (Schmale et al 1990). VEG-P accumulates in the clefts of taste papilla, as demonstrated by immunocytochemistry using antibodies against the rat VEG-P (right panel) (Kock et al 1992). Scale bars = 100 μm.

150 different proteins (Beidler 1990), some of which are unique, may be more relevant for gustatory function than the proteins of the major salivary glands. A role for VEGs in taste perception has been implicated by the fact that their secretions significantly reduce neurophysiological taste responses when applied in the presence of gustatory stimuli (Gurkan & Bradley 1988). We have characterized an 18 kDa secretory protein abundantly expressed in rat and human VEGs that has significant sequence identity with the odorant-binding proteins (OBPs) present in the mucus that covers olfactory receptor neurons. OBP and VEG

proteins are members of a large family of lipophilic ligand carrier proteins, called lipocalins (Bignetti et al 1985, Pevsner et al 1988, Schmale et al 1990). Surprisingly, despite the different chemical nature of odorants (highly volatile, hydrophobic) and taste stimulants (mostly water-soluble, hydrophilic), both chemoreception systems express similar proteins in their perireceptor environment. Here we summarize and extend the existing information on rat and human VEG proteins in an attempt to clarify the proposed function of the 18 kDa VEG protein as a carrier molecule for taste stimuli such as lipophilic bitter substances.

Molecular cloning of rat and human VEG protein cDNAs

Cell-free translation of mRNA prepared from rat or human VEG revealed a prominent protein of about 20 kDa that was not expressed in other salivary glands of the rat (Schmale et al 1990, Bläker et al 1993). Initially, the remarkable abundance of the mRNA encoding the secretory protein termed von Ebner's gland protein (VEG-P) allowed us to identify clones in cDNA libraries of rat VEG simply from their restriction fragment pattern. cDNA clones encoding the human VEG-P were later isolated by homology screening using rat hybridization probes. The rat and human VEG-P pre-proteins consist of 177 and 176 amino acids, respectively, including the N-terminal signal peptide typical of secretory proteins. The sizes of mature proteins were deduced from the cDNA sequences and from protein sequences of the N-termini of purified VEG-Ps. The mature rat VEG-P consists of 157 amino acids (M_r 17 670), whereas the human VEG-P has 158 residues (M_r 17 446).

Comparison of the predicted amino acid sequences reveals 60% identity between rat and human VEG-Ps. Both proteins share conserved sequence motifs with members of the lipocalin superfamily of proteins that serve as carriers for hydrophobic molecules such as retinol and odorants (Pervaiz & Brew 1985, Godovac-Zimmermann 1988, Pevsner et al 1990, Pelosi & Tirindelli 1989). Although members of the lipocalin family generally have only about 20% sequence identity with one another, at least three of them—retinol-binding protein, bilin-binding protein and β-lactoglobulin—have remarkably similar tertiary structures (Huber et al 1987), consisting of eight anti-parallel β strands forming a β barrel core that encapsulates the hydrophobic ligand. Of the 20 or so lipocalins known by sequence, VEG-Ps show the highest similarity (45%) to the odorant-binding protein II (OBP II) identified in the rat lateral nasal glands (Dear et al 1991). OBP II and the rat and human VEG-Ps have several identical stretches of conserved amino acids and clearly constitute a new lipocalin subfamily (Fig. 3).

In addition to the rat VEG-P shown in Fig. 3, now termed VEG-P I, we have isolated cDNA clones for another rat VEG protein, VEG-P II, exhibiting 95% sequence identity with VEG-P I. The rat VEG-P II is encoded by a separate gene, as demonstrated by Southern blotting and a PCR-based genomic

```
VEGh   HHLLASDEEIQDVSGTWYLKAMTVDREFPEMNLE--SVTPMT-L    41
           # ####### ########## # #####    ##### * #
VEGr   FPTTEENQDVSGTWYLKAAAWDKEIPDKKFGSVSVTPMK-I
       *       * *   *   *         *  *    *  * *
OBPII  EAPPDDQEDFSGKWYTKATVCDRNHTDGKRP-MKVFPMT-V

VEGh   TTLEGGNLEAKVTMLISGRCQEVKAVLEKTDEPGKYTADG------GKHV    85
       ## ##### ##  *    *   *   **########     #    ## #
VEGr   KTLEGGNLQVKFTVLIAGRCKEMSTVLEKTDEPAKYTAYS------GKQV
       * ***** *   *        *     * ******* **       ** *
OBPII  TALEGGDLEVRITFRGKGHCHLRRITMHKTDEPGKYTFK-------GKKT

VEGh   AYIIRSHVKDHYIFYCEGELHG---KPVRGVKLVGRDPK-NNLEALEDFEK   132
         *    ###### #     #    #    ##### ## ##### ## ##
VEGr   LYIIPSSVEDHYIFYYEGKIHR---HHFQIAKLVGRDPE-INQEALEDFQS
        ***    ****** **  *            ***** *   ***** *
OBPII  FYTKEIPVKDHYIFYIKGQRHG---KSYLKGKLVGRDSK-DNPEAMEEKKF

VEGh   AAGARGLST-ESILIPRQSET-CSPGSD    158
        #  # ### *     ##### ####
VEGr   VVRAGGLNP-DNIFIPKQSET-CPLGSN
          *              ***  * *
OBPII  VKSSKGFRE-ENITVPELLDE-CVPGSD
```

FIG. 3. Alignment of the sequence of human VEG-P (*VEGh*) with rat VEG-P I (*VEGr*) and odorant-binding protein (*OBP II*). Amino acids that are identical in the human and rat VEG-Ps are indicated by #. Comparison of the two VEG-Ps with OBP II (identical residues in all three proteins are indicated by asterisks) reveals conserved sequences (boxes).

analysis (Kock et al 1993). A second class of cDNA clones was also observed in the course of characterizing human VEG protein cDNAs. These clones differ only in their 3' untranslated regions, without alteration of the predicted protein sequence (M. Bläker, unpublished results).

Gene structure

Classification of proteins as members of the lipocalin family is based primarily on their amino acid sequence similarities. These are restricted to a few conserved motifs. Because of the low overall sequence similarity, the membership of some proteins (including VEG-P) of this superfamily has been questioned (Flower et al 1991). Comparison of gene organization among a group of proteins provides another means of determining their relationship and evolutionary origin. We have isolated and sequenced the gene for the rat VEG-P I from a genomic library and compared its organization with that of other known lipocalin genes (Kock et al 1993). The gene spans about 5 kb and contains seven exons ranging in size from 27 bp (exon 6) to 164 bp (exon 7). The organization of the VEG-P I gene is remarkably similar to that of the β-lactoglobulin gene, having an almost identical pattern of exons and introns (Fig. 4). As with all other known lipocalin genes, the conserved tryptophan residue of the motif Gly-X-Trp in the second exon is encoded by the fifth codon (Igarashi et al 1992). The genomic organization therefore provides clear evidence that the VEG-Ps are indeed members of the lipocalin superfamily.

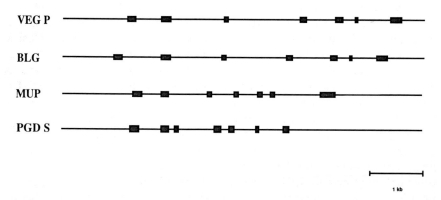

FIG. 4. Comparison of lipocalin gene structures. Exons are shown as boxes. The four genes are aligned at their second protein-coding exon. VEG-P, rat von Ebner's gland protein (Kock et al 1993); BLG, ovine β-lactoglobulin (Ali & Clark 1988); MUP, murine major urinary protein (Clark et al 1985); PGD S, human prostaglandin D synthase (Igarashi et al 1992).

Localization of VEG protein and mRNA

We localized VEG-P mRNA by *in situ* hybridization using RNA probes on sections of rat and human tongue containing VEGs (Schmale et al 1990, Bläker et al 1993). Antisense RNA probes label the cytoplasm of all acinar cells in the adult glands intensely (Fig. 2). Expression of the VEG-Ps begins at postnatal Day 2 in cells differentiating and branching off from VEG ducts and sharply increases with further enlargement and maturation of the gland. Expression of the protein itself was demonstrated in rat and human VEGs by immunocytochemistry, using polyclonal antibodies raised against rat VEG-Ps (Kock et al 1992, Bläker et al 1993). VEG-P is restricted to the serous VEGs and neighbouring mucous salivary glands are not stained. Secreted VEG-P is transported through salivary ducts to the trenches surrounding the foliate and vallate papillae, where it accumulates (Fig. 2). In human saliva collected from the cleft surrounding the circumvallate and foliate papillae, VEG-P is by far the most abundant protein (Bläker et al 1993, Spielman et al 1993). However, there is no apparent difference in secretion of VEG-P upon stimulation with sour, sweet, salty or bitter stimuli (Fig. 5).

In the oral cavity of rats and humans, VEGs are the only site of VEG-P expression. Recently, it has been shown that, at least in humans, tear fluid contains a prealbumin identical in sequence to the human VEG-P (Redl et al 1992). Whether nasal mucus and sweat proteins reacting with an antibody against tear prealbumin are identical or closely related to VEG-P remains to be determined.

Protein purification and ligand-binding studies

The ligand-binding properties of the members of the lipocalin family vary considerably. OBP can bind virtually any of the 80-plus odorants tested, with an apparent affinity in the micromolar range (Pevsner et al 1990, Pelosi & Tirindelli 1989), whereas retinol-binding protein is more specific for its ligand. The true physiological role of the majority of lipocalins remains unknown. As a first step to understanding the ligand-binding specificity of VEG-Ps, particularly with respect to their proposed role in gustation, we purified the rat VEG-P to apparent homogeneity from VEG extracts, using a combination of Sephadex G 75 gel filtration and Mono Q ion-exchange chromatography, or electroelution (Kock et al 1993). For a preliminary characterization, we measured binding of the ^{3}H- or ^{14}C-labelled compounds (Table 1) in assays using polyethyleneimine-treated glass fibre filters (Bruns et al 1983). The purified rat VEG-P does not bind retinol, cholesterol or odorants (pyrazine, 2-phenylethanol, 1-hexanol), or any of the bitter-tasting substances (strychnine, sucrose octaacetate, aloin, cholic acid). Surprisingly, only fatty acids such as oleate and palmitate show measurable binding activity. Thus, none of the ligands active in gustation or olfaction examined here appears to have affinity for rat VEG-P.

TABLE 1 Binding of various radioactively labelled compounds to purified rat VEG-P

Ligand	Taste	Binding activity
Retinol	Tasteless	−
Cholesterol	Tasteless	−
Cholic acid	Bitter	−
Strychnine	Bitter	−
Sucrose octaacetate	Bitter	−
Aloin	Bitter	−
Oleic acid	Tasteless	+
Palmitic acid	Tasteless	+
2-Isobutyl-3-methoxypyrazine	Tasteless	−
2-Phenylethanol	Tasteless	−
1-Hexanol	Tasteless	−

Filter-binding assays were performed according to Bruns et al (1983). Ligands, each at a concentration of 500 nM, were incubated with 1 μM VEG-P in 20 mM Tris-HCl, pH 7.5, for 1 h at 20 °C. All ligands were ^{3}H-labelled, except for ^{14}C-labelled sucrose octaacetate.

Species-specific expression of the VEG proteins

Knowledge of the distribution of VEG-Ps in different species may help us understand their function in taste. Immunoblot analysis of VEG extracts with antibodies to VEG-P indicated that in addition to rat and human, pig (domestic and wild) VEGs contain considerable amounts of an immunoreactive protein (Fig. 6a). Immunocytochemistry and *in situ* hybridization studies confirm the presence and localization of a VEG-related protein in pig VEGs. Surprisingly, in two different strains of mice, and in cow and guinea pig VEG extracts, no immunoreactivity was visible (Fig. 6a) (Spielman et al 1993). Furthermore, mouse and cow VEGs do not contain the mRNA that codes for the abundant 18 kDa protein found in rat and human VEGs. Southern blot analysis indicates that mice do not even have a homologous gene to this that could be detected by hybridization under reduced stringency (Fig. 6b), despite their close evolutionary relationship to rats. Under the same conditions, the human VEG-P gene, with about 60% identity to the rat gene, could be demonstrated easily. Therefore, at least for mice, expression of a VEG-P-like gene in salivary glands can be excluded. Species specificity is not an uncommon feature in the lipocalin superfamily, where, for example, β-lactoglobulin is found in ruminants, but not in rodents (Pervaiz & Brew 1985), and OBP expression is confined to a subset of animal species (Pelosi & Tirindelli 1989). The absence of VEG-Ps in several species questions the importance of their function in gustation.

Conclusions

What is the role of VEG-P in tasting? It has been suggested that VEG-P is a carrier for lipophilic molecules (e.g. bitter-tasting substances) and delivers them to, or concentrates them at, taste receptor sites (Schmale et al 1990).

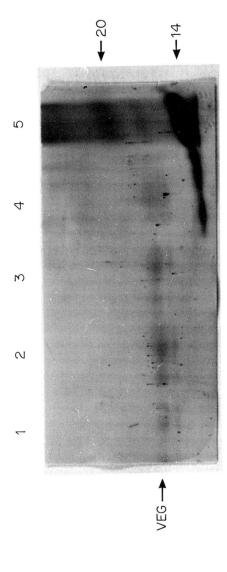

FIG. 5. Western blot analysis of human von Ebner saliva produced in response to sour (lane 1), sweet (lane 2), salt (lane 3) and bitter (lane 4) stimuli. Proteins were separated on an 11% SDS–polyacrylamide gel, transferred onto nitrocellulose paper and probed with antibodies to rat VEG-P. The apparent molecular mass (kDa) of markers (lane 5) is shown to the right.

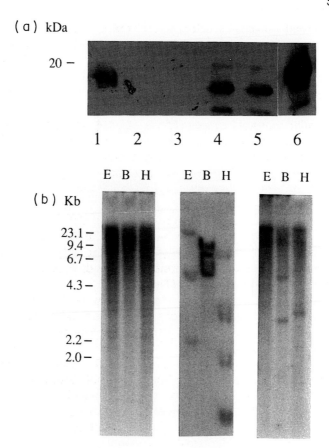

FIG. 6. Species-specific expression of VEG-P demonstrated by immunoblot (a) and Southern blot (b) analysis. (a) Protein extracts of VEG tissue of human (lane 1), mouse (lane 2), cow (lane 3), pig (domestic, lane 4; wild, lane 5) and rat (lane 6) were separated on a 15% SDS-polyacrylamide gel then subjected to Western blotting using antibodies against rat VEG-P (Bläker et al 1993). (b) DNA blot hybridization of mouse, rat and human genomic DNA (*left to right*) after digestion with the restriction enzymes indicated (E, *Eco*RI; B, *Bam*HI; H, *Hind*III). Each lane was loaded with 10 μg DNA. Mouse and human blots were hybridized with a ^{32}P-labelled rat VEG-P cDNA probe under reduced stringency (4×SSPE, 10% formamide, 42 °C), whereas the rat DNA blot was hybridized and washed at high stringency (Schmale et al 1990).

Several observations are not compatible with the role of VEG-P as a general carrier of gustatory and, in particular, bitter-tasting molecules to taste receptors. (1) Rat VEG-P apparently does not bind several of the bitter substances tested so far. If it had a general function in bitter taste, VEG-P would be expected to exhibit a broad range of specificity, like OBP. The limited number of VEG-Ps (two genes in rat) would each have to recognize a wide variety of structurally diverse bitter-tasting substances. Not all bitter compounds are hydrophobic: ammonium ions and arginine-containing peptides are highly charged bitter molecules. (2) VEG-P is expressed in a species-specific fashion. Closely related rodents—rat and mouse—differ in VEG-P expression, but respond to similar threshold concentrations of bitter substances. (3) VEG-P expression, at least in human, is not restricted to the oral cavity, but is also found in lacrimal glands, and there appears to be no difference in the type or amount of VEG-P secreted in response to different gustatory stimuli (Fig. 5). In light of these observations, other roles of VEG-P must be considered.

VEG-P might protect receptors from excessively high concentrations of gustatory stimuli or noxious substances that are likely to accumulate, especially in the clefts of vallate and foliate papillae. Remarkably, many lipocalin superfamily members are found in exocrine secretions that protect the epithelia against potentially harmful chemicals (Table 2).

The function of those lipocalins may initially have been associated with maintenance of epithelial integrity. Their abundance, broad ligand-binding specificity and binding characteristics make them ideally suited for clearance of lipophilic, potentially harmful substances. Their affinity for ligands is in the micromolar range and their dissociation kinetics are unusual. For example, odorants, once bound, are not readily released, even if the ligand concentration falls by two orders of magnitude (Pevsner et al 1990). Likewise, the purification of lipocalin proteins with their ligands still bound, as reported for mouse urinary proteins (Bacchini et al 1992) or aphrodisin (Singer & Macrides 1990), can be explained by this unusual behaviour.

In some tissues, lipocalins may have gained additional functions, unrelated to their primary ones. During the course of evolution, selective pressure on conservation of particular lipocalin genes differed among animal species, giving rise to the species-specific pattern of expression and sequence diversity. Basic

TABLE 2   Lipocalin proteins found in protective exocrine secretions

Protein	Secretion	Epithelium
Tear prealbumin	Tears	Cornea
OBPs	Olfactory mucus	Olfactory epithelium
VEG-Ps	von Ebner saliva	Gustatory epithelium
Mouse urinary proteins	Urine	Kidney and bladder epithelia

house-keeping functions common to many species, such as transport of retinol in the blood, are fulfilled by highly conserved proteins, for example retinol-binding proteins, present in all vertebrates studied so far (Blaner 1989). In contrast, species-specific functions, such as social interaction involving pheromones, are mediated by unique lipocalins, such as aphrodisin in female hamsters (Singer & Macrides 1990). Along these lines, our observations on VEG-P expression point to a special species-specific function, which may include pheromone transport. An example of pheromone transport by salivary gland proteins can be found in pigs. In submaxillary glands of the male pig, the secretory protein pheromaxein binds 16-androstene steroids and thus provides a mechanism by which first to concentrate the steroids from the blood and then transport them in the saliva for release of pheromones (Booth & White 1988).

One possibility has to be considered—that VEG-P has no specific role in tasting, but rather, similar to other salivary proteins such as amylase, it has a digestive function. In this regard, VEG-P could be a lipid-binding protein, facilitating the action of the von Ebner-specific lingual lipase. This function is consistent with the binding of VEG-P to fatty acids (Table 1). However, not all species express lingual lipase. Although rats possess both VEG-P and lipase, mice have only lipase, humans secrete VEG-P but apparently have very low levels of lipolytic activity (Moreau et al 1988) and guinea pigs have neither VEG-P nor lingual lipase (Spielman et al 1993). In order to clarify this, it would be interesting to extend VEG-P-binding experiments to lipids other than fatty acids.

If VEG-P turns out not to be taste associated, this in no way minimizes the role of the von Ebner's salivary glands in tasting, associated with the circumvallate and foliate taste papillae. After all, von Ebner's saliva has additional proteins and peptides, as well as numerous other components (such as lipids and electrolytes) that could be important in tasting. Very little is known about the ionic composition of von Ebner's saliva and the changes that occur during stimulation. Evidence suggests that von Ebner's saliva has a modulatory (diminishing) effect on tasting (Gurkan & Bradley 1988). The factors associated with such activity remain to be identified.

Acknowledgements

We thank Dr P. Pelosi (University of Pisa, Italy) and Dr J. Brand (Monell Chemical Senses Center, Philadelphia) for providing the radioactively labelled compounds pyrazine, aloin, 1-hexanol, 2-phenylethanol and sucrose octaacetate, and Dr F. Buck for protein sequence analysis. We also thank H. Christiansen, D. Schneider and G. Ellinghausen for technical assistance. This work was supported by Deutsche Forschungsgemeinschaft, National Institutes of Health and New York University.

References

Ali S, Clark AJ 1988 Characterization of the gene encoding ovine β-lactoglobulin. J Mol Biol 199:415–426

Azen EA, Lush IE, Taylor BA 1986 Close linkage of mouse genes for salivary proline-rich proteins (PRPs) and taste. Trends Genet 2:199–200

Bacchini A, Gaetani E, Cavaggioni A 1992 Pheromone binding protein of the mouse, *Mus musculus*. Experientia 48:419–421

Beidler JL 1990 Protein composition of the von Ebner gland secretions. Chem Senses 15:552(abstr)

Bignetti E, Cavaggioni A, Pelosi P, Persaud KC, Sorbi RT, Tirindelli R 1985 Purification and characterization of odorant-binding protein from cow nasal tissue. Eur J Biochem 149:227–231

Blaner WS 1989 Retinol-binding protein: the serum transport protein for vitamin A. Endocr Rev 10:308–316

Bläker M, Kock K, Ahlers C, Buck F, Schmale H 1993 Molecular cloning of human von Ebner's gland protein, a member of the lipocalin superfamily highly expressed in lingual salivary glands. Biochim Biophys Acta 1172:131–137

Booth WD, White CA 1988 The isolation, purification and some properties of pheromaxein, the pheromonal steroid-binding protein, in porcine submaxillary glands and saliva. J Endocrinol 118:47–57

Bruns RF, Lawson-Wendling K, Pugsley TA 1983 A rapid filtration assay for soluble receptors using polyethylenimine-treated filters. Anal Biochem 132:74–81

Clark AJ, Ghazal P, Bingham RW, Barrett D, Bishop JO 1985 Sequence structures of a mouse urinary protein gene and pseudogene compared. EMBO (Eur Mol Biol Organ) J 4:3159–3165

Dawes C 1984 Stimulus effects on protein and electrolyte concentrations in parotid saliva. J Physiol 346:579–588

Dear TN, Campbell K, Rabbitts TH 1991 Molecular cloning of putative odorant-binding and odorant-metabolizing proteins. Biochemistry 30:10376–10382

Flower DR, North ACT, Attwood TK 1991 Mouse oncogene protein 24p3 is a member of the lipocalin protein family. Biochem Biophys Res Commun 180:69–74

Glendinning JI 1992 Effect of salivary proline-rich proteins on ingestive responses to tannic acid in mice. Chem Senses 17:1–12

Godovac-Zimmermann J 1988 The structural motif of β-lactoglobulin and retinol-binding protein: a basic framework for binding and transport of small hydrophobic molecules? Trends Biochem Sci 13:64–66

Gurkan S, Bradley RM 1988 Secretions of von Ebner's glands influence responses from taste buds in rat circumvallate papilla. Chem Senses 13:655–661

Huber R, Schneider M, Mayr I et al 1987 Molecular structure of the bilin binding protein (BBP) from *Pieris brassica* after refinement at 2.0 Å resolution. J Mol Biol 198:499–513

Igarashi M, Nagata A, Toh H, Urade Y, Hayaishi O 1992 Structural organization of the gene for prostaglandin D synthase in the rat brain. Proc Natl Acad Sci USA 89:5376–5380

Kinnamon SC, Cummings TA 1992 Chemosensory transduction mechanisms in taste. Annu Rev Physiol 54:715–731

Kock K, Bläker M, Schmale H 1992 Postnatal development of von Ebner's glands: accumulation of a protein of the lipocalin superfamily in taste papillae of rat tongue. Cell Tissue Res 267:313–320

Kock K, Ahlers C, Schmale H 1993 Genes of rat von Ebner's gland proteins: genomic organization reveals close relationship to the lipocalins, submitted

Mogi M, Hiraoka BY, Harada M, Kage T, Chino T 1986 Analysis and identification of human parotid salivary proteins by micro two-dimensional electrophoresis and Western blot techniques. Arch Oral Biol 31:337–339

Moreau H, Laugier R, Gargouri Y, Ferrato F, Verger R 1988 Human preduodenal lipase is entirely of gastric fundic origin. Gastroenterology 95:1221–1226

Pelosi P, Tirindelli R 1989 Structure/activity studies and characterization of an odorant-binding protein. In: Brand JG, Teeter JH, Cagan RH, Kare MR (eds) Chemical senses: receptor events and transduction in taste and olfaction. Marcel Dekker, New York, p 207–226

Pervaiz S, Brew K 1985 Homology of β-lactoglobulin, serum retinol-binding protein, and protein HC. Science 228:335–337

Pevsner J, Reed RR, Feinstein PG, Snyder SH 1988 Molecular cloning of odorant-binding protein: member of a ligand carrier family. Science 241:336–339

Pevsner J, Hou V, Snowman AM, Snyder SH 1990 Odorant-binding protein. Characterization of ligand binding. J Biol Chem 265:6118–6125

Redl B, Holzfeind P, Lottspeich F 1992 cDNA cloning and sequencing reveals human tear prealbumin to be a member of the lipophilic-ligand carrier protein superfamily. J Biol Chem 267:20282–20287

Rehnberg BG, Hettinger TP, Frank ME 1992 Salivary ions and neural responses in the hamster. Chem Senses 17:179–190

Schmale H, Holtgreve-Grez H, Christiansen H 1990 Possible role for salivary gland protein in taste reception indicated by homology to lipophilic-ligand carrier proteins. Nature 343:366–369

Singer AG, Macrides F 1990 Aphrodisin: pheromone or transducer? Chem Senses 15:199–203

Spielman AI 1990 Interaction of saliva and taste. J Dent Res 69:838–843

Spielman AI, Brand JG, Kare MR 1991 Taste and tongue. In: Dulbecco R (ed) Encyclopedia of human biology. Academic Press, San Diego, CA, p 527–535

Spielman AI, D'Abundo S, Field RB, Schmale H 1993 Protein analysis of human von Ebner saliva and a method for its collection from the foliate papillae. J Dent Res 72, in press

Witt M, Reutter K 1988 Lectin histochemistry on mucous substances of the taste buds and adjacent epithelia of different vertebrates. Histochemistry 88:453–461

DISCUSSION

Getchell: Have you done binding studies of VEG protein with aphrodisin (Henzel et al 1988) or pheromaxein (Booth & White 1988)?

Schmale: No, so far we have only tried the ligands shown in Table 1 of my paper. For a while, people thought that the whole protein was the ligand, but recombinant expression in bacteria has shown that the pure protein isn't active: it picks up the natural ligand after synthesis in the vaginal gland (Singer & Macrides 1990).

Getchell: What is the sequence homology of VEG-P compared with vomeromodulin, another putative pheromone transporter?

Schmale: Vomeromodulin is a glycoprotein of 70 kDa, so it is much larger than VEG-P, which doesn't have an N-glycosylation site and is therefore probably not a glycoprotein. In addition, we did mass spectroscopy measurements on purified VEG-P; it has exactly the molecular mass you would predict from its sequence which suggests it doesn't have a carbohydrate component.

Perireceptor events in taste 181

Margolis: How do you know that VEG-P doesn't occur in mouse salivary secretions, other than by its absence in electrophoretic studies?

Schmale: First, our antibody against the rat VEG-P does not detect a similar protein in mouse von Ebner glands in immunocytochemical studies. Second, on mouse genomic Southern blots we were not able to demonstrate a homologous gene by hybridization under reduced stringency with a rat VEG-P gene probe. The smear you saw in the lanes containing mouse DNA (Fig. 6b) most likely represents cross-hybridization to other distantly related genes of the lipocalin family. And, finally, when we did cell-free translation of mRNA from mouse salivary glands, we couldn't find the abundant 20 kDa protein we found after translation of human and rat von Ebner mRNA.

Hwang: In our binding experiments, there were two other bitter substances which didn't bind to rat VEG-P: one is radioactive hydroquinine and the other is radioactive denatonium.

Lancet: What is the lower limit of the affinity when you say that no binding is occurring?

Schmale: The highest ligand concentration that we used in our preliminary binding studies was $0.5\,\mu M$. This is in the range used in binding studies with mouse urinary proteins (Bacchini et al 1992) and odorant-binding proteins (Pelosi & Tirindelli 1989, Pevsner et al 1990).

Reed: Hartwig, your experiments were looking at binding in purified biochemical preparations of protein. Paul Hwang was talking about *in situ* binding.

Hwang: I actually did binding experiments with crude soluble extracts from rat von Ebner's gland, as well as autoradiography on frozen tissue sections.

Reed: Do you see any binding of bitter substances at all on saliva from rat?

Schmale: We haven't tested total saliva for binding of bitter substances. We identified the VEG-P in the course of screening of taste tissue cDNA libraries and suggested a possible function on the basis of its relationship to the lipocalin superfamily.

Reed: I don't know even if there's supposed to be any binding for bitter compounds in the saliva to start with, contributed by any other proteins.

Schmale: There are other proteins that bind tastants, for example, the proline-rich proteins (PRPs) are known at least to bind tannic acid, which is bitter and astringent. There is a study by John Glendinning (1992) that shows that these PRPs bind tannic acid and that their presence has an influence on the diet of mice. The PRPs are expressed in parotid and von Ebner's glands, but in rodents their secretion has to be stimulated by tannic acid or β-adrenergic agents. In humans they are expressed constitutively, so they are present all the time.

Lancet: Can you please clarify the question of binding constants. Are you saying that $0.5\,\mu M$ was the highest concentration you tested? Because OBP binds at micromolar and higher concentrations for different odorants.

Schmale: We used that concentration because studies not only of OBPs, but also of other members of the lipocalin family, including retinol-binding protein (Cogan et al 1976) and the mouse urinary protein I mentioned before, indicated binding constants in the micromolar range.

However, it is not clear to me why filter-binding assays used for odorant-binding studies work with such low binding constants. In general, dilution of the ligand during washing steps should lead to dissociation from the binding protein in less than seconds, so that one could not work fast enough without losing the bound ligand. The only explanation may be the very low dissociation rates of at least some of the lipocalin proteins: they bind but don't release their ligand in a typical way (Pevsner et al 1990). This behaviour is also illustrated by the fact the several members of the lipocalin family, such as aphrodisin (Singer & Macrides 1990) and mouse urinary proteins, have been isolated with the ligands still bound after extensive purification.

Lancet: It doesn't make sense to look at proteins that are putative carriers for ligands without attempting to reach higher ligand concentrations in the millimolar range. If necessary, one should use additional methodologies.

Schmale: But many bitter substances have much lower thresholds—quinine has a threshold of about $10\,\mu M$ and denatonium has a threshold of about $0.01\,\mu M$ (Saroli 1984).

Hwang: What dissociation constants did you calculate from your binding experiments?

Schmale: These are preliminary experiments and we have had technical problems doing competition experiments with the very lipophilic fatty acids that showed binding activity. Let me add another point regarding the dissociation kinetics of lipocalins. Pevsner et al (1990) reported that if you do the binding with an odour at 10 nM, dilute this assay mixture 50-fold and take samples an hour later, 90% of the odorant is still bound. This is unusual behaviour for a ligand with this binding constant.

Lancet: Actually it's quite natural behaviour, because off constants can be in the range of minutes or hours for association constants in the micromolar and nanomolar ranges.

Pelosi: We also tried to measure the binding of the VEG protein isolated from the pig tongue, with odorants like 2-isobutyl-3-methoxypyrazine, 2-phenylethanol, 1-hexanol and borneol, all labelled with tritium. All the experiments have been unsuccessful, although both the protein and the ligand were used at a concentration ($100\,\mu M$) much higher than the expected dissociation constant. The conditions were the same as those used routinely for odorant-binding proteins (OBPs), known to give positive and reproducible results.

It is strange that, according to the experiments of Pevsner et al (1990), the dissociation of ligands from OBP is extremely low, given a dissociation constant in the micromolar range. In general, it is assumed that the velocity of the association is diffusion limited and therefore the 'on' kinetic constant is about

Perireceptor events in taste

the same for most binding proteins, about 10^7 mol/sec. Therefore, differences in the values of the thermodynamic association constants are generally related to the 'off' kinetic constants. With OBPs, this model seems not to be working: either the 'on' rate is very slow, or the kinetics of association and dissociation follow a more complicated path.

Interestingly, some proteins, called OBPs because they are similar to other OBPs in their amino acid sequence or other characteristics, do not bind 2-isobutyl-3-methoxypyrazine. One example is the so-called BG protein, isolated from the Bowman's glands of the frog (Lee et al 1987). OBP candidates isolated from the nasal mucosa of some mammals have so far failed to bind radioactive 2-isobutyl-3-methoxypyrazine (P. Pelosi, M. Garibotti & M. Ganni, unpublished results).

Lancet: Is is strange that we assume that just because protein A binds ligand X with a high affinity that protein B should bind ligand Y in the same way. There is diversity. To me, Paolo Pelosi's experiments, where he tested 10^{-5} M concentrations of both counterparts and found no binding, involve high enough concentrations to be a good indication of the absence of binding. If he had tested any substances at 10^{-7} M, I wouldn't have been convinced.

Pelosi: We have also tried to measure binding of moth pheromones to their relevant pheromone-binding proteins (PBPs). Using both a crude antennal extract and the purified PBP from *Bombyx mori*, we could not measure any binding of tritiated bombykol, except when the protein–ligand mixture was seperated by ion-exchange chromatography or native gel electrophoresis (Maida et al 1993). There are no reports that binding of pheromones to PBP has been measured in solution.

Margolis: This sounds strange to me. What are the actual protein concentrations in solution and in the gel bands? Do you think that this observation is related to the reports that some of these proteins function as dimers to bind single molecules of ligand (Pelosi & Maida 1990)? It is possible that they need to be at high concentration for dimer formation and ligand binding.

Pelosi: This could be one explanation. In fact, during both ion-exchange chromatography and electrophoresis, the proteins become several orders of magnitude more concentrated than in the initial solution, possibly reaching the very high levels (10 mM) measured *in vivo* for the PBP of *Antheraea polymphemus* (Vogt et al 1990). Although the diluted extract of the *Bombyx minor* PBP appears to be a monomer (Maida et al 1993), in the sensillum lymph it could well be present as a dimer or even a tetramer.

Reed: What's the evidence that PBPs really bind pheromones? I assume that someone has demonstrated pheromone binding?

Breer: Binding of pheromones to native proteins has been shown by several groups using different approaches (Vogt & Riddiford 1981, Kaissling 1987). We have cloned several genes encoding PBPs from different insect species (Krieger et al 1991). The moth *Antheraea perny* expresses two

PBPs and has two sex pheromones. We have expressed a cDNA clone encoding one of these PBPs. This heterologously expressed protein bound one of these pheromones but not the other (Krieger at al 1992). So these proteins do bind pheromones and can even discriminate between different pheromone molecules. Nevertheless, insect pheromones are usually very hydrophobic molecules and as a consequence it is not easy to perform straightforward binding assays.

Margolis: Aren't pheromones soluble at the very low concentrations that we are talking about?

Breer: The hydrophobic molecules tend to stick to everything—including glass, pipette tips, filters—so it is very difficult to control the exact concentration of ligand.

van Houten: Is there any action of saliva on these taste stimulants? For example, it is not known whether OBP binds modified odorants and therefore whether it might be acting in a protective manner in the clearance mechanism. Could saliva be transforming any of these tastants so that your protein is not binding the pristine tastant but a modified form?

Schmale: I can't exclude that possibility. I don't know anything about the enzymic activity of saliva on taste stimulants. However, there is one member of the lipocalin family that exhibits isomerase activity—prostaglandin D synthase (Nagata et al 1991).

DeSimone: Along those lines, what conditions of pH, ion composition and ionic strength did you employ in your binding assay? Particularly where binding may be weak, factors such as these may be important because they can influence the shape of protein binding sites.

Schmale: I used 20 mM Tris buffer at pH 7.5.

Kinnamon: Gurkan & Bradley (1988) stimulated VEGs to increase output of the gland and obtained modified responses to taste stimuli.

Schmale: They exposed both glossopharyngeal nerves in the rat and then on one side they stimulated salivary secretion from VEG (Gurkan & Bradley 1988). Although the gland is only activated on one side, saliva flows from the whole papilla accessing the entire continuous cleft. From the contralateral glossopharyngeal nerve they recorded neural activity after application of the different chemical stimuli to the circumvallate papilla. Various chemicals (NaCl, KCl, NH_4Cl, citric acid, saccharin) were used and VEG saliva diminished taste responses to all of them. However, the magnitude of reduction varied significantly among stimuli and was most pronounced for NH_4Cl.

References

Bacchini A, Gaetani E, Cavaggioni A 1992 Pheromone-binding protein of the mouse, *Mus musculus*. Experientia 48:419–421

Booth WD, White CA 1988 The isolation, purification and some properties of pheromaxein, the pheromonal steroid-binding protein, in porcine submaxillary glands and saliva. J Endrocrinol 118:47–57

Cogan U, Kopelman M, Mokady S, Shinitzky M 1976 Binding affinities of retinol and related compounds to retinol-binding proteins. Eur J Biochem 65:61–78

Glendinning JI 1992 Effect of salivary proline-rich proteins on ingestive responses to tannic acid in mice. Chem Senses 17:1–12

Gurkan S, Bradley RM 1988 Secretions of von Ebner's glands influence responses from taste buds in rat circumvallate papilla. Chem Senses 13:655–661

Henzel WJ, Rodriguez H, Singer AG et al 1988 The primary structure of aphrodisin. J Biol Chem 263:16682–16687

Kaissling K-E 1987 RH Wright lectures on insect olfaction. Simon Fraser University, Burnaby, Canada

Krieger J, Raming K, Breer H 1991 Cloning of genomic and complementary DNA encoding pheromone-binding proteins: evidence for microdiversity. Biochim Biophys Acta 1088:277–284

Krieger J, Raming K, Prestwich GD, Frith D, Stabel S, Breer H 1992 Expression of a pheromone-binding protein in insect cells using a baculovirus vector. Eur J Biochem 203:161–166

Lee HK, Wells RG, Reed RR 1987 Isolation of an olfactory cDNA: similarity to retinol-binding protein suggests a role in olfaction. Science 253:1053–1056

Maida R, Steinbrecht A, Ziegelberger G, Pelosi P 1993 The pheromone-binding protein of *Bombyx mori*: purification, characterization and immunocytochemical localization. Insect Biochem Molec Biol 23:243–253

Nagata A, Suzuki Y, Igarashi M, Eguchi N, Toh H, Urade Y, Hayaishi O 1991 Human brain prostaglandin D synthase has been evolutionary differentiated from lipophilic-ligand carrier proteins. Proc Natl Acad Sci USA 88:4020–4024

Pelosi P, Maida R 1990 Odorant-binding proteins in vertebrates and insects: similarities and possible common function. Chem Senses 15:205–215

Pelosi P, Tirindelli R 1989 Structure/activity studies and characterization of an odorant-binding protein. In: Brand JG, Teeter JH, Cagan RH, Kare MR (eds) Chemical senses: receptor events and transduction in taste and olfaction. Marcel Dekker, NY, p 207–226

Pevsner J, Hou V, Snowman AM, Snyder SH 1990 Odorant-binding protein. Characterization of ligand binding. J Biol Chem 265:6118–6125

Saroli A 1984 Structure-activity relationship of a bitter compound: denatonium chloride. Naturwissenschaften 71:428–429

Singer AG, Macrides F 1990 Aphrodisin: pheromone or transducer? Chem Senses 15:199–203

Vogt RG, Riddiford LM 1981 Pheromone binding and inactivation by moth antennae. Nature 293:161–163

Vogt RG, Rybczynski R, Lerner MR 1990 The biochemistry of odorant reception and transduction. In: Schild D (ed) Chemosensory information processing (NATO ASI series H, vol 39). Springer-Verlag, Berlin, p 33–76

Gustducin and transducin: a tale of two G proteins

Susan K. McLaughlin, Peter J. McKinnon, Alain Robichon, Nancy Spickofsky and Robert F. Margolskee

Roche Research Center, Roche Institute of Molecular Biology, 340 Kingsland Street, Nutley, NJ 07110-1199, USA

Abstract. In the vertebrate taste cell, heterotrimeric guanine nucleotide-binding proteins (G proteins) are involved in the transduction of both bitter and sweet taste stimulants. The bitter compound denatonium raises the intracellular Ca^{2+} concentration in rat taste cells, apparently via G protein-mediated increases in inositol trisphosphate. Sucrose causes a G protein-dependent generation of cAMP in rat taste bud membranes; elevation of cAMP levels leads to taste cell depolarization. To identify and characterize those proteins involved in the taste transduction process, we have cloned G protein α subunit (Gα) cDNAs from rat taste cells. Using degenerate primers corresponding to conserved regions of G proteins, we used the polymerase chain reaction to amplify and clone taste cell Gα cDNAs. Eight distinct Gα cDNAs were isolated, cloned and sequenced from a taste cell library. Among these clones was α gustducin, a novel taste Gα closely related to the transducins. In addition to α gustducin, we cloned rod and cone transducins from taste cells. This is the first identification of transducin expression outside photoreceptor cells. The primary sequence of α gustducin shows similarities to the transducins in the receptor interaction domain and the phosphodiesterase activation site. These sequence similarities suggest that gustducin and transducin regulate taste cell phosphodiesterase, probably in bitter taste transduction.

1993 The molecular basis of smell and taste transduction. Wiley, Chichester (Ciba Foundation Symposium 179) p 186-200

The molecular mechanisms underlying taste transduction have been only partially elucidated. Each taste submodality has its own specific mechanism: salty taste is due to Na^+ flux through apical Na^+ channels; sour is mediated by H^+ blockade of K^+ or Na^+ channels; bitter and sweet are mediated by heterotrimeric guanine nucleotide-binding protein (G protein)-dependent mechanisms (Roper 1992, Kinnamon 1988, Avenet & Lindemann 1989, Teeter & Cagan 1989). Sweet compounds cause a GTP-dependent generation of cAMP in rat tongue membranes (Striem et al 1989, 1990). External application or microinjection of cAMP inactivates K^+ channels in vertebrate taste cells and leads to their depolarization (Avenet & Lindemann 1987, Tonosaki & Funakoshi 1988,

Avenet et al 1988). It has been known for some time that taste tissue contains very high levels of adenylate cyclase (Kurihara & Koyama 1972, Nomura 1978, Asanuma & Nomura 1982). These observations suggest that sweet transduction involves sweet receptor activation of a G_s-like protein, which leads to adenylate cyclase-mediated elevation of cAMP concentrations. Bitter taste, at least in part, is also transduced via a G protein-coupled receptor: the bitter compound denatonium leads to Ca^{2+} release from internal stores (Akabas et al 1988), presumably mediated by G protein activation of phospholipase C to generate inositol trisphosphate ($InsP_3$).

The heterotrimeric G proteins couple membrane-bound receptors to effectors in many diverse signal transduction processes. Most of the G protein's specificity with regard to receptor and effector interactions resides within the α subunit. Presently, 16 different Gα genes have been identified (Wilkie et al 1992). Four G proteins are apparently restricted in their expression to specific sensory cells: G_{olf} (olfactory neurons) (Jones & Reed 1989), gustducin (taste cells) (McLaughlin et al 1992) and the rod and cone transducins (retinal photoreceptor cells) (Lochrie et al 1985, Medynski et al 1985, Tanabe et al 1985, Yatsunami & Khorana 1985). The other Gα subunits are, for the most part, more widely expressed.

We set out to identify and clone proteins involved in vertebrate taste transduction. Our initial focus was on Gα subunits. The experimental evidence cited above implicates multiple G proteins in the transduction of both sweet and bitter taste. Furthermore, Gα subunits are highly conserved, making it

TABLE 1 Isolates of Gα subunit clones from polymerase chain reaction-amplified rat taste tissue cDNA

G_α subunit	5'KWIHCF 3'FLNKKD	5'DVGGQR 3'FLNKKD	5'HLFNSIC 3'VFDAVTD	5'TIVKQM 3'FLNKQD
α_{i-2}	4	–	–	–
α_{i-3}	5	–	–	–
α_{12}	–	1	–	–
α_{14}	–	1	–	–
α_s	–	–	–	1
α_{gust}	5	–	4	1
α_{t-rod}	3	–	2	1
α_{t-cone}	–	–	2	–

Degenerate PCR primers corresponding to conserved amino acids of G proteins were made and used pairwise in the polymerase chain reaction (PCR) to amplify DNA encoding G proteins. The Gα subunit isolates cloned from PCR-amplified rat taste tissue cDNA are listed in the left hand column. The heading above each column lists the upstream (5') and downstream (3') primers. The numbers in each column represent the number of independent clonal isolates from the particular PCR amplification.

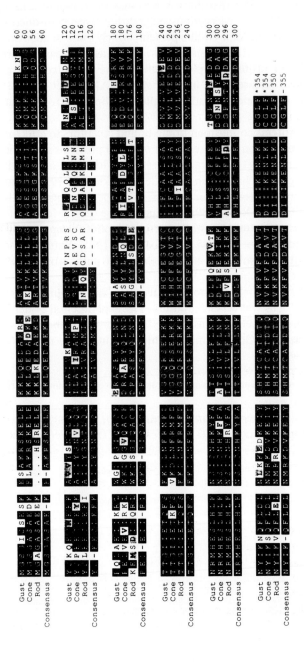

FIG. 1. Alignment of amino acid sequences of α subunits of rat gustducin (Gust), rat rod transducin (Rod) and bovine cone transducin (Cone). Amino acid sequence alignments were produced iteratively by the 'BestFit' routine of the Wisconsin GCG software package. Consensus sequence matches (i.e. at least two out of the three proteins match) are denoted by white letters within black boxes. Conservative changes are denoted by black letters within shaded boxes. Non-conservative changes are depicted by black letters in unboxed regions. The consensus line shows positions where two or three of the three proteins match; dashes in the consensus sequence correspond to non-conserved positions. Dots in the rod transducin sequence correspond to gaps to align it with the other sequences. Note the high degree of conservation of the three sequences throughout their length. The last 38 amino acids of all three proteins are identical: this region has previously been implicated in receptor interaction.

possible to design specific primers that can be used in the polymerase chain reaction (PCR) to amplify and then clone their DNAs. The identification of G proteins with elevated or specific expression in gustatory tissue could provide insight into the mechanisms of the taste pathways. We have cloned eight different Gα subunits from rat taste tissue. Among these clones is α gustducin, a novel Gα expressed specifically in taste cells (McLaughlin et al 1992). In addition to α gustducin, we cloned rod α transducin ($α_{t\text{-rod}}$) and cone α transducin ($α_{t\text{-cone}}$) from taste cells. Previously, the transducins had only been found in photoreceptor cells of the retina. The close relatedness of gustducin to transducin and the presence of both proteins in taste cells suggests that they play similar roles in taste transduction.

Cloning of Gα subunits from taste tissue

To identify G proteins present in taste tissue, we used degenerate oligonucleotide primers from conserved regions of the Gα subunits in the PCR, with taste tissue cDNA as the template. In this way, we isolated eight different Gα clones from the taste cDNA library (Table 1). Five of the Gα clones ($α_{i\text{-}2}$, $α_{i\text{-}3}$, $α_{12}$, $α_{14}$ and $α_s$) had been previously identified and are expressed in several tissues other than lingual epithelium; two clones ($α_{t\text{-rod}}$ and $α_{t\text{-cone}}$) had been previously identified only in retina. In addition to these previously known α subunits, we isolated a novel clone, α gustducin, which is taste cell specific and closely related to the transducins.

The entire coding sequence of α gustducin and $α_{t\text{-rod}}$ was obtained by PCR and screening of a rat taste tissue cDNA library. The gustducin cDNA encodes a protein of 354 amino acids and the $α_{t\text{-rod}}$ cDNA encodes a protein of 351 amino acids. An alignment of rat α gustducin, rat $α_{t\text{-rod}}$ and bovine $α_{t\text{-cone}}$ is presented in Fig. 1. These three proteins share extensive sequence similarity (79–80% amino acid identity, 89–90% similarity), suggesting that gustducin and the transducins are evolutionarily related members of a subgroup of G proteins. In contrast, gustducin and the transducins are only 65–68% identical to the $α_i$ subunits and only 42–46% identical to $α_s$.

Alignment of α gustducin with $α_{t\text{-rod}}$ and $α_{t\text{-cone}}$ shows that the general structure of all three α subunits is highly conserved (Fig. 1). In fact, the C-terminal 38 amino acids of α gustducin and both α transducins are identical. This C-terminal identity is of particular importance, since this region has been implicated in G protein–receptor interactions. These comparisons suggest that the guanine nucleotide-binding properties and GTPase activities of these three α subunits are likely to be similar. All three α subunits contain a potential N-myristoylation site (MGXXXS) at their N-terminus, which, if utilized, may anchor these α subunits to the inner face of the plasma membrane.

Expression of α gustducin and α transducin

By RNase protection, we demonstrated that α gustducin RNA only occurs in taste tissue-enriched preparations (Fig. 2A). No α gustducin RNA was detected in the other tissues we examined: non-taste lingual tissue, olfactory epithelium, retina, brain, liver, heart or kidney. We also used RNase protection to demonstrate that $\alpha_{t\text{-rod}}$ mRNA only occurs in retina and taste tissue (Fig. 2B). $\alpha_{t\text{-rod}}$ mRNA was not found in the other tissues we examined: non-taste lingual tissue,

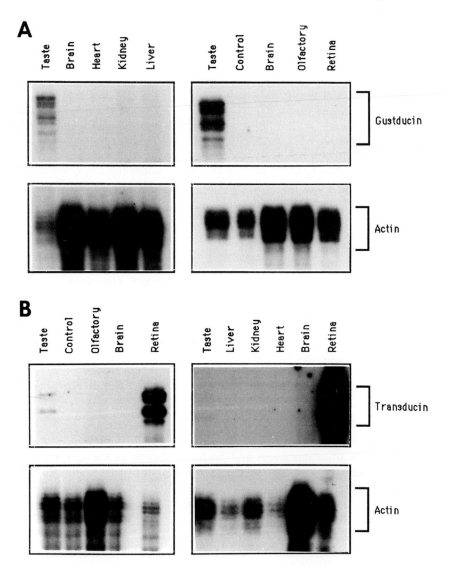

olfactory epithelium, brain, liver, heart or kidney. This is consistent with previous Northern results which demonstrated that $\alpha_{t\text{-}rod}$ is not expressed in heart, liver and brain (Medynski et al 1985). However, previous workers had not looked for the expression of transducin in taste tissue. Furthermore, the levels of $\alpha_{t\text{-}rod}$ mRNA are several hundred-fold higher in retina than in taste tissue.

We also examined the expression of gustducin and $\alpha_{t\text{-}rod}$ protein in taste tissue. Immunoprecipitation of Gα subunits which had been ADP-ribosylated by pertussis toxin demonstrated the presence of transducin protein in taste bud-containing tissue and its absence from non-sensory lingual epithelium (Fig. 3). Pre-incubation of the antiserum with a transducin-specific peptide blocked the immunoprecipitation of ADP-ribosylated transducin.

The RNase protection results demonstrate that expression of gustducin mRNA is restricted to taste bud-containing tissue and absent from non-taste portions of the lingual epithelium. To determine directly if α gustducin mRNA expression is confined to the taste buds, we did *in situ* hybridization and immunohistochemistry on tongue sections containing taste papillae. *In situ* hybridization demonstrated the presence of α gustducin mRNA in the taste buds of circumvallate, foliate and fungiform papillae (McLaughlin et al 1992) (Fig. 4a,b). Immunohistochemistry demonstrated the presence of gustducin protein in taste receptor cells of the foliate papillae (Fig. 4e). Gustducin mRNA and protein are absent from the adjacent lingual epithelium, muscle, connective tissue and von Ebner's glands. The pattern of gustducin protein expression is consistent with membrane association. *In situ* hybridization (Fig. 4c,d) and immunohistochemistry (Fig. 4f) also show that $\alpha_{t\text{-}rod}$ is expressed specifically within the taste buds of the circumvallate and foliate papillae.

FIG. 2 (*opposite*). (A) Tissue-specific expression of α gustducin transcripts as assayed by RNase protection. cDNA libraries from various tissues were prepared, linearized with restriction endonuclease, then *in vitro* run-off transcripts were generated (right panels). RNase protection was done simultaneously with α gustducin (upper panels) and actin probes (lower panels) to normalize for expression. α Gustducin transcripts are only present in RNA from the taste cell-enriched library (Taste) and are absent from libraries derived from non-taste lingual epithelium (Control), retina, brain and olfactory epithelium. RNase protection with total RNA (15 μg) from various tissues demonstrates that α gustducin is absent from brain, heart, kidney and liver (left panels). (B) Tissue-specific expression of rod transducin transcripts assayed by RNase protection of *in vitro*-generated RNA (left panels). RNase protection was done simultaneously with rod α transducin (upper panels) and actin probes (lower panels) to normalize for expression. Transducin transcripts are only present in RNA from the taste cell-enriched library (Taste) and the retinal library (Retina); transducin transcripts are absent from non-taste lingual epithelium (Control), olfactory epithelium and brain. Note that expression of transducin is several hundred-fold higher in retina than in taste tissue. RNase protection with total RNA (15 μg) from various tissues demonstrates that rod transducin is absent from liver, kidney, heart and brain, but present in retina and taste tissue RNA (right panels).

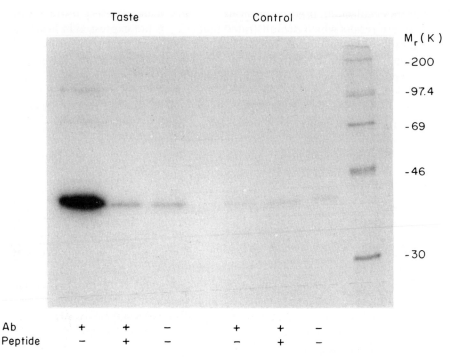

FIG. 3. Expression of rod α transducin analysed by ADP ribosylation and immunoprecipitation. Membrane extracts from bovine taste bud-containing tissues (Taste) and non-sensory lingual tissues (Control) were radioactively labelled by pertussis toxin-mediated ADP ribosylation. The products of these labelling reactions were immunoprecipitated by a transducin-specific antibody (Ab + lanes). Pre-incubation of the antibody with the cognate transducin peptide (peptide + lanes) blocks the immunoprecipitation, demonstrating that the immunoprecipitated product from taste tissue is transducin. In the absence of anti-transducin antibody (Ab − lanes), some non-specific trapping of ADP-ribosylated α gustducin (Taste) and $α_i$ (Taste and Control) occurs.

Discussion

We have identified and cloned α gustducin and the α transducins from rat taste cells. Gustducin is only expressed within the taste buds of the tongue: its mRNA is present in apparently all of the taste buds examined. Gustducin mRNA is not expressed in retina, olfactory epithelium, brain, liver, kidney, muscle or heart. Transducin had previously been found only in retina; we have shown that $α_{t\text{-rod}}$ mRNA and protein are expressed specifically in rat taste cells. Gustducin and transducin are evolutionarily related members of an α subtype gene family; presumably functional constraints have maintained a high degree of structural conservation in these two proteins.

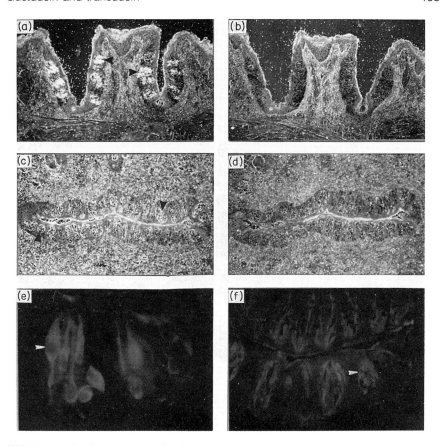

FIG. 4. a–d, photomicrographs (200 × magnification) of frozen sections of rat taste papillae hybridized with ^{33}P-labelled α gustducin or α rod transducin RNA probes and then stained with haematoxylin–eosin. (a) Dark field view of cross section of foliate papillae hybridized to α gustducin antisense probe; (b) dark field of foliate papillae, gustducin sense probe control; (c) dark field of circumvallate papillae, α rod transducin antisense probe; (d) dark field of circumvallate papillae, α rod transducin sense probe control. Arrowheads indicate areas of hybridization within taste papillae to gustducin (a) or transducin (c). e–f, immunofluorescence photomicrographs of frozen sections of rat taste papillae treated with anti-α gustducin or anti-α rod transducin antibodies and rhodamine-conjugated secondary antibody. (e) Foliate papillae reacted with anti-gustducin antibody (1000 ×); (f) circumvallate papillae reacted with anti-rod transducin antibody (400 ×). Arrowheads indicate taste receptor cells immunoreactive for gustducin (e) or transducin (f).

We propose that gustducin and transducin play roles in taste transduction similar to that of transducin in visual transduction. In the retina, the transducins relay activation of a receptor (rhodopsin or the colour opsins) into activation of a cytoplasmic effector enzyme (cGMP phosphodiesterase [PDE]).

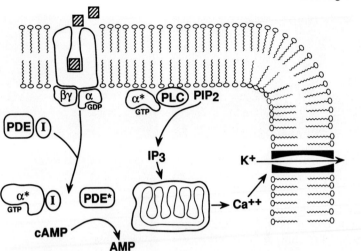

FIG. 5. Proposed mechanisms for bitter taste transduction. Bitter compounds bind to and activate specific receptors which activate taste cell-specific heterotrimeric G proteins. The activated Gα subunit may modulate the activity of phospholipase C (PLC) leading to inositol trisphosphate (IP_3) generation. This would lead to release of Ca^{2+} from internal stores, which in turn may activate basolateral K^+ channels (leading to hyperpolarization). Ca^{2+} may also lead to neurotransmitter release from taste cells. Transducin and gustducin are also proposed to play roles in bitter taste transduction: stimulated bitter receptors may activate transducin and/or gustducin to bind to an inhibitory subunit (I) of taste cell phosphodiesterase (PDE). This action would remove the inhibition of PDE, leading to breakdown of taste cell cAMP. The net effect of lower cAMP levels would be to hyperpolarize the taste cell. PDE activation would be antagonistic to sweet-activated adenylate cyclase, leading to bitter opposition to sweet at the taste cell or taste bud level.

Transducin removes the inhibition of cGMP PDE by binding to the inhibitory γ subunits; this leads to breakdown of cGMP and causes closure of cyclic nucleotide-gated channels. We propose that α gustducin and α transducin transduce taste receptor activation into activation of taste cell PDEs (Fig. 5). Consistent with this proposal, extremely high levels of cAMP PDE are present in taste tissue (Nomura 1978, Asanuma & Nomura 1982, Law & Henkin 1982). We propose that gustducin and/or transducin bind to an inhibitory subunit of a taste cell cAMP PDE to activate this enzyme and thereby decrease cAMP levels. At the level of the taste receptor we expect to find some functional and structural homology to the opsins. The C-terminal end of transducin is known to interact with the opsins (Ui et al 1984, VanDop et al 1984); the C-terminal 38 amino acids of α gustducin and the α transducins are identical, suggesting that gustducin can interact with opsin and opsin-like G protein-coupled receptors and conversely that transducin can interact with taste receptors. The third cytoplasmic loop of rhodopsin is a key interaction site for transducin (Kuhn 1984, Findlay &

Pappin 1986); we anticipate that the analogous region of the taste receptor will show structural and functional similarity to the opsins.

Several biochemical and electrophysiological studies argue for sweet pathway involvement of G_s (or a G_s-like protein), which activates adenylate cyclase to raise taste cell cAMP concentrations. Transducin does not activate adenylate cyclase and gustducin, by homology, is unlikely to do so. Gustducin and transducin may inhibit the sweet response (PDE activation would decrease sweet-induced cAMP) or they may play a role in bitter transduction. In support of these proposals, previous workers have correlated bitter compounds with PDE activation (Schiffman et al 1985) and several PDE inhibitors have been shown to enhance sweet perception (Price 1973, Schiffman et al 1986). The biochemical and molecular biological characterization of taste transduction has previously been hindered because taste receptor cells constitute only a small, inaccessible fraction of the lingual epithelium. Using PCR, we have cloned α gustducin and α transducins from rat taste cells. It should be possible to clone other components of the taste transduction pathways in a similar manner.

References

Akabas MH, Dodd J, Al-awqati Q 1988 A bitter substance induces a rise in intracellular calcium in a subpopulation of rat taste cells. Science 242:1047–1050

Asanuma N, Nomura H 1982 Histochemical localization of adenylate cyclase and phosphodiesterase in the foliate papillae of the rabbit. II: Electron microscopic observation. Chem Senses 7:1–9

Avenet P, Lindemann B 1987 Patch-clamp study of isolated taste receptor cells of frog. J Membr Biol 97:223–239

Avenet P, Lindemann B 1989 Perspectives of taste reception. J Membr Biol 112:1–8

Avenet P, Hofmann F, Lindemann B 1988 Transduction in taste receptor cells requires cAMP-dependent protein kinase. Nature 331:351–354

Findlay JBC, Pappin DJC 1986 The opsin family of proteins. Biochem J 238:625–642

Jones DT, Reed RR 1989 G_{olf}: an olfactory neuron specific-G protein involved in odorant signal transduction. Science 244:790–795

Kinnamon SC 1988 Taste transduction: a diversity of mechanisms. Trends Neurosci 11:491–496

Kuhn H 1984 Interactions between photoexcited rhodopsin and light-activated enzymes in rods. In: Osborne N, Chader J (eds) Progress in retinal research. Pergamon Press, New York, p 123–153

Kurihara K, Koyama N 1972 High activity of adenyl cyclase in olfactory and gustatory organs. Biophys Res Commun 48:30–34

Law JS, Henkin RI 1982 Taste bud adenosine-3′ 5′-monophosphate phosphodiesterase: activity, subcellular distribution and kinetic parameters. Res Commun Chem Pathol Pharmacol 38:439–452

Lochrie MA, Hurley JB, Simon MI 1985 Sequence of the α subunit of photoreceptor G protein: homologies between transducin, ras, and elongation factors. Science 228:96–99

McLaughlin SK, McKinnon PJ, Margolskee RF 1992 Gustducin is a taste-cell specific G protein closely related to the transducins. Nature 357:563–569

Medynski DC, Sullivan K, Smith D et al 1985 Amino acid sequence of the α subunit of transducin deduced from the cDNA sequence. Proc Natl Acad Sci USA 82:4311–4315

Nomura H 1978 Histochemical localization of adenylate cyclase and phosphodiesterase activities in the foliate papillae of the rabbit. I: Light microscopic observations. Chem Senses Flavor 3:319–374

Price S 1973 Phosphodiesterase in tongue epithelium: activation by bitter taste stimuli. Nature 241:54–55

Roper SD 1992 The microphysiology of peripheral taste organs. J Neurosci 12:1127–1134

Schiffman SS, Gill JM, Diaz C 1985 Methylxanthines enhance taste: evidence for modulation of taste by adenosine receptor. Pharmacol Biochem Behav 22:195–203

Schiffman SS, Diaz C, Beeker TG 1986 Caffeine intensifies taste of certain sweeteners: role of adenosine receptor. Pharmacol Biochem Behav 24:429–432

Striem BJ, Pace U, Zehavi U, Naim M, Lancet D 1989 Sweet tastants stimulate adenylate cyclase coupled to GTP binding protein in rat tongue membranes. Biochem J 260:121–126

Striem BJ, Yamamoto T, Naim M, Lancet D, Jakinovich W, Zehavi Y 1990 The sweet taste inhibitor methyl 4,6-dichloro-4,6-dideoxy-α-D-galactopyranoside inhibits sucrose stimulation of the chorda tympani nerve and of the adenylate cyclase in anterior lingual membranes of rats. Chem Senses 15:529–536

Tanabe T, Nukada T, Nishikawa Y et al 1985 Primary structure of the α-subunit of transducin and its relationship to *ras* proteins. Nature 315:242–245

Teeter JH, Cagan RH 1989 Mechanisms of taste transduction. In: Cagan RH (ed) Neural mechanisms in taste. CRC Press, Boca Raton, FL, p 1–20

Tonosaki K, Funakoshi M 1988 Cyclic nucleotides may mediate taste transduction. Nature 331:354–356

Ui M, Katada T, Murayama T et al 1984 Islet-activating protein, pertussis toxin: a specific uncoupler of receptor-mediated inhibition of adenylate cyclase. Adv Cyclic Nucleotide Protein Phosphorylation Res 17:145–151

VanDop C, Yamanaka G, Steinberg F, Sekura RD, Manclark CR, Stryer L 1984 ADP-ribosylation of transducin by pertussis toxin blocks the light-stimulated hydrolysis of GTP and cGMP in retinal photoreceptors. J Biol Chem 259:23–26

Wilkie TM, Gilbert DJ, Olsen AS et al 1992 Evolution of the mammalian G protein α subunit multigene family. Nature Genet 1:85–91

Yatsunami K, Khorana HG 1985 GTPase of bovine rod outer segments: the amino acid sequence of the α subunit as derived from the cDNA sequence. Proc Natl Acad Sci USA 82:4316–4320

DISCUSSION

Carlson: Are there any knockout mutants for transducin?

Margolskee: Transducin knockout mutants are being made in Mel Simon's laboratory (personal communication). They have made transgenes where the transducin promoter is driving T antigen and one would presume, depending on the particular promoter construct, that we should also find T antigen expression in taste buds when we look at the tongues from the transgenic mice.

Carlson: Are there any human blindnesses associated with transducin defects? Are any blindnesses associated with taste defects?

Margolskee: I don't know of any. Most congenital blindnesses where the molecular defect is known involve rhodopsin or phosphodiesterase (PDE) mutations.

Buck: Do you see gustducin and transducin expression in all of the taste receptor cells, or in only a subpopulation?

Margolskee: We haven't done the experiments very often, nor have we done them quantitatively, so we can't give a definitive answer. However, I would say the majority of taste receptor cells in a given taste bud express gustducin and a similar number also express transducin. Expression is certainly not restricted to just a small number of cells within the taste buds.

Buck: Is this true of the fungiform papillae?

Margolskee: We haven't done immunocytochemistry on fungiform papillae, so let me restrict this conclusion to the foliate and circumvallate papillae.

Reed: You must have been surprised at the distribution of gustducin within the taste cells. It's unusual among the sensory α subunits to see them apparently uniformly distributed throughout the whole cell body. Do you have any explanation?

Margolskee: Shigeru Takami in Tom and Marilyn Getchell's group and Susan McLaughlin in my group (unpublished work) have both seen differences in gustducin expression patterns. Depending on how much antibody we use, we can see gustducin either spread throughout the entire cell or more apically distributed in a membrane-associated pattern.

Getchell: We have also observed two expression zones in human circumvallate papillae. There is a very clear pattern of intense immunoreactivity located in the apical membranes of the taste cells subjacent to the taste pore and there is also a cytosolic distribution pattern which we have seen in one or two cells in 15 μm sections of taste papilla. How do you explain this?

Reed: You should try using a gluteraldehyde-based fixative. At least in the visual system, of all the Gα subunits, these α transducins are probably among the most loosely coupled to the plasma membrane. It may be that by using fixatives that don't cross-link proteins tightly, you are getting diffusion. With better, gluteraldehyde-based fixatives you may find that transducin has an intense localization somewhere else. We see differences in the relative distribution depending on whether we use paraffin or frozen sections, or with different fixatives. We still get strong cilia localization under all conditions, but with more cross-linking it is easier to see specific cilial localization.

Lancet: I would emphasize this point. In its active GTP-bound form, transducin is soluble. In the rod outer segment there is no way out for the soluble protein because it is stuck among the discs. But here we have a completely different geometry: there is a very small apex and a cell body of large volume, so it's not surprising that you find the protein mostly in the cytoplasm.

Hwang: I would be very cautious about interpreting immunofluorescence micrographs which show labelling at the apical end of taste bud cells, because these cells have apical vesicles or other granular substances that will often take up the fluorescent tag non-specifically. I have had problems with this phenomenon before in immunofluorescence experiments.

Margolskee: To show definitively that transducin is expressed in taste or any other tissue, it is very important to use multiple independent techniques. We were quite concerned, when we first cloned transducin from taste tissue, that it was a PCR artifact. We have therefore used multiple independent assays to demonstrate that it really is present in taste tissue or taste buds. Specifically, we have used cloning, reverse transcriptase PCR, RNase protection, Western blotting, immunoprecipitation, *in situ* hybridization and immunohistochemistry to demonstrate its presence.

Lindemann: Micha Naim and his colleagues from Israel have shown that sweet compounds stimulate adenylate cyclase isolated from the tongue of the rat (Striem et al 1989). They also showed that intact taste buds from the circumvallate papilla show an increase in cAMP concentrations when challenged with sucrose (Striem et al 1991). This implicates the cAMP pathway in transduction of sweet taste. For bitter taste, most people agree that there is a variety of transduction pathways. Denatonium, a widely used bitter-tasting compound, does not change the cAMP signal (Y. Barush, personal communication). Is there any more recent evidence for cAMP being involved in bitter taste transduction?

Margolskee: Based on the work of Akabas et al (1988) and Hwang et al (1990), we think that denatonium activates a pathway involving a G_q-like protein. G_{14} is elevated in taste buds and is G_q-like, therefore it is a good candidate for involvement in this pathway. Activation of G_{14} is unlikely to alter cAMP levels. We don't yet have direct evidence implicating cAMP or PDE in a bitter transduction pathway. Price (1973) found a correlation between several bitter compounds and their ability to activate lingual PDE. However, Kurihara (1972) found that the bitter compounds he examined were PDE inhibitors. These results may both be true: perhaps there is specialization of receptor cells within the taste bud, such that some receptor cells would read an increase in cAMP as bitter and some would read a decrease in cAMP as bitter. There is certainly room for multiple mechanisms in bitter taste transduction. Furthermore, diverse mechanisms may have been selected by evolution. One level of diversity may involve a gene family of bitter receptors, perhaps 10–20 of them. Another mechanism to generate diversity may involve bitter receptor-coupling to multiple G proteins. Transducin, gustducin, G_{i-3} and G_{14} are all candidate G proteins for involvement in bitter transduction. There may also be diversity at the effector level; there may be multiple PDEs in these taste buds.

I feel that a transgenic approach using gustducin knockout mutants will give us specific information: perhaps gustducin knockouts will display discrete deficits in the detection of particular bitter compounds. These transgenic mice should also show electrophysiological changes that correlate with any behavioural changes.

Hwang: What do you think the end effect of lowering the cAMP in taste transduction is, since there are no electrophysiological data implicating cyclic nucleotide-gated channels in taste receptor cell signal transduction?

Margolskee: I would predict that lowered cAMP levels would lead to taste cell hyperpolarization. If there is an interneuron equivalent interposed, then the hyperpolarization could lead to the depolarization of the afferent neuron. We hope to pursue these studies when we have biologically active gustducin. Perhaps by microinjecting GTPγS-activated gustducin into taste receptor cells, we will be able to induce hyperpolarization of the cell.

Kinnamon: There have not been any behavioural or electrophysiological studies using bitter compounds that might activate PDE. I've looked for evidence in the literature of hyperpolarization in response to bitter compounds and I can't find a report where anyone has tested anything other than the classic bitter taste stimuli like quinine and NH_4Cl.

Bartoshuk: What would you want tested?

Kinnamon: Steve Price (1973) reported that, in addition to quinine, thioacetamide, propylthiouracil, urea and acetamide activated PDE in lingual epithelium. I would have expected these compounds to hyperpolarize taste cells and antagonize sweet taste.

Bartoshuk: John DeSimone and I have been talking about water taste for a long time. One of the things about human water taste that's so interesting is that if you put bitter substances on the tongue and then rinse them off, they taste sweet; if you put sweet substances on the tongue and then rinse them off, you get a bitter taste (Bartoshuk 1968, McBurney & Bartoshuk 1973).

Kinnamon: Is that true of all bitter compounds?

Bartoshuk: It's not true equally of all, but it's certainly true for quinine, urea and caffeine, and it's true of all the sugars.

Margolskee: One fact that is relevant to this is that when many of the high potency sweeteners are altered slightly, they become high potency bitter compounds (DuBois et al 1993). In effect, you convert something that is a sweet agonist into an antagonist.

Kinnamon: When I add the PDE inhibitor IBMX (3-isobutyl-1-methyl-xanthine) to isolated sweet-responsive taste cells in the absence of a sweet ligand, I see a slow depolarization, apparently mediated by a block of K^+ channels.

Margolskee: This result indicates that there is a tonic level of PDE activity in taste cells that can be inhibited by IBMX. Perhaps this same PDE is activated by transducin or gustducin during bitter transduction.

Consistent with this model for a bitter transduction pathway, we have cloned two cAMP-type PDEs from taste tissue (Spickofsky et al 1992). One of these PDEs is elevated for expression in taste tissue versus non-sensory portions of the tongue; the other is very highly expressed in taste tissue but not detectable in control non-sensory lingual tissue. This finding is consistent with these PDEs playing a role in initiation of bitter transduction or termination of sweet transduction.

Ronnett: You used IBMX as a PDE inhibitor. However, there are a whole range of PDE inhibitors which are specific for high and low affinities and cAMP versus cGMP. They are useful if you are trying to dissect out function and are fairly readily available from drug companies.

Lancet: I'm a little worried about the fact that gustducin occurs in every one of the taste papillae, as these are expected to be cells with different specificities. *A priori* I would expect if a G protein is related to one of the four taste modalities, it should show a little more cellular specificity.

Margolskee: Let me clarify some of the comments about expression of gustducin. In the experiment where we did *in situ* hybridization in the fungiform papillae, it looked as if there were two or three cells positive for gustducin expression. However, on the periphery of the taste bud, there were many cells which were negative for expression. Susan McLaughlin and I have used *in situ* hybridization to study gustducin expression in many fungiform, circumvallate and foliate papillae (McLaughlin et al 1992). It is our qualitative impression that the highest level of gustducin mRNA expression is in the circumvallate papillae and the lowest expression is in the fungiform papillae. In our immunological studies, we have not yet looked at the fungiform papillae. However, we know that not every cell in the foliate and circumvallate papillae expresses gustducin.

References

Akabas MH, Dodd J, Al-aqwati Q 1988 A bitter substance induces a rise in intracellular calcium in a subpopulation of rat taste cells. Science 242:1047–1050

Bartoshuk LM 1968 Water taste in man. Percept Psychophys 3:69–72

DuBois GE, Walters DE, Kellog MS 1993 Mechanism of human sweet taste and implications for rational sweetener design. In: Ho C-T, Manley CH (eds) Flavor Measurement. Marcel Dekker, NY, p 239–266

Hwang PM, Verma A, Bredt DS, Snyder SH 1990 Localization of phosphatidylinositol signaling components in rat taste cells: role in bitter taste transduction. Proc Natl Acad Sci USA 87:7395–7399

Kurihara K 1972 Inhibition of cyclic 3′,5′-nucleotide phosphodiesterase in bovine taste papillae by bitter taste stimuli. FEBS (Fed Eur Biochem Soc) Lett 27:279–281

McBurney DH, Bartoshuk LM 1973 Interactions between stimuli with different taste qualities. Physiol & Behav 10:1101–1106

McLaughlin SK, McKinnon PJ, Margolskee RF 1992 Gustducin is a taste-cell-specific G protein closely related to the transducins. Nature 357:563–569

Price S 1973 Phosphodiesterase in tongue epithelium: activation by bitter taste stimuli. Nature 241:54–55

Spickofsky N, McLaughlin SK, McKinnon PJ, Margolskee RF 1992 Molecular cloning of taste transduction proteins. Chem Senses 17:701 (abstr)

Striem BJ, Pace U, Zehavi U, Naim M, Lancet D 1989 Sweet tastants stimulate adenylate cyclase coupled to GTP-binding protein in rat tongue membranes. Biochem J 260:121–126

Striem BJ, Naim M, Lindemann B 1991 Generation of cyclic AMP in taste buds of the rat circumvallate papilla in response to sucrose. Cellular Physiol Biochem 1:46–54

Role of apical ion channels in sour taste transduction

Sue C. Kinnamon

Department of Anatomy and Neurobiology, Colorado State University, Ft Collins, CO 80523 and the Rocky Mountain Taste and Smell Center, University of Colorado Health Sciences Center, Denver, CO 80262, USA

Abstract. Sour taste perception depends primarily on the concentration of H^+ in the taste stimulus. Acid stimuli elicit concentration-dependent action potentials in taste cells. Recent patch-clamp studies suggest that protons depolarize taste cells by direct interaction with apically located ion channels. In *Necturus maculosus*, the voltage-dependent K^+ conductance is restricted to the apical membrane of taste cells. The current flows through a variety of K^+ channels with unitary conductances ranging from 30 to 175 pS, all of which are blocked directly by citric acid applied to outside-out or perfused cell-attached patches. In contrast, hamster fungiform taste cells appear to utilize the amiloride-sensitive Na^+ channel for acid transduction. Amiloride completely inhibits H^+ currents elicited by acid stimuli in isolated taste cells, with an inhibition constant similar to that for amiloride-sensitive Na^+ currents ($K_i = 0.2\ \mu M$). Treatment of isolated taste cells with the bioactive peptide arginine^8-vasopressin results in similar increases in both the amiloride-sensitive Na^+ and H^+ currents; the effect is mimicked by 8-bromocyclic AMP. These results suggest that H^+ can permeate amiloride-sensitive Na^+ channels in hamster fungiform taste cells, contributing to the transduction of sour stimuli.

1993 The molecular basis of smell and taste transduction. Wiley, Chichester (Ciba Foundation Symposium 179) p 201–217

The application of modern biophysical techniques to isolated taste cells, as well as taste cells *in situ*, has resulted in tremendous advances in our understanding of the initial events in taste transduction. It now appears that taste cells use a variety of mechanisms for transduction. Ionic stimuli, such as salts and acids, modulate apically located ion channels directly, while sweeteners, amino acids and many bitter compounds utilize specific membrane receptors. Both ligand-gated channels and G protein-coupled receptors appear to be involved (for reviews, see Kinnamon & Cummings 1992, Roper 1992).

The perception of sour taste depends primarily on the concentration of protons in the taste stimulus. Protons interact with the apical membrane of taste receptor cells, resulting in membrane depolarization, action potential initiation and release of transmitter onto gustatory afferent neurons. The results of many studies suggest that several mechanisms may contribute to sour transduction, all involving direct H^+ modulation of apically located ion channels.

This paper describes recent experiments in our laboratory on mechanisms of sour transduction in mudpuppy (*Necturus*) and hamster taste cells. In mudpuppy, sour transduction involves a direct proton block of K^+ channels localized to the apical membrane of the taste cells. The protons block a resting K^+ efflux, depolarizing the taste cells. Sour taste in hamster fungiform taste cells appears to use the amiloride-sensitive Na^+ channel, the channel also responsible for Na^+ salt taste. When Na^+ concentrations are low, protons permeate the channel to depolarize taste cells. These mechanisms probably contribute to the perception of sour taste.

Methods

Necturus taste receptor cells were isolated by first removing the non-gustatory epithelial layer with collagenase, leaving the taste buds attached to their connective tissue papillae. Individual cells were then dissociated in Ca^{2+}-free saline and plated onto coverslips coated with Cell-Tak. The method is described in detail elsewhere (Kinnamon et al 1988a). For hamster, entire taste buds are isolated, plated onto Cell-Tak-coated coverslips and individual cells within the bud are selected for recording. These methods are described by Béhé et al (1990) and Gilbertson et al (1993). Standard giga-seal whole-cell and single-channel recording techniques (Hamill et al 1981) were used in most studies (Kinnamon & Roper 1988b, Cummings & Kinnamon 1992), except that perforated-patch recording (Korn et al 1991) was used in preference to whole-cell recording for most hamster taste cells (Gilbertson et al 1993). A combination of whole-cell and loose-patch recording was used to map the distribution of ion channels on *Necturus* taste cells. These procedures are described by Kinnamon et al (1988b). The loose-patch technique for recording from mammalian taste buds *in situ* is described in detail by Avenet & Lindemann (1991) and Gilbertson et al (1992a).

Results and discussion

Proton block of apical K^+ channels

A variety of studies have shown that, unlike non-gustatory epithelial cells, taste receptor cells are electrically excitable and possess a variety of voltage-gated ion channels including tetrodotoxin-sensitive Na^+ channels, Ca^{2+} channels and a variety of K^+ channels (for review, see Kinnamon & Cummings 1992). Although the role of these voltage-sensitive channels remains uncertain, trains of action potentials are elicited in response to taste stimuli in amphibians (Avenet

& Lindemann 1987) as well as in mammals (Avenet & Lindemann 1991, Avenet et al 1991, Gilbertson et al 1992a).

Studies in our laboratory on *Necturus* taste cells have shown that in addition to repolarizing the action potential, voltage-sensitive K^+ channels participate in the transduction of sour stimuli. The role of K^+ channels in sour taste was first hypothesized following microelectrode studies showing that the K^+ channel blocker tetraethylammonium (TEA) completely inhibited the response to sour stimuli (Kinnamon & Roper 1988a). In addition, whole-cell voltage-clamp recordings from isolated taste cells showed that citric acid (1 mM; pH 3) reduced the voltage-dependent K^+ current (Kinnamon & Roper 1988b). However, citric acid reduced other currents, including Na^+ and Ca^{2+} currents, as well. We concluded from these experiments that K^+ channel block could be an effective transduction mechanism for sour taste, but only if K^+ channels were localized selectively to the apical membrane.

The distribution of ion channels was determined by a combination of whole-cell and loose-patch recording from isolated taste cells. The whole-cell pipette was used to voltage-clamp the taste cell and record whole-cell currents. The loose-patch pipette was used to record that fraction of whole-cell current that flowed through patches of membrane in different regions of the cell. The results showed a striking localization of K^+ current to the apical membrane, with little or no K^+ current in the basolateral membrane. In contrast, Na^+ and Ca^{2+} currents were distributed uniformly over the cell surface (Kinnamon et al 1988b, 1989). Similar results were obtained by Roper & McBride (1989) in experiments involving intracellular recordings from *Necturus* taste cells in an Ussing chamber. The localization of K^+ channels to the apical membrane clearly suggests a prominent role of K^+ channels in transduction, since the entire K^+ conductance would be blocked in response to acid stimulation.

To determine if sour stimuli block K^+ channels directly (i.e. without the need for second messengers), we studied apical K^+ channels with single-channel recording techniques. Patches from the apical membrane typically contained many channels with unitary conductances ranging from 30 to 175 pS in symmetrical K^+ solutions (Cummings & Kinnamon 1992). Channels showed a small but significant probability of opening at the resting potential. Although the single channels varied in their response to intracellular factors such as Ca^{2+} and ATP, all channels were blocked by citric acid applied to the outside of the channel, either in outside-out patches (Fig. 1A) or in perfused cell-attached patches. In contrast, citric acid applied to the rest of the cell failed to block channels in cell-attached patches when citric acid was omitted from the pipette solution (Fig. 1B). These results indicate that acid blocks K^+ channels directly, not as a consequence of changes in intracellular pH or second messengers.

It is important to note that these apical K^+ channels are blocked by a variety of bitter-tasting compounds as well as the sour-tasting acids (Bigiani & Roper 1991, Cummings & Kinnamon 1992). Thus, this transduction mechanism is not

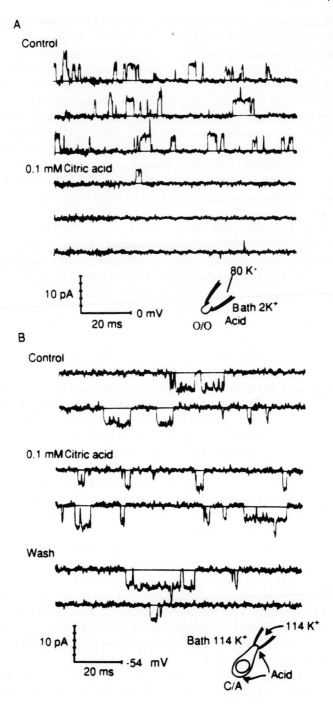

limited to sour taste. Recent behavioural experiments suggest that compounds which block the apical K^+ conductance are all repellent to mudpuppies (Bowerman & Kinnamon 1992).

Proton permeability of the amiloride-sensitive Na^+ conductance

It was obviously of interest to determine if apical K^+ channels are involved in the mammalian response to sour stimuli. To study mechanisms of sour transduction in hamster fungiform taste buds, we used a technique developed by Avenet & Lindemann (1991) for recording responses to taste stimuli from taste receptor cells maintained in the tongue *in situ*. The technique involves recording action currents from taste receptor cells with an extracellular loose-patch pipette placed over a taste pore. Because stimuli are limited to the apical membrane, many non-specific and potentially damaging effects of acids could be avoided using this technique. Perfusion of taste pores with citric acid (3 mM; pH 2.6) elicited fast transient currents, reflecting action potentials in taste receptor cells (Gilbertson et al 1992a). Action current frequency increased in a dose-dependent manner with decreasing pH, from a threshold of approximately pH 5. These action currents were usually smaller in amplitude than action currents elicited by the salty stimulus, NaCl.

The potential role of apical K^+ channels in the sour response was examined by recording responses to citric acid in the presence of the K^+ channel blocker, TEA. There was no significant effect of TEA on the sour-induced response, suggesting that K^+ channels were not involved. Surprisingly, the sour response was completely blocked by amiloride, the inhibitor of the epithelial Na^+ channel responsible for the transduction of Na^+ salt taste (Fig. 2). The concentration of amiloride required for half-maximal block of the acid response (K_i) was 2.4 µM, a value almost identical to the K_i for NaCl-induced action currents in the same taste buds. When NaCl and citric acid were applied together, the large action currents typical of the NaCl response were reduced and the response often resembled a sour response alone. Because the K_i for amiloride-block of the acid and NaCl responses was similar, and because there was an interaction when both stimuli were applied together, we hypothesized that protons were permeating the amiloride-sensitive Na^+ channel to mediate the sour response.

FIG. 1. (*opposite*) (A) Results from an outside-out patch from the apical membrane of an isolated *Necturus* taste cell. Patch contained many K^+ channels which were blocked by 0.1 mM citric acid (pH 3.8). The pipette contained 80 mM potassium gluconate and the bath contained 2 mM K^+. (B) Data recorded from a cell-attached apical patch. When citric acid (0.1 mM) was bath-applied to the cell, with no direct access to channels in the patch, it had no effect on K^+ channel activity. The pipette contained 114 mM potassium gluconate and the bath 114 mM KCl to eliminate the resting potential of the cell. Reproduced with permission from Cummings & Kinnamon (1992).

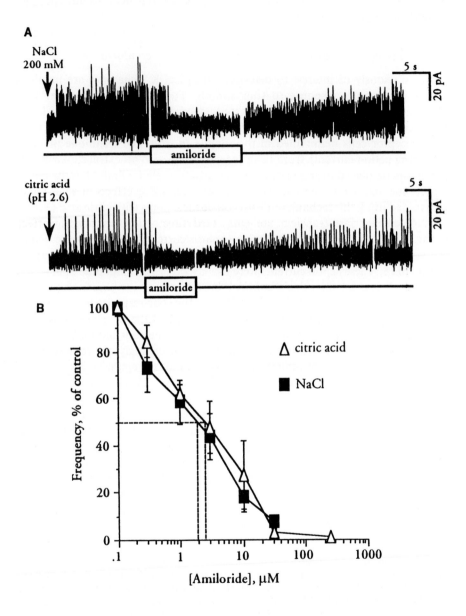

Whole-cell and perforated-patch recordings were made from taste cells in isolated taste buds to test directly the role of the amiloride-sensitive Na^+ channel in the sour response (Gilbertson et al 1992b, 1993). First, taste cells were tested for the presence of an amiloride-sensitive Na^+ current by examining the effect of amiloride (30 μM) on steady-state inward currents. In approximately 50% of the cells, amiloride inhibited a steady inward current and caused a decrease in membrane conductance. Cells were then tested for the presence of a proton current by replacing Na^+ in the bath solution with the impermeant cation N-methyl-D-glucamine (NMDG) and applying citric acid focally to taste cells from a puffer pipette positioned close to the apical portion of the taste bud. Citric acid elicited an inward current accompanied by an increase in membrane conductance; the proton-induced current was completely inhibited by amiloride. Amiloride-sensitive proton currents were found only in taste cells that also exhibited an amiloride-sensitive Na^+ current, suggesting that the same channel was mediating both the Na^+ and H^+ responses.

In recent experiments, this model was tested directly by examining the effect of the diuretic hormone arginine[8]-vasopressin on amiloride-sensitive Na^+ and H^+ currents in isolated taste buds. This hormone has been shown to increase the incorporation of amiloride-sensitive Na^+ channels into transporting epithelia via a cAMP-dependent mechanism (Garty & Benos 1988). Incubation with vasopressin for 10–15 min resulted in a threefold potentiation of both the amiloride-sensitive Na^+ current and the amiloride-sensitive H^+ current in taste cells. This effect was mimicked by 8-bromocyclic AMP, a membrane-permeant analogue of cAMP. These results provide strong evidence for a role of amiloride-sensitive Na^+ channels in sour as well as salt transduction.

Hypothesis of sour transduction

Since mammals are capable of distinguishing sour and salty, as well as sour and bitter compounds, neither of the above mechanisms can explain sour taste completely. It is our working hypothesis that several mechanisms may contribute to the perception of sour taste. The amiloride-sensitive Na^+ channel appears to be important in fungiform taste cells, at least in hamster. Indeed, afferent nerve responses to sour stimuli are partially inhibited by amiloride in the hamster (Hettinger & Frank 1990). There appears to be little or no effect of amiloride

---

FIG. 2 (*opposite*). (A) *In situ* recording from a hamster fungiform taste bud, showing that action currents in response to both NaCl (200 mM) and citric acid (pH 2.6) were inhibited by amiloride (30 μM). Citric acid was dissolved in a carrier solution of 30 mM NMDG-Cl. (B) Dose-dependence of the amiloride block for both acid- and NaCl-induced responses. The concentration of amiloride producing a half-maximal block was 1.9 μM for NaCl responses and 2.4 μM for citric acid responses. Reproduced with permission from Gilbertson et al (1992a).

on sour responses in the rat (Giza & Scott 1991) or human (Schiffman et al 1983), so other mechanisms are likely in these cases. Apical K^+ channels may be involved in the rat, since rat fungiform taste buds respond more vigorously to KCl than hamster taste buds (D. Harris & S. C. Kinnamon, unpublished results). Circumvallate taste buds appear to lack the amiloride-sensitive Na^+ channel (Formaker & Hill 1991, Gilbertson et al 1993), so an alternative mechanism is necessary to explain sour transduction in these cells. Apical K^+ channels may contribute to sour taste transduction in circumvallate taste buds, because glossopharyngeal nerve recordings show strong responses to K^+ salts and compounds known to block K^+ channels, including acids (Frank 1991).

Additional mechanisms for sour transduction are likely, since protons have been shown to modulate most ion channels (Hille 1992). In frog taste cells, protons are thought to gate a Ca^{2+} conductance that is responsible for taste cell depolarization (Miyamoto et al 1988). In dog, protons induce an anionic conductance in lingual epithelial preparations (Simon & Garvin 1985). Basolateral ion channels may also contribute to sour transduction, since paracellular pathways have been shown to be permeable to a number of small ions (Ye et al 1991). Thus, it is likely that sour perception depends on a complex interaction of mechanisms, reflecting the wide range of possible pH effects on ion channels.

Summary and conclusions

Two mechanisms of sour taste transduction have been described: proton block of apically located K^+ channels and proton permeability of amiloride-sensitive Na^+ channels. Proton block of apical K^+ channels depolarizes taste cells by preventing a resting efflux of K^+, whereas protons themselves depolarize taste cells directly by flux through the amiloride-sensitive Na^+ channel. It is likely that both of these mechanisms contribute to the perception of sour taste.

Acknowledgements

I thank Drs Timothy Gilbertson and Thomas Cummings for reading the manuscript and for assistance with figures. Supported by National Institutes of Health grants DC00766 and DC00244.

References

Avenet P, Lindemann B 1987 Action potentials in epithelial taste receptor cells induced by mucosal calcium. J Membr Biol 95:265–269
Avenet P, Lindemann B 1991 Non-invasive recording of receptor cell action potentials and sustained currents from single taste buds maintained in the tongue: the response to mucosal NaCl and amiloride. J Membr Biol 124:33–41
Avenet P, Kinnamon SC, Roper SD 1991 In situ recording from hamster taste cells: response to salt, sweet and sour. Chem Senses 16:498(abstr)

Béhé P, DeSimone JA, Avenet P, Lindemann B 1990 Membrane currents in taste cells of the rat fungiform papilla: evidence for two types of Ca^{2+} currents and inhibition of K^+ currents by saccharin. J Gen Physiol 96:1061-1084

Bigiani AR, Roper SD 1991 Mediation of responses to calcium in taste cells by modulation of a potassium conductance. Science 252:126-128

Bowerman AG, Kinnamon SC 1992 The effect of K^+ channel blockers on mudpuppy feeding behavior. Chem Senses 17:597-598

Cummings TA, Kinnamon SC 1992 Apical K^+ channels in *Necturus* taste cells: modulation by intracellular factors and taste stimuli. J Gen Physiol 99:591-613

Formaker BK, Hill DL 1991 Lack of amiloride sensitivity in SHR and WKY glossopharyngeal taste responses to NaCl. Physiol & Behav 50:765-769

Frank M 1991 Taste-responsive neurons of the glossopharyngeal nerve of the rat. J Neurophysiol 65:1452-1463

Garty H, Benos DJ 1988 Characteristics and regulatory mechanisms of the amiloride-blockable Na^+ channel. Physiol Rev 68:309-373

Gilbertson TA, Avenet P, Kinnamon SC, Roper SD 1992a Proton currents through amiloride-sensitive Na^+ channels in hamster taste cells: role in acid transduction. J Gen Physiol 100:803-824

Gilbertson TA, Roper SD, Kinnamon SC 1992b Effects of acid stimuli on isolated hamster taste cells. Soc Neurosci Abstr 18:844

Gilbertson TA, Roper SD, Kinnamon SC 1993 Proton currents through amiloride-sensitive Na^+ channels in isolated hamster taste cells: enhancement by vasopressin and cAMP. Neuron 10:931-942

Giza BK, Scott TR 1991 The effect of amiloride on taste-evoked activity in the nucleus tractus solitarius of the rat. Brain Res 550:247-256

Hamill OP, Marty A, Neher E, Sakmann B, Sigworth FJ 1981 Improved patch-clamp techniques for high-resolution current recordings from cells and cell-free membrane patches. Pfluegers Arch Eur J Physiol 391:85-100

Hettinger TP, Frank ME 1990 Specificity of amiloride inhibition of hamster taste responses. Brain Res 513:24-34

Hille B 1992 Ionic channels of excitable membranes, 2nd edn. Sinauer, Sunderland, MA

Kinnamon SC, Cummings TA 1992 Chemosensory transduction mechanisms in taste. Annu Rev Physiol 54:715-731

Kinnamon SC, Roper SD 1988a Evidence for a role of voltage-sensitive apical K^+ channels in sour and salt taste transduction. Chem Senses 13:115-121

Kinnamon SC, Roper SD 1988b Membrane properties of isolated mudpuppy taste cells. J Gen Physiol 91:351-371

Kinnamon SC, Cummings TA, Roper SD 1988a Isolation of single taste cells from lingual epithelium. Chem Senses 13:355-366

Kinnamon SC, Dionne VE, Beam KG 1988b Apical localization of K^+ channels in taste cells provides the basis for sour taste transduction. Proc Natl Acad Sci 85:7023-7027

Kinnamon SC, Cummings TA, Roper SD, Beam KG 1989 Calcium currents in isolated taste receptor cells of the mudpuppy. Ann NY Acad Sci 560:112-115

Korn SJ, Marty A, Connor JA, Horn R 1991 Perforated patch recording. In: Conn PM (ed) Electrophysiology and microinjection. Academic Press, New York (Methods Neurosci 4) p 364-373

Miyamoto T, Okada Y, Sato T 1988 Ionic basis of receptor potential of frog taste cells induced by acid stimuli. J Physiol 405:699-711

Roper SD 1992 The microphysiology of peripheral taste organs. J Neurosci 12:1127-1134

Roper SD, McBride DW Jr 1989 Distribution of ion channels on taste cells and its relationship to chemosensory transduction. J Membr Biol 109:29-39

Schiffman SS, Lockhead E, Maes FW 1983 Amiloride reduces the taste intensity of Na$^+$ and Li$^+$ salts and sweeteners. Proc Natl Acad Sci USA 80:6136–6140

Simon SA, Garvin JL 1985 Salt and acid studies on canine lingual epithelium. Am J Physiol 249:C398–C408

Ye Q, Heck GL, DeSimone JA 1991 The anion paradox in sodium taste reception: resolution by voltage-clamp studies. Science 254:724–726

DISCUSSION

Lindemann: The evidence pointing to the identity of proton and Na$^+$ channels is still somewhat circumstantial. In addition to showing that vasopressin regulates them both, you might compare blocking effects of amiloride analogues on Na$^+$ and H$^+$.

Kinnamon: Tim Gilbertson, Steve Roper and I (unpublished results) have shown that phenamil, an analogue specific for the channel, completely inhibits the response to citric acid recorded with the *in situ* recording technique. The citric acid response is not inhibited by hexamethylene amiloride, an analogue specific for the Na$^+$/H$^+$ antiporter. We have also tried using a fluorescent analogue of amiloride recently developed by Molecular Probes that is supposed to be relatively specific for the channel. We thought that by using fluorescent amiloride we would be able to distinguish exactly where it binds and where it doesn't, but the results were confusing: it seems to bind everywhere, including circumvallate taste buds, which do not appear to express the channel in the hamster. To determine if fluorescent amiloride was binding to the antiporter as well as to the channel, we pretreated the cells with hexamethylene amiloride before adding the fluorescent amiloride. It decreased the binding a bit, but not much. It's clear that amiloride is a pretty dirty drug and that we should probably work with channel inhibitors. We would also like to do the single-channel recording, but it's going to be hard to see single-channel proton currents. We will need to use noise analysis to be able to distinguish conductance through these channels. We predict that it would be a much lower conductance than for Na$^+$, which is already a very small conductance according to your results.

Lindemann: Concerning the fluorescent amiloride conjugate, amiloride itself is fluorescent at 380 nM excitation (Brigmann et al 1983) and can completely ruin experiments with the calcium dye, fura-2.

Lancet: At what wavelength is the molecular probe fluorescent?

Kinnamon: At about the same wavelength as fluorescein.

Lancet: That's much larger than the amiloride wavelength, which is probably why you've not noticed it.

DeSimone: I found the vasopressin results to be particularly exciting, but in a different context. You're familiar with the work of David Hill and his colleagues on Na$^+$-restricted rats. These animals give a poor chorda tympani response to NaCl (Hill 1987). There is reason to think that the cause is a deficit in taste cell apical membrane Na⁺ channels that mediate Na⁺ salt taste transduction. If the rats are given a Na⁺-replete diet, they recover a normal chorda tympani response, which is amiloride sensitive, suggesting that the Na⁺ channels are once again functional. Your data suggest a cellular mechanism by which this recovery could occur. Na⁺-restricted rats probably have high aldosterone levels and basal, perhaps depressed, vasopressin levels. An increase in antidiuretic hormone levels, which might be expected when the rats are permitted free access to salt, may act on taste cells by triggering Na⁺ channel activation. Your data suggest that this is a hypothesis worth pursuing.

Kinnamon: Yes, it would be very interesting.

Schmale: What concentration of vasopressin did you use?

Kinnamon: It's the same concentration that other people have used for other transporting epithelia—10 milliunits. I haven't figured out what it is in terms of micromoles.

DeSimone: Do you know how it compares with normal circulating levels?

Kinnamon: No, I don't.

Schmale: When vasopressin is acting as a physiological agent, what side does it approach the cell from?

Kinnamon: Most likely the basolateral membrane, if it reaches the cells via the circulation.

Schmale: Could it be from secretory vesicles that are transported in the glossopharyngeal nerve towards the synapses with taste receptor cells?

Kinnamon: Yes, it could. There is immunoreactivity to other peptides in the nerve fibres supplying the taste buds, but no one has looked for vasopressin.

Getchell: Isn't vasopressin a circulating hormone?

Kinnamon: Vasopressin has been found to exist as a neurotransmitter as well as a circulating hormone, but its role as a neurotransmitter is not well understood (Caffe et al 1991).

Lindemann: The hormone arginine-vasopressin, which acts on the distal tubules and collecting ducts of the kidney and on other Na⁺-absorbing epithelia, is released from the neural pituitary. The targets respond with a delay of 2–5 minutes (e.g. Li et al 1982). When Okada et al (1991) showed that vasopressin improved NaCl taste in the frog, the delay was about 30 min and the stimulation increased up to 150 min. I was surprised that this hormone should be so slow to have its effect on taste.

Kinnamon: The effect we studied didn't begin until after about 5 min, but then we got a steady increase to saturation at an average of 10 min.

I was also interested in the paper by Okada et al (1991) for a different reason: they found potentiation of acid as well as NaCl responses, but they never tried using amiloride. This suggested that vasopressin acts in the frog in the same way as it does in our system; that the channel can be permeable to protons in the frog as well. They thought that the vasopressin effect was working through the inositol trisphosphate system, so it may be that their receptor is a different receptor than ours.

DeSimone: The effect of vasopressin in mammalian kidney is mainly on the regulation of water channels inserted into the apical membranes of cells in the collecting duct (Fushimi et al 1993). In the case of mammals, I do not recall if vasopressin also regulates renal tubular Na^+ channel density. Is this also an established effect?

Lindemann: In the distal nephron, vasopressin induces not only a hydroosmotic but also a natriferic response mediated by apical Na^+ channels (e.g. Reif et al 1986).

Margolis: Apart from direct release from the central nervous system, isn't there evidence for peptidergic innervation of various taste receptor cells by vasopressin-releasing neurons?

Kinnamon: We don't know, but Tom Firger will be looking at this question.

Getchell: Certainly, in the olfactory mucosa there are vasoactive intestinal peptide-innervating fibres.

Schmale: There are, of course, other locations in the central nervous system, apart from the supraoptic and paraventricular nuclei, where vasopressin is expressed. Also there are a few peripheral tissues, like the adrenal medulla, where vasopressin is made (Weindl & Sofroniew 1985).

Bartoshuk: In the human there is an interesting overlap between the sour and salty taste. If you do blind tests, so that the subjects have no idea of what stimuli you are giving them, salt is described as partly sour and acids are described as partly salty. The salty taste of citric acid is so powerful it's even used as a salt substitute in Kosher cuisine—Kosher sour salt is powdered citric acid. There's another stimulus that's intermediate between the two—LiCl. We talk about LiCl as if it's a substitute for the salty taste of NaCl, but in fact LiCl tastes more sour than NaCl does. Might LiCl be of some value in your experiments?

Kinnamon: Yes. It would be interesting to look at Li^+ permeability of the channel in taste cells.

Lindemann: It is interesting that Li^+ tastes more sour than Na^+. The epithelial Na^+ channel is most permeable to H^+, less permeable to Li^+ and least permeable to Na^+. Then comes a long gap and it is 1000-fold less permeable to K^+ (Palmer 1987).

Hatt: You have shown that K^+ channels are blocked by citric acid and quinine and that these channels have several conductances. Do you think these conductances are all subconductances or are there several different types of channel?

Kinnamon: I would like to know the answer to that. I know that some of them are separate channels, because they respond differentially to some intracellular factors. For example, there's a channel that operates in the 95–105 pS range that's blocked by intracellular ATP (Cummings & Kinnamon 1992). That channel, at least, is different from another group of channels which have conductances larger than 135 pS and are Ca^{2+} dependent. I know there's more than one channel, but I think your idea of substates is really interesting. There are so many channels in the patch you can't really do the kinetic analysis necessary to see the transitions that are expected of substate behaviour.

Bargmann: Do you know whether mudpuppies discriminate between sour and bitter taste?

Kinnamon: No; we just know they spit out gelatin cubes containing either bitter or sour chemicals. I think this is a generalized aversive mechanism, but we haven't done the behavioural studies that would be needed to tell if they are discriminating between these compounds. Nor have we done the afferent nerve single-fibre studies which might suggest that one fibre responded much more strongly to sour than to bitter.

Firestein: Is it possible that the Na^+/H^+ channel is simply an electrolyte detector? Its role might be just to get a record of the electrolyte concentration in the saliva.

Kinnamon: These channels aren't going to be affected by divalent cations, so I don't think they are general electrolyte detectors. We can clearly distinguish between Na^+ and H^+, though.

Lindemann: At the Na^+ channels of frog skin and toad bladder, Ca^{2+} has a blocking effect that is dependent on the membrane potential. If you hyperpolarize the apical membrane potential by about 50–60 mV, then the Ca^{2+} block becomes apparent and at more extreme potentials, Ca^{2+} can block completely (Lindemann 1989). Amiloride-blockable channels of frog taste cells are also somewhat inhibited by extracellular divalent cations (Avenet & Lindemann 1989).

DeSimone: We shouldn't forget that there is good reason to believe that amiloride-blockable Na^+ channels in taste cells mediate responses in the taste nerves to Na^+ salt stimuli. Whole-nerve and single-fibre responses to Na^+ salts are blocked by amiloride. Also, rats behave as if these channels function as salt detectors (Hill et al 1990).

Kinnamon: In rat, amiloride blocks the afferent nerve response to NaCl, and in hamster, it blocks the response to both NaCl and HCl.

Bartoshuk: The effect of amiloride is very controversial in humans.

DeSimone: We think that this is the result of an anion effect. As I showed in my paper (this volume: DeSimone et al 1993), the Cl^- ion adds a voltage-independent component to the salt response in rats. Hill's lab has shown that this Cl^--dependent response is also amiloride insensitive (Formaker & Hill 1988). The fact that both amiloride and voltage insensitivity emerge together when Cl^- is the anion suggest that Cl^- ions allow Na^+ ions access to sites in the taste bud below the tight junctions that exclude the larger amiloride molecule.

Kurahashi: Is the amiloride-sensitive Na^+ channel localized apically?

Kinnamon: We don't know the answer to that because we are bathing the whole cell in amiloride.

Lindemann: If you take a perfused pipette and put it on a fungiform papilla, you get a response to NaCl comprising both action potentials and sustained current. If you add amiloride to the solution, which only reaches the apical membrane, it immediately blocks both the action potential and the sustained current (Avenet & Lindemann 1991). The conclusion is that these amiloride-blockable channels are present in the apical membrane. If they were also present and functional in the basolateral membrane in sizeable numbers, then the cell would depolarize and it could not fire these action potentials.

Kurahashi: Dr Kinnamon, you find K^+ channels in the apical membrane and that bitter substances block almost all ion channels. Is it possible that some of these channels in the apical membrane are amiloride-sensitive Na^+ channels that are also blocked by bitter substances?

Kinnamon: Surprisingly, in the mudpuppy, there is no evidence for amiloride-sensitive Na^+ channels. There's probably a weak Na^+ conductance that is unaffected by amiloride. There's also some question about amiloride-sensitive Na^+ channels in mammalian circumvallate taste buds. Afferent nerve recordings have suggested that in some cases the circumvallate taste buds may not express this channel (Formaker & Hill 1991).

DeSimone: I would like to comment further on the species-dependent expression of the salt-transducing cation channel. Single-unit recordings in rats by Erickson (1963), in rats and hamsters by Frank et al (1983) and in the geniculate ganglion by Boudreau et al (1983), point strongly to a special Na^+ taste system in these animals that does not react to K^+ salts. On the other hand, in the case of dogs (Mierson et al 1988) and frogs (Herness 1987), amiloride blocks the responses of both Na^+ and K^+ salt stimuli. Both dogs and frogs are primarily carnivorous, whereas rodents tend to be mainly herbivorous. Boudreau et al (1983) speculate that the Na^+-specific salt taste system is a specialization of herbivores and omnivores, whereas the non-specific Na^+/K^+ salt system is a property of carnivores, i.e. animals that do not have to forage for Na^+ in their environment, but like herbivores, must avoid brackish water. As more species are tested, it will be interesting to see if the amiloride-blockable channel in herbivores and omnivores remains Na^+ specific while the channel in carnivores generally accepts K^+ as well as Na^+.

Lancet: I knew there was overlap between sweet and bitter in the sense that bitter compounds and sweet compounds may be very analogous to each other, but I didn't know about the 'after-taste' effect, similar to shining bright red light and getting an after-image that is green. Linda Bartoshuk suggested that there is something like that between bitter and sweet. Likewise, there seems to be a relationship between salty and sour that transcends what I had always suspected. It leads me to think that taste is actually quite a peculiar sense. In olfaction, no matter what we sense—if it's a pure compound or if it's a complex mixture of compounds—in the end a pattern of neuronal activity is generated. We don't need to classify beyond the actual recognition of the pattern. In taste, we are trying to do something much more complicated: we are trying to classify a vast array of compounds into only four categories. If you were to ask a computer scientist how to do something like this, the answer would be to take several relatively non-specific detectors (indeed, no taste receptors appear to be really specific for H^+, Na^+ or sweet and bitter compounds) and do a subtraction of their signals (or some other linear combination), so you can 'purify' one of the classification components. The concept I would like to put forward is that, whilst in olfaction there may be integration in the sense that we add up several factors and get a pattern, in taste we are subtracting. For example, there may be two sour detectors, one that has a stronger sour component and the other that has a weaker one. If you subtract one signal from the other you get a pure signal. This might explain our inability to look at individual cells or axons in the gustatory epithelium and say what each responds to.

DeSimone: Single-unit electrophysiological studies in rodents at the level of the chorda tympani and the nucleus of the solitary tract indicate the presence of inhibitory taste responses (McPheeters et al 1990). These may play a role in quality coding as you suggest. Some of the coding possibilities may also reside in the structure of the taste bud itself. It is not clear why the taste receptor cells exist in clusters—the taste buds. The chemoreceptor cells in the lower gastrointestinal tract, for example the acid receptors of the duodenum (secretin-releasing cells), exist as isolated cells, never as chemoreceptor cell clusters (this volume: DeSimone et al 1993). Yet they seem to function as simple detectors. Grouping the cells in clusters, however, allows them to communicate and permits some cells to be primary receptors and others to react secondarily. The modes of communication could be synaptic or via gap junctions (Roper 1992), or by stimulus diffusion through the paracellular pathway (this volume: DeSimone et al 1993). This richness of interaction could provide either augmentation or subtraction, i.e. a first order integration of a receptor output at the taste bud level.

Bartoshuk: What about your anionic inhibition?

DeSimone: This would be an example of a parallelogram mode of information transfer in the taste bud. In this case, the anion mobility in the weak cation exchanger tight junctional complexes determines the extent to which Na^+ has access to cells in the taste bud interior. Anion mobility also influences field potentials sensed by receptor cells. Without the unique topology of the taste bud, anion effects of this sort could not occur.

Lancet: Anions are interesting, because we can often recognize Na^+ irrespective of its anion. Maybe there is an anion detector that then gets subtracted, so you get pure Na^+ perception.

Bartoshuk: Schulkin (1986) has argued that if animals develop a deficiency of almost any cation, it's manifested behaviourally as a Na^+ craving; they go out and look for it. Apparently, this is how they find all their cations, because they tend to occur together in their environment.

DeSimone: Denton (1967) describes some studies on herbivores that nicely illustrate the specificity of Na^+ taste detection in these animals. Na^+-deficient wild rabbits were given the opportunity to sample licks of NaCl, $NaHCO_3$, KCl, $MgCl_2$ and other salts. In short order, the Na^+ salt licks were exhausted, while the other salt licks were virtually ignored after a few probe licks.

Firestein: What if they were K^+ deprived: would they still go for the Na^+ lick?

DeSimone: That does not normally occur in wild herbivores because of the high K^+ levels in plants. However, experimental rats that have been K^+ depleted still prefer to ingest Na^+ over K^+ (Adam & Dawson 1972).

References

Adam WR, Dawson JK 1972 Effect of potassium depletion as mineral appetite in the rat. J Comp Physiol Psych 78:51–58

Avenet P, Lindemann B 1989 Chemoreception of salt taste. The blockage of stationary sodium currents by amiloride in isolated receptor cells and excised membrane patches. In: Brand JG, Teeter JH, Cagan RH, Kare MR (eds) Chemical senses: receptor events and transduction in taste and olfaction. Marcel Dekker, NY, vol 1:171–182

Avenet P, Lindemann B 1991 Noninvasive recording of receptor cell action potentials and sustained currents from single taste buds maintained in the tongue: the response to mucosal NaCl and amiloride. J Membrane Biol 124:33–41

Boudreau JC, Hoang NK, Oraveck J, Do TL 1983 Rat neurophysiological taste responses to salt solutions. Chem Senses 8:131–150

Briggman JV, Graves JS, Spicer SS, Cragoe EJ 1983 The intracellular localization of amiloride in frog skin. Histochem J 15:239–255

Caffe AR, Holstege JC, van Leeuwen FW 1991 Vasopressin immunoreactive fibers and neurons in the dorsal pontine tegmentum of the rat, monkey and human. Prog Brain Res 88:227–240

Cummings TA, Kinnamon SC 1992 Apical K^+ channels in *Necturus* taste cells: modulation by intracellular factors and taste stimuli. J Gen Physiol 99:591–613

Denton DA 1967 Salt appetite. In: Code CF (ed) Handbook of physiology, Vol I, Section 6: Alimentary canal. Amer Physiological Soc, Washington DC, p 433–459

DeSimone JA, Ye Q, Geck GE 1993 Ion pathways in the taste bud and their significance for transduction. In: The molecular basis of smell and taste transduction. Wiley, Chichester (Ciba Found Symp 179) p 218–234

Erickson RP 1963 Sensory neural patterns and gustation. In: Zotterman Y (ed) Olfaction and taste. Pergamon Press, Oxford, p 205–213

Formaker BK, Hill DL 1988 An analysis of residual NaCl taste response after amiloride. Am J Physiol 255:R1002–R1007

Formaker BK, Hill DL 1991 Lack of amiloride sensitivity in SHR and WKY glossopharyngeal taste responses to NaCl. Physiol & Behav 50:765–769

Frank ME, Contreras RJ, Hettinger TP 1983 Nerve fibers sensitive to ionic taste stimuli in the chorda tympani of the rat. J Neurophysiol 50:941–960

Fushimi K, Uchida S, Hara Y, Hirata Y, Marumo F, Sasaki S 1993 Cloning and expression of apical membrane water channel of rat kidney collecting tubule. Nature 361:549–552

Herness S 1987 Are apical ion channels involved in frog taste transduction? Anns NY Acad Sci 510:362–367

Hill DL 1987 Susceptibility of the developing rat gustatory system to the physiological effects of dietary sodium deprivation. J Physiol 393:413–424

Hill DL, Formaker BK, White KS 1990 Perceptual characteristics of the amiloride-suppressed sodium chloride response in the rat. Behav Neurosci 104:734–741

Li JH-Y, Palmer LG, Edelman IS, Lindemann B 1982 The role of sodium channel density in the natriferic response of the toad urinary bladder to an antidiuretic hormone. J Membrane Biol 64:77–89

Lindemann B 1969 Sodium and calcium dependence of threshold potential in frog skin excitation. Biochim Biophys Acta 163:424–426

McPheeters M, Hettinger TP, Nuding SC, Savoy LD, Whitehead MC, Frank ME 1990 Taste-responsive neurons and their locations in the solitary nucleus of the hamster. Neuroscience 34:745–758

Mierson S, DeSimone SK, Heck GL, DeSimone JA 1988 Sugar-activated ion transport in canine lingual epithelium. J Gen Physiol 92:87–111

Okada Y, Miyamoto T, Sato T 1991 Vasopressin increases frog gustatory neural responses elicited by NaCl and HCl. Comp Biochem Physiol A Comp Physiol 100:693–696

Palmer L 1987 Ion selectivity of epithelial Na^+ channels. J Membrane Biol 96:97–106

Roper SD 1992 The microphysiology of peripheral taste organs. J Neurosci 12:1127–1134

Reif MC, Troutman SL, Schafer JA 1986 Sodium transport by rat cortical collecting tubule. J Clin Invest 77:1291–1298

Schulkin J 1986 Behavioral dynamics in the appetite for salt in rats. In: de Caro ANEB, Massi M (eds) The physiology of thirst and sodium appetite. Plenum, New York, p 497–502

Weindl A, Sofroniew M 1985 Neuroanatomical pathways related to vasopressin. Curr Top Neuroendocrinol 4:137–196

Ion pathways in the taste bud and their significance for transduction

John A. DeSimone, Qing Ye and Gerard L. Heck

Department of Physiology, Virginia Commonwealth University, Richmond, VA 23298-0551, USA

Abstract. Taste buds share a topology with ion-transporting epithelia and evidence now indicates that neural responses in rats to Na^+ salts of differing anion are mediated by both transcellular and paracellular ion transport. Na^+ exerts its effects mainly on the transcellular pathway. Neural responses to Na^+ salts are enhanced by negative voltage clamp and suppressed by positive clamp in a manner indicating modulation of the apical membrane potential of receptor cells. Anion effects are mainly paracellular. Under zero current clamp increasing anion size reduces the neural response at constant Na^+ concentration. Below about 50 mM this difference is entirely eliminated under voltage clamp. This suggests that paracellular transepithelial potentials normally create an anion difference. At higher concentrations the relatively high permeability of the paracellular shunt to Cl^- permits sufficient electroneutral diffusion of NaCl below the tight junctions to stimulate cells that do not make direct contact with the oral cavity. In general, the sensitivity of a response to perturbations in the apical membrane potential indicates that some phase of Na^+ salt taste transduction is accompanied by changes in an apical membrane channel conductance.

1993 The molecular basis of smell and taste transduction. Wiley, Chichester (Ciba Foundation Symposium 179) p 218-234

Mature taste-bud cells share the characteristic polar structure of the epithelial cells that line the gastrointestinal tract. Like the taste-bud cells, some of the intestinal cells are also chemosensory—most notably the endocrine cells of the gut that function as specialized chemosensory cells (Fugita 1991). The chemosensory cells of the mouth and gut make contact with a luminal environment of variable composition (in which the chemical stimulus is usually presented) and a submucosal environment that has a plasma-like composition. Their apical cell membranes are in direct contact with the luminal environment, which in the case of taste cells is the oral cavity. It has been generally assumed that this membrane contains the molecular receptors that interact selectively with substances to be detected. This would present the simplest route of interaction between stimulus and receptor for both taste and intestinal chemoreceptors. In support of this view, it is known that taste cells detect certain proteins as sweet-tasting stimuli (Yamashita et al 1990). Because of their high molecular masses (27 800 Da for sweet-tasting curculin, Yamashita et al 1990), these proteins are unlikely to be membrane permeable. It is therefore probable that receptors for these sweet-tasting stimuli reside in the apical membranes of the chemosensory cells. However, tastants administered intravascularly can also evoke responses in the taste nerves (Bradley 1973).

The existence of intravascular taste suggests a more complex pattern of functional organization among taste-bud cells. One possibility is that molecular receptors for taste stimuli are normally present on the basolateral, as well as the apical, membranes of taste-bud cells (Bradley 1973). Another possibility is that there are developing taste cells that do not initially make direct contact with the oral cavity. They may, none the less, be innervated and be stimulated by taste stimuli that reach them intravascularly. A third possibility is that low molecular mass taste stimuli administered intravascularly diffuse through the tight junctions that form the structural barrier between the apical and basolateral membranes, and thus stimulate the taste receptors on the apical membrane (Holland et al 1991). To complicate matters further, these possibilities are not mutually exclusive.

The possible physiological importance of the movement of taste stimuli across the tight junctions surrounding taste cells (especially when it occurs in the usual mucosal to submucosal direction) suggests that the topology of the taste bud itself may significantly determine its response to a stimulus. Ultrastructural and electrophysiological studies place the paracellular shunt barrier in the tight junctional complexes that bind lingual epithelial cells together (Holland et al 1989). The paracellular shunt pathways in the lingual epithelium can transport ions and sugars (DeSimone et al 1984, Simon & Garvin 1985, Mierson et al 1988). The non-gustatory cells of the dorsal lingual epithelium are coupled electrically by gap junctions (Holland et al 1989). This is good evidence that, like many other transporting epithelia, the dorsal lingual epithelium is a functional syncytium. Dye-coupling studies indicate that some taste receptor cells may also be coupled by gap junctions (Yang & Roper 1987). In addition, taste receptor cells appear to be coupled synaptically to basal cells, which, in turn, may be coupled synaptically to taste nerves (Roper 1992). The implications of such coupling for the transduction and coding of the various taste submodalities are beginning to be appreciated. It is clear that in their potential for cell–cell coupling, taste and intestinal chemoreceptors differ greatly. The latter exist exclusively as isolated cells and therefore lack the multicellular response possibilities inherent in the topological organization of the taste bud.

The paracellular pathway and the salt anion effects

Active ion transport across the lingual epithelium is an electrogenic process which generates a standing transepithelial electrical potential (DeSimone et al 1981, 1984,

Mierson et al 1985, 1988). The size of the potential is, in part, determined by the ion-transporting properties of the paracellular shunts. These are passive, leaky and moderately cation-selective pathways that obey the laws of electro-diffusion (DeSimone et al 1984, Simon & Garvin 1985, Ye et al 1991). Because the shunt pathways exist in parallel with the transcellular ion pathways in the taste bud, the transepithelial potential dropped across the tight junctions (the major transport barriers in the paracellular pathway) equals the sum of the potentials across the taste cell apical membranes and basolateral cell membranes. This topological relation places constraints on the instantaneous receptor potential (the stimulus-evoked change in potential across the receptor cell basolateral membrane before adaptation processes have occurred), as shown in equation 1. Here V_r is the receptor potential (referenced to the submucosa),

$$V_r = V_a - V_t \qquad (1)$$

V_a is the potential across the apical membrane (referenced to the mucosa) and V_t is the transepithelial potential (referenced to the mucosa). In general, changes in apical membrane potential are not independent of changes in transepithelial potential. However, if two salt stimuli are identical in all respects except that the first gives rise to a more electropositive transepithelial potential than the second, then the first is the weaker of the stimuli. That is, it produces the smaller instantaneous receptor potential.

This topological restriction exerts a major influence on the responses of taste receptors to Na^+ salts of differing anions. It is well established that the response of rat chorda tympani to Na^+ salts is anion dependent (Beidler 1954). Among the anions tested, Cl^- always results in the largest response for a given Na^+ concentration—larger anions usually produce lower responses. Harper (1987) and Elliott & Simon (1990), using different approaches, each proposed that electrodiffusion of the salts across the junctional complexes between receptor cells might be the source of the anion influence. If the larger anions have lower shunt permeability, then Na^+ salts with large anions would produce larger values of V_t than smaller anion salts. By equation (1), this would result in more hyperpolarized receptor potentials and, therefore, smaller neural responses.

We have used a lingual epithelial voltage clamp to demonstrate this mechanism (Heck et al 1989). The voltage clamp essentially allows for the decoupling of V_t from anion influences (Ye et al 1991, 1993). In zero current clamp mode, there is an inverse relation between the transepithelial potential produced by a given concentration of Na^+ salt and the neural response that it evokes (Ye et al 1991). This is clearly a necessary condition, but it is not sufficient to prove that anion-induced changes in V_t cause the anion effect. However, under voltage clamp of V_t, paracellular influences should, in principle, be removed. Figure 1 shows integrated chorda tympani responses to NaCl, sodium acetate

(NaAc) and sodium gluconate (NaGlu). The results confirm that anion-related differences disappear when V_t is held fixed. The process of fixing V_t has a predictable effect on V_a as well. As shown in Fig. 1, clamping V_t at 0 mV and at -60 mV enhanced responses to all the salts, whereas clamping V_t at $+60$ mV suppressed the responses. Voltage sensitivity of this type is what would be expected if the Na$^+$ sensor is an apical membrane passive channel (see below). Because of the topological constraints on the system, a fraction of the fixed value of V_t will be sensed also at the apical and basolateral membranes of the receptor cells. Therefore, increasing electronegative values of V_t serve to steepen the electrical gradient at the apical cell membrane, that is, to increase the driving force for depolarizing Na$^+$ influx into the receptor cells. Increasing electropositive values of V_t have the opposite effect and therefore diminish responses to the Na$^+$ salts.

The paracellular pathway between receptor cells, although cation selective, is far from a perfect cation-exchanger membrane. *In vitro* electrophysiological studies show that the potential across the lingual epithelium is much less than an ideal Nernst potential (DeSimone et al 1984). Because of the leakiness of the shunts, a co-ion-limited electroneutral diffusion of Na$^+$ salt can be expected to occur. Given the low volumes of the intercellular spaces, even low paracellular fluxes could result in large changes in salt concentration in the submucosal microenvironment. Provided that Na$^+$-sensing channels exist on the submucosal side (and intravascular taste studies suggest they may), a second shunt-mediated anion effect could emerge. The evidence is that a second anion effect does exist. It is Cl$^-$ dependent, amiloride insensitive, voltage independent and is observed only at concentrations above about 50 mM (Ye et al 1993). If penetration of the shunt is required for the second effect, each of the observed properties can be explained as follows. The anion effect is Cl$^-$ dependent because only a relatively mobile anion will permit much Na$^+$ penetration. It is amiloride insensitive because amiloride cannot penetrate below the tight junctions to block a submucosal event. It is voltage independent because most of the clamp voltage is sensed at the apical membranes and tight junctions; the intercellular spaces are all at roughly the same potential. A threshold concentration exists because passive influx requires a favourable Na$^+$ gradient across the tight junctions. For NaCl, this second anion effect may be of greater magnitude than the first anion effect (on V_t).

Apical membrane channels and the Na$^+$ electrochemical concentration

Because the transcellular and paracellular ion pathways in a transporting epithelium are arranged in parallel, changes in the transepithelial potential, ΔV_t, will also be sensed at the apical and basolateral cell membranes. A simple equivalent circuit model (Reuss 1991) is shown in equation 2, where ΔV_a is the

$$\Delta V_a = \delta \Delta V_t \qquad (2)$$

change in potential across the apical cell membrane, ΔV_t is the change in transepithelial potential and δ is the fraction of ΔV_t dropped across the apical membrane, shown in equation 3, where R_a and R_b are the equivalent resistances

$$\delta = R_a/(R_a + R_b) \tag{3}$$

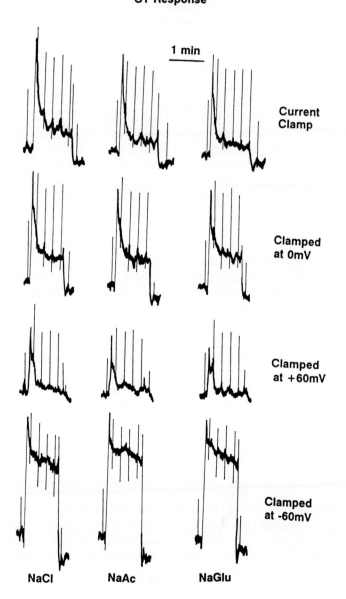

of the apical and basolateral membranes, respectively. Again, these relations are expected to apply immediately following the application of a stimulus and before compensating adaptation processes have come into play.

If the key initial event in transduction in the case of the Na$^+$ ion is the influx of Na$^+$ into receptor cells through specific Na$^+$ channels (Avenet & Lindemann 1988, 1991), the chorda tympani response, R_{ct}, will be proportional to J, the influx of Na$^+$. The influx, J, will in turn be proportional to the electrochemical concentration of Na$^+$, C_e (Ye et al 1993) given by equation 4.

$$C_e = Ce^{-\delta\phi} \quad (4)$$

C is the stimulus Na$^+$ concentration, δ is as defined in equation 3 and ϕ is the dimensionless potential, $F\Delta V_t/RT$, where F, R and T have their usual thermodynamic significance. The initial influx of Na$^+$ through the channels will be given by equation 5 (Fidelman & Mierson 1989), where J_m is the maximum

$$J = \frac{J_m C_e}{C_e + K_m} \quad (5)$$

flux and K_m is the electrochemical concentration at which J is half maximal. If the influx, J is causal to the neural response, then a generalization of the taste equation follows equation 6 (Beidler 1954, Ye et al 1993) where R_{ct} is the

$$R_{ct}(C, \Delta V_t) = \frac{R_{ctm} C_e}{C_e + K_m} \quad (6)$$

FIG. 1. (*opposite*) Integrated chorda tympani responses to 25 mM NaCl, sodium acetate (NaAc) and sodium gluconate (NaGlu). In each case the rinse and adapting solution was 10 mM KHCO$_3$. Recordings were made from a section of rat lingual epithelium, lying within a flow chamber affixed by vacuum to the tongue. Stimuli were applied by forcing solutions into the centre of the chamber, allowing them to flow over the tongue and then to exit along the periphery. Current-passing and voltage-sensing electrodes were also in the chamber, enabling monitoring of the neural response with the stimulated receptive field under current or voltage clamp (Heck et al 1989, Ye et al 1991). The top row shows the responses recorded under zero current clamp (equivalent to open circuit measurements). Typically, NaCl gives larger responses than NaAc or NaGlu. Under voltage clamp (next three rows), all anion differences are eliminated and the neural responses are seen to be strong functions of voltage. Responses are enhanced by more electronegative clamping voltages and suppressed by more positive clamping voltages.

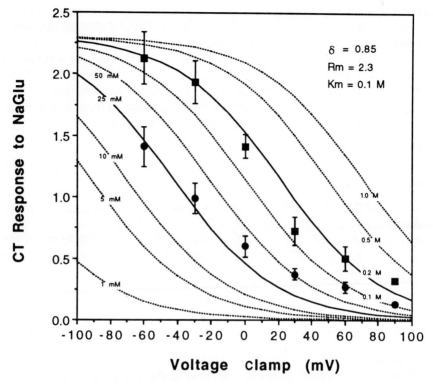

FIG. 2. The voltage dependence of the response of chorda tympani (CT) to 25 mM NaGlu (circles) and 200 mM NaGlu (squares). Each point represents the mean ± SEM ($n=4$). The solid lines represent the best fit of the data to the model expressed in equation (6) with the best-fit parameters given (*top right*). The broken curves present the predicted dependence of CT response to voltage at other NaGlu concentrations. It is clear that responses are *not* constant for a given [Na$^+$]; they are in fact constant across equivalent *electrochemical* concentrations: $[Na^+]e\{-\delta(F\Delta V_t/RT)\}$.

chorda tympani response and R_{ctm} is its maximum value. From equation 6 we would expect the neural response to be a function of the instantaneous voltage drop across the receptor cell apical membrane ($\Delta V_a = \delta \Delta V_t$) as well as concentration of Na$^+$ salt.

Figure 2 shows the voltage dependence of the chorda tympani response to 25 mM and 200 mM NaGlu. These data were fit to equation 6. Best fit was achieved with the following parameters: $\delta = 0.85$, $R_{ctm} = 2.3$ and $K_m = 100$ mM. With these parameters, the predicted chorda tympani responses can be plotted as a function of clamping voltage for NaGlu concentrations between 1 mM and 1 M. As expected for a channel, the results show that apical membrane voltage as well as Na$^+$ concentration determines the intensity of the response. Figure 2 shows that a given response can be achieved by pairing concentration

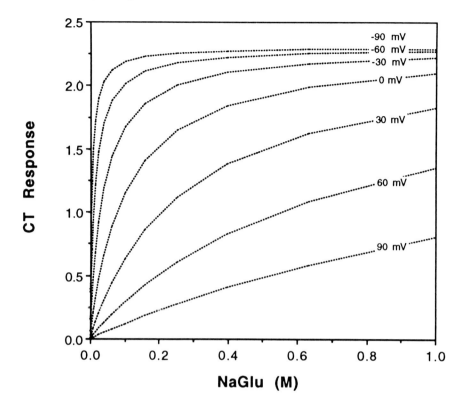

FIG. 3. Values of the CT response as a function of NaGlu concentration, predicted by equation (6), using the parameters found from fitting the data in Fig. 2. The dose-response curves shift to the left along the NaGlu axis (relative to the zero voltage curve) when voltages are electronegative and to the right when they are positive. Electronegative voltages decrease the apparent K_m (enhance the response), whereas positive voltages increase K_m (suppress the response).

and potential so as to achieve constant electrochemical concentration (equation 4).

Figure 3 shows the predicted response (equation 6) as a function of concentration for fixed values of the voltage. This form produces the more familiar saturating isotherms. The effect of voltage is clearly seen as a shift in the apparent value of K_m. Increasing the driving force for Na^+ by making ΔV_t more negative shifts the curve to the left on the concentration axis. It is clear that for increasingly negative values of ΔV_t, the maximum response can be achieved at lower NaGlu concentrations.

(a) Paracellular Pathway

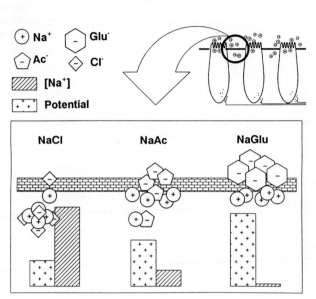

(b) Transcellular Pathway

$$\text{Flux} \propto [Na^+]e^{-F\Delta V/RT}$$

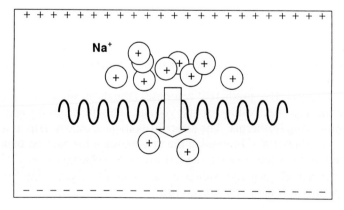

Discussion

The early events surrounding the detection of Na$^+$ salts in the taste buds are remarkably similar to those that take place in ion-transporting epithelia. The detecting elements involve apical membrane Na$^+$ channels, probably within a subset of taste-bud cells. Na$^+$ ions enter taste cells carrying the depolarizing current necessary to trigger the subsequent transduction events leading to the release of neurotransmitters that can excite the afferent taste nerves. There are also, of course, paracellular ion pathways. The principal barriers to paracellular ion transport are the tight junctions that bind the apical poles of the receptor cells together. These act as weakly cation-selective barriers mediating electrodiffusion into the intercellular microenvironment.

Transport through paracellular pathways is responsible for the significant influence of anions on the neural response to Na$^+$ salts. Figure 4a illustrates the two effects. In the case of NaGlu, the anion is relatively large. Its limited mobility through the tight junctions results in a large electropositive transepithelial potential and very little submucosal penetration of Na$^+$. The more electropositive the transepithelial potential, the lower the receptor potential. The limited penetration of Na$^+$ prevents stimulation of cells that are not in direct contact with the oral cavity. In the case of NaAc, the anion-mediated transepithelial potential is less electropositive and somewhat more submucosal penetration of Na$^+$ can be expected, both effects producing a measurably larger neural response to a given Na$^+$ concentration. At the other extreme, the highly mobile Cl$^-$ produces the least electropositive transepithelial potential; it mediates significant penetration of Na$^+$ into submucosal regions where cells lacking direct contact with the oral cavity can be stimulated. This accounts for the extra amiloride-insensitive component in the NaCl chorda tympani response (Formaker & Hill 1988).

FIG. 4. (*opposite*) (a) A schematic illustration of the effects of the paracellular pathway on the microenvironment in a taste bud. The principal barrier in the paracellular pathway is in the tight junctional complex. This acts as a weak cation exchanger membrane. For a series of sodium salts, the rate of diffusion of the salt through the tight junctions will be anion limited. The relatively mobile Cl$^-$ ion permits substantial permeation of NaCl below the tight junctions, allowing Na$^+$ access to additional sensory cells. Because of the high Cl$^-$ mobility, the anion-dependent transepithelial potential will be relatively small. This also promotes a larger cell receptor potential (see equation 1). In the cases of NaAc and NaGlu, limitations on anion mobility result in lower Na$^+$ salt penetration of the tight junctions and higher electropositive transepithelial potentials. Both effects reduce the stimulus power of these salts. (b) For Na$^+$ salts, the major transduction sites on taste cells are Na$^+$ channels in the apical membranes. The influx of depolarizing Na$^+$ ions is driven by differences in both the Na$^+$ concentration and the apical membrane potential, ΔV. The stimulus intensity is therefore the electrochemical concentration: $[Na^+]e\{-\delta(F\Delta V/RT)\}$.

If the Na^+ sensors are apical membrane Na^+ channels, the actual Na^+ stimulus intensity should be the apical membrane electrochemical concentration (Fig. 4b). This is borne out by recording under voltage clamp. Although the initial conditions of stimulus concentration and apical membrane voltage might appear to determine only the initial response of the system, subsequent adaptation processes appear to be tightly coupled to the magnitude of the initial depolarizing current. This has been demonstrated experimentally by showing that responses at constant electrochemical concentration are not only identical in initial peak magnitude but also adapt at the same rate (Ye et al 1993). The electrochemical concentration is therefore the true intensity dimension for the Na^+ 'salty' taste submodality. This fact is definitive proof that in this particular instance, chemical and electric taste are mechanistically identical phenomena. In the first case, changes in the electrochemical concentration are effected by altering its chemical part, while in the second case the changes are effected by altering its electrical part. Figure 3 shows the great potential that exists for enhancement of the response to a given Na^+ concentration. Anything that results in steepening of the electrical gradient across the apical membrane of the taste cell will enhance the response to salt. Enhancement takes the form of achievement of maximal response at lower concentration, not elevation of the maximal response itself.

The mechanisms responsible for the expression and control of the taste response to Na^+ salts are not fully understood. Our results show that there are several key loci where control could be exerted. These include the apical membrane channels (number or conductance properties), tight junction ion exchanger permeability and selectivity, and metabolic control over cell ion pumps. As seen in Figs. 2 and 3, regulation of the apical membrane potential alone would substantially up- or down-regulate the sensitivity to a given concentration of Na^+. As in the case of other Na^+ transporting epithelia, such regulation might be controlled by hormones in response to body Na^+ needs. The voltage-clamp recording method provides a means of probing these regulatory mechanisms along the electrochemical concentration dimension that defines Na^+ taste intensity.

Acknowledgement

This work was supported by National Institutes of Health grant DC00122.

References

Avenet P, Lindemann B 1988 Amiloride-blockable sodium currents in isolated taste receptor cells. J Membr Biol 105:245–255

Avenet P, Lindemann B 1991 Non-invasive recording of receptor cell action potentials and sustained currents from single taste buds maintained in the tongue: the response to mucosal NaCl and amiloride. J Membr Biol 124:33–41

Beidler LM 1954 A theory of taste stimulation. J Gen Physiol 38:133–139

Bradley RM 1973 Electrophysiological investigations of intravascular taste using perfused rat tongue. Am J Physiol 224:300–304

DeSimone JA, Heck GL, DeSimone SK 1981 Active ion transport in dog tongue: a possible role in taste. Science 214:1039–1041

DeSimone JA, Heck GL, Mierson S, DeSimone SK 1984 The active ion transport properties of canine lingual epithelia in vitro: implications for gustatory transduction. J Gen Physiol 83:633–656

Elliott EJ, Simon SA 1990 The anion in salt taste: a possible role for paracellular pathways. Brain Res 535:9–17

Fidelman ML, Mierson S 1989 Network thermodynamic model of rat lingual epithelium: effects of hyperosmotic NaCl. Am J Physiol 257:G475–G487

Formaker BK, Hill DL 1988 An analysis of residual NaCl taste response after amiloride. Am J Physiol 255:R1002–R1007

Fugita T 1991 Taste cells in the gut and on the tongue. Their common, paraneuronal features. Physiol & Behav 49:883–885

Harper HW 1987 A diffusion potential model of salt taste receptors. Ann NY Acad Sci 510:349–351

Heck GL, Persaud KC, DeSimone JA 1989 Direct measurement of translingual epithelial NaCl and KCl currents during the chorda tympani taste response. Biophys J 55:843–857

Holland VF, Zampighi GA, Simon SA 1989 Morphology of fungiform papillae in canine lingual epithelium: location of intercellular junctions in the epithelium. J Comp Neurol 279:13–27

Holland VG, Zampighi GA, Simon SA 1991 Tight junctions in taste buds: possible role in perception of intravascular gustatory stimuli. Chem Senses 16:69–79

Mierson S, Heck GL, DeSimone SK, Biber TUL, DeSimone JA 1985 The identity of the current carriers in canine lingual epithelium in vitro. Biochim Biophys Acta 816:283–293

Mierson S, DeSimone SK, Heck GL, DeSimone JA 1988 Sugar-activated ion transport in canine lingual epithelium. Implication for sugar taste transduction. J Gen Physiol 92:87–111

Reuss L 1991 Tight junction permeability to ions and water. In: Cereijido M (ed) Tight junctions. CRC Press, Boca Raton, FL, p 49–66

Roper SD 1992 The microphysiology of peripheral taste organs. J Neurosci 12:1127–1134

Simon SA, Garvin JL 1985 Salt and acid studies on canine lingual epithelium. Am J Physiol 249:C398–C408

Yamashita H, Theerasilp S, Aiuchi T, Nakaya K, Nahamura Y, Kurihara Y 1990 Purification and complete amino acid sequence of a new type of sweet protein with taste-modifying activity: curculin. J Biol Chem 265:15770–15775

Yang J, Roper SD 1987 Dye-coupling in taste buds in the mudpuppy, *Necturus maculosus*. J Neurosci 7:3561–3565

Ye Q, Heck GL, DeSimone JA 1991 The anion paradox in sodium taste reception: resolution by voltage-clamp studies. Science 254:724–726

Ye Q, Heck GL, DeSimone JA 1993 Voltage dependence of the rat chorda tympani response to Na^+ salts: implications for the functional organization of taste receptor cells. J Neurophysiol 70:167–178

DISCUSSION

Lindemann: Have you tried studying the localization of Na^+ channels in taste receptor membranes?

DeSimone: We have done immunofluorescence experiments using the anti-Na$^+$ channel antibody prepared by Dale Benos and his colleagues. In these experiments, the staining seemed to be uniform between both the apical and basolateral membranes. This has been found consistently using this antibody. Dale has been quite generous in making it available to our group and to others interested in taste reception.

Smith & Benos (1991) have recently reviewed the literature on the epithelial Na$^+$ channel, including the methods of Na$^+$ channel isolation and antibody production. Sid Simon and his colleagues were the first to report that this antibody labels taste buds; they used canine circumvallate papillae (Simon et al 1993). The antibody labelling methods have been adapted and refined for rat fungiform papillae by Robert Stewart working in David Hill's lab (Stewart & Hill 1993).

Lindemann: Patrick Avenet and I did some work on frog taste cells, where, under whole-cell conditions, we recorded 100 pA of amiloride-blockable current. We then pulled an outside-out patch from the basolateral membrane and got 10% of the amiloride-blockable current from much less than 10% of the total membrane area. We pulled another patch from the same cell and this patch also had 10%, suggesting that there may be a cytosolic inhibitor (Avenet & Lindemann 1989a). Indeed, there must be some factor present that keeps basolateral Na$^+$ channels turned off, because if they were functioning, the cell would be depolarized. Consequently, it is interesting that you have shown that amiloride-blockable channels are present in the basolateral membranes of taste receptor cells.

Kinnamon: It was Patrick Avenet's idea that some factor in the intact cell kept these basolateral channels from opening, so that the probability of them opening was very low. When patches of this membrane were pulled, he thought that this factor was somehow lost and the probability of opening was greatly increased. I think that this low probability of opening would allow the cell to have basolateral channels and might also explain the amiloride-insensitive component of the afferent nerve responses to NaCl.

DeSimone: Another valid interpretation is that the basolateral membrane channels are not exactly the same as the apical ones, but they are in the process of becoming the same channels, which they do when they get to the apical side. The antibody might not recognize that subtle difference.

It's still not clear how stimulation of taste cells below tight junctions occurs. It may be that some cells already have a differentiated apical-like region—a kind of 'pro-apical membrane'. It seems that Cl$^-$ allows Na$^+$ to diffuse into the intracellular spaces below the tight junctions, but precisely which cells and which membranes are then stimulated is as yet unknown. As you point out, if the amiloride-blockable channels on the basolateral side were all in a functional state, the cells would be very depolarized. This is clearly not the case, so the antibody may be binding to functional and non-functional channels alike. Using the antibody to determine which are in a conducting state may then not be possible.

Lancet: What do you know about this antibody?

DeSimone: It is a polyclonal antibody raised against the purified epithelial Na^+ channel from bovine renal papilla. The channel itself was obtained by Dale Benos and his colleagues employing standard biochemical methods including enrichment by amiloride affinity chromatography (Benos et al 1986). The polyclonal antibodies against the purified channel bind two of the Na^+ channel subunits on Western blots (the 300 kDa and 110 kDa polypeptides) and cross react with the channel protein isolated from the amphibian A6 renal epithelial cell line. They can also be used to immunoaffinity purify the channel (Sorscher et al 1988). Immunochemical localization studies show that in the cells of both the intact bovine collecting tubule and the A6 confluent monolayers, only the apical membranes stain (Tousson et al 1989). Taste cells clearly show antibody binding but not localization of this type.

Lancet: There is a lot of controversy about the molecular nature of the amiloride channel. There is a claim that it has been cloned, but at present this is not clear. Does this antibody cross react with the putative cloned channel? The antibody is still not very well defined in terms of its molecular target, so we have to be rather careful about inference with respect to taste cells.

Hwang: The antibody recognizes five to six different protein bands on Western analysis. When Dale Benos and his co-workers did immunocytochemistry, they demonstrated apical membrane localization in the kidney. It is strange that the whole taste receptor cell should light up without any subcellular compartmentalization.

Lancet: Or the antibody may be recognizing five to six different proteins that happen to be apical because that's the way the antibody was generated. In the taste cells, the cross reactivities that are generated by these five or six different specificities in the polyclonal antibody could recognize almost anything on the membrane, including basolateral components.

DeSimone: Yes; perhaps even Na^+ pumps. The amiloride-sensitive Na^+ channel and the α subunit of the Na^+/K^+ ATPase appear to have a common antigenic site. Antibody raised against the α subunit also binds the channel (Smith & Benos 1991). It may be the case that in taste cells, the anti-Na^+ channel antibody also binds to pump sites, which are localized on the basolateral membranes.

Margolis: It's also important to bear in mind that when you do immunocytochemistry, the proteins have been treated in a very different way from how they are handled for Western blots. The cells have been aldehyde-fixed for immunocytochemistry and on Western blots they have probably been treated with SDS. You may have modified the nature of the epitopes that are being presented for detection by the two different techniques.

DeSimone: Yes; all I'm saying is that a major component of the labelling seems to be throughout the taste cells.

Lancet: Avenet & Lindemann (1989b) have shown very nicely that it is likely that the amiloride-sensitive channel that mediates salt taste is not the same molecular species as the kidney one.

DeSimone: I agree that there could be some important differences, but the structures may be sufficiently homologous so that there is some cross reaction—obviously there is.

Getchell: So that the model doesn't get ahead of the experimental results, I think it is necessary to distinguish carefully between the voltage-sensitive and voltage-insensitive amiloride channels, because there is always a residual component of Na^+ channels that are not amiloride sensitive. Also, in terms of interpreting the immunocytochemical results, a number of controls are necessary to obtain better resolution of the localization of the immunoreactivity along the membranes or within the cells. I suggest that you label the antibody with the fluorescent probe and then do confocal laser scanning microscopy of sections of the taste bud. With this technique I think you will observe better resolution at the membrane level.

DeSimone: We may also get more selective antibodies, because one of the subunits of the amiloride-sensitive Na^+ channel has been recently cloned, sequenced and expressed (Canessa et al 1993). But even if this proves impossible, I want to emphasize once more that, from the electrophysiology that I presented, the density of *functional* apical membrane Na^+ channels can be deduced. Our results show that the Na^+-transducing channels are for the most part in the apical membrane. The absence of voltage sensitivity in a response means that the channel is not located in the apical membrane. If the channel exists at all, it must be located on basolateral membranes.

Hwang: Are there any electrophysiological data concerning these Na^+ channels in the nerve fibres innervating the taste buds?

DeSimone: Yes, in some preparations you can see staining that appears to be following tracts in the taste bud. It has occurred to us that this fibre-like pattern of staining in the fungiform papillae could be nerve fibres entering the taste bud. This is certainly one interpretation and if it is true, it would be exciting. The other possibility is that the pattern represents the branching microcirculation of the papillae.

Reed: I have always thought that the electrical potential is all generated out in the cilia. Is it possible that there are additional currents that are generated in the soma of the receptor cells?

DeSimone: Like most epithelial cells, the taste cell potential depends upon ion transport accross the the apical and basolateral cell membranes. Paracellular shunt electrodiffusion also modulates the cell potential. Na^+ pumps and K^+ channels are primarily concentrated on the basolateral membrane. Simon et al (1991) have used immunohistochemical methods to demonstrate the localization of Na^+/K^+ ATPase on the basolateral membranes of taste bud cells. There is also evidence for a Ca^{2+}-activated Cl^- channel in taste cells (Taylor & Roper 1992), but we think this may have more to do with adaptation than with transduction.

Firestein: Why?

DeSimone: First, the anion effect in Na^+ salt taste is not a dichotomous one, with NaCl contributing one class of response and all other Na^+ salts another class. If that were the case, one would expect a Cl^- channel, dedicated to transduction, to react to NaCl, but not to all other Na^+ salt stimuli. We observe, however, a graded effect. Among the anions Cl^-, acetate and gluconate, the neural responses rank: $Cl^- >$ acetate $>$ gluconate. Second, if an apical membrane Cl^- channel were involved, it would in most cases be conducting a hyperpolarizing current, so NaCl would be expected to give the smallest neural responses among Na^+ salts. In fact, it gives the largest. However, if the initial phases of transduction result in elevated Ca^{2+} levels that subsequently activate Cl^- channels, the latter may be dominant in directing the time-course of taste cell adaptation.

Getchell: One of the points that has come from the results that you have reported is that one needs to consider more than simply the concentration of the stimulus (particularly for salty and sour stimuli) as the driving factor for the current, which presumably is related to the magnitude of the response. One needs to consider those factors that regulate the transmucosal potential as an integral component of the driving force on the ions.

DeSimone: Yes; transmucosal potentials can modulate transcellular potentials, because, as I showed in my paper, changes in the transmucosal potential are divided between the taste cell apical membrane and the basolateral membrane. This influence derives, in part, from the topological relations existing between the transcellular and the paracellular ion pathways. The transmucosal potential might be subject to hormonal control. One way this could occur is through the actions of aldosterone and vasopressin on Na^+ transport (Palmer 1992). We can see that even if the affected cells were non-gustatory, they could still influence the taste cells by their influence on the transmucosal potential.

Siddiqi: I would like to comment on the difference between the insect and vertebrate taste chemoreceptors that supports your interpretation of the anion effect. In *Drosophila* chemoreceptors, the threshold for salt sensitivity is between 10 and 100 μM—rather low. I don't think that at this concentration there will be much inflow of Na^+. In the fly, there is no amiloride effect on the salt cell L1 and there is no strong anion effect either. Therefore, it may be that you have to have these basolateral channels to produce both effects.

Firestein: In both olfactory channels and taste-transducing channels, these epithelia, with their tight junctions, have the ability to utilize chloride in an important way: they can actually maintain two reversal potentials for Cl^-. The intracellular Cl^- concentrations are the same, say 50 μM. In the extracellular solution, the Cl^- concentration may be 160 mM, but in the mucus it could be considerably lower, say 25 mM. So, if you have Cl^- channels in different places, they will do different things even though it's the same cell. This is because the Cl^- reversal potential will be different in the two compartments. On one side the Cl^- channels may be depolarizing, on another they may be hyperpolarizing.

Kinnamon: You assume that the intracellular Cl^- concentration is 50 mM, whereas it is probably closer to 12–14 mM. The only way the cell would depolarize would be if there was a high internal Cl^- concentration and I think that most cells regulate their Cl^- at fairly low concentrations.

References

Avenet P, Lindemann B 1989a Chemoreception of salt taste. The blockage of stationary sodium currents by amiloride in isolated receptor cells and excised membrane patches. In: Brand JG, Teeter JH, Cagan RH, Kare MR (eds) Chemical senses, vol. 1: Receptor events and transduction in taste and olfaction. Marcel Dekker, NY, p 171–182

Avenet P, Lindemann B 1989b Perspectives of taste reception. J Membr Biol 112:1–8

Benos DJ, Saccomani G, Brenner BM, Sariban-Sohraby S 1986 Purification and characterization of amiloride-sensitive sodium channel from A6 cultured cells and bovine renal papilla. Proc Natl Acad Sci USA 83:8525–8529

Canessa CM, Horisberger J-D, Rossier BC 1993 Epithelial sodium channel related to proteins involved in neurodegeneration. Nature 361:467–470

Palmer LC 1992 Epithelial Na^+ channels: function and diversity. Annu Rev Physiol 54:51–56

Simon SA, Holland VF, Zampighi GA 1991 Localization of Na^+,K^+-ATPase in lingual epithelia. Chem Senses 16:283–293

Simon SA, Holland VF, Benos DJ, Zampighi GA 1993 Transcellular and paracellular pathways in lingual epithelia and their influence in taste transduction. J Electron Microsc Tech 25, in press

Smith PR, Benos DJ 1991 Epithelial Na^+ channels. Annu Rev Physiol 53:509–530

Sorscher EJ, Accavitti MA, Keeton D, Steadman E, Frizzell RA, Benos DJ 1988 Antibodies against purified epithelial sodium channel protein from bovine renal papilla. Am J Physiol 255:835–843

Stewart RE, Hill DL 1993 The developing gustatory system: functional, morphological, and behavioural perspectives. In: Simon S, Roper S (eds) Mechanisms of taste transduction. CRC Press, Boca Raton, FL, p 127–158

Taylor RS, Roper SD 1992 A calcium-dependent anion conductance in *Necturus* taste receptor cells. Soc Neurosci Abstr 354:8

Tousson A, Alley C, Sorscher EJ, Brinkley BR, Benos DJ 1989 Immunochemical localization of amiloride-sensitive sodium channels in sodium-transporting epithelia. J Cell Sci 93:349–362

The cellular and genetic basis of olfactory responses in *Caenorhabditis elegans*

Piali Sengupta, Heather A. Colbert, Bruce E. Kimmel, Noelle Dwyer and Cornelia I. Bargmann

Programs in Developmental Biology, Neuroscience and Genetics, The University of California, San Francisco, CA 94143-0452, USA

Abstract. The small soil nematode *Caenorhabditis elegans* has only 302 neurons in its entire nervous system, so it is possible to analyse the functions of individual neurons in the animal's behaviour. We are using behavioural, cellular and genetic analyses of chemotactic responses to find out how olfactory behaviour patterns are generated and regulated. Single chemosensory neurons in *C. elegans* can recognize several different attractive odorants that are distinguished by the animal. Distinct sets of chemosensory neurons detect high and low concentrations of a single odorant. Odorant responses adapt after prolonged exposure to an odorant; this adaptation is odorant specific and reversible. Mutants with defects in odorant responses have been identified. Some genes appear to be necessary for the development or function of particular kinds of sensory neurons. Other genes have effects that suggest that they participate in odorant reception or signal transduction.

1993 The molecular basis of smell and taste transduction. Wiley, Chichester (Ciba Foundation Symposium 179) p 235–250

To understand the genes that affect a behaviour, it is necessary to understand the neurons that generate the behaviour. The relationship between individual cells and behaviour is understood in only a few cases (e.g. Selverston & Miller 1980, Chiel et al 1988). In vertebrates, the activities of individual neurons can be correlated with perception of particular stimuli and in some cases it can be shown that direct stimulation of a group of neurons can influence the behaviour of an animal (Newsome et al 1990). However, the enormous complexity of the vertebrate brain limits the situations in which such experiments can be conducted.

We are using genetic, cellular and behavioural assays to study the chemotactic behaviour of the nematode *Caenorhabditis elegans*, which has a small and well-described nervous system. *C. elegans* detects chemical and mechanical stimuli, temperature and electrical fields. Every aspect of the animal's

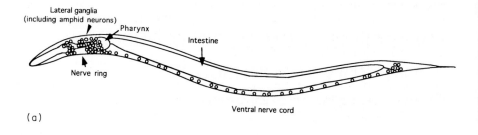

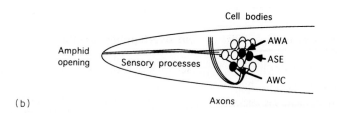

FIG. 1. Neuroanatomy of *C. elegans*. (a) Schematic anatomy of *C. elegans*. Cell bodies of some of the 302 neurons in the adult hermaphrodite are shown (small circles). The lateral ganglia of the head contain the cell bodies of all chemosensory neurons of the amphid discussed here. Their axons run into a neuropil called the nerve ring. Additional neurons are found in the body and tail. Anterior is at left, dorsal at top. For further detail, see White et al (1986), Ward et al (1975), Ware et al (1975). (b) Anatomy of chemosensory neurons in the head. The cell bodies of the twelve amphid neurons on the left side and a representative group of their axons and dendrites are shown. The chemosensory neurons are bipolar; each neuron has one sensory process and one axon. An opening at the anterior tip of the nose allows the ciliated endings of the sensory processes (dendrites) access to the environment. The axons extend into the neuropil of the nerve ring, where they form synapses with other neurons. The AWA, AWC and ASE neurons sense chemoattractants. Anterior is at left, dorsal at top.

biology is influenced by chemical stimuli: its development, movement, feeding and egg-laying are all regulated by chemical signals from bacteria, its natural food source (Wood 1988). In addition, *C. elegans* will track towards some substances and avoid others.

Chemotaxis and chemical avoidance behaviour can be elicited by single, defined chemicals as well as by complex food sources (Ward 1973, Dusenbery 1974, Culotti & Russell 1978). The chemicals to which *C. elegans* is attracted include molecules that are known to be produced by bacteria, such as cAMP, biotin, amino acids, alcohols, ketones and esters (Ward 1973, Dusenbery 1974, Bargmann & Horvitz 1991a, Bargmann et al 1993). Repulsive compounds include acid, any compound at high osmotic strength and a different set of alcohols and ketones (Dusenbery 1974, Culotti & Russell 1978, Bargmann et al 1993). Interestingly, the same compound (for example, benzaldehyde) can be attractive at one concentration and repulsive at another (Bargmann et al 1993).

We have identified neurons required for responses to different chemicals and characterized mutant animals in which those responses are abnormal (Bargmann et al 1990, 1993, Bargmann & Horvitz 1991a,b). The properties of individual neurons and genes are surprisingly complex, but we have been able to understand some of the mechanisms of odorant recognition and discrimination in *C. elegans* by comparing results obtained by cellular and genetic approaches.

Individual sensory neurons have distinct but overlapping functions

Figure 1 illustrates the anatomy of chemosensory neurons in *C. elegans*. An adult hermaphrodite *C. elegans* has 302 neurons in its nervous system, of which about 30 appear to be chemosensory (White et al 1986). All of the chemosensory neurons for which functions have been identified are associated with a sensory structure called the amphid. The ciliated sensory endings of the amphid neurons are presented to the environment through a small opening in the cuticle at the tip of the animal's nose (Ward et al 1975, Ware et al 1975). Because *C. elegans* is transparent, the cell bodies of these neurons can be identified in live animals based on their morphology and position. The functions of the neurons have been examined by using laser microsurgery to kill individual neurons or groups of neurons, and analysing the effects of these lesions on the animal's olfactory behaviour.

Chemical signals can initiate a variety of different responses in *C. elegans*— altered egg-laying, developmental changes, chemotaxis and chemical avoidance. Each of these behaviours also depends on the presence of particular subsets of neurons (Bargmann & Horvitz 1991a,b, Bargmann et al 1993, Thomas & Horvitz 1993, E. Sawin & H. R. Horvitz, personal communication). We infer that these neurons sense environmental conditions to regulate these responses; however, more complex models are also consistent with these data.

Two general observations can be made about the chemosensory neurons. First, each type of neuron seems to have a unique set of functions (defined by laser killing), although these functions may overlap in part with those of other neurons. Second, while a single neuron may participate in several kinds of responses, functional subgroups of neurons can be identified. For example, the neurons that sense attractive compounds are largely distinct from those that sense repellents. The most important neurons for attraction are ASE, AWC and AWA, though at least four other classes of sensory neurons also participate

TABLE 1 Chemosensory neurons required for normal chemotactic responses to volatile or water-soluble attractants

Attractant	Sensory neuron
Benzaldehyde	AWC
2-Butanone	AWC
Isoamyl alcohol	AWC > AWA
Diacetyl (1 nl)	AWA
Pyrazine	AWA
Trimethylthiazole	AWC & AWA
cAMP, biotin, Na^+ and Cl^- ions	ASE > ASG, ASI, ADF
Lysine	ASE, ASG, ASI, ASK

The six volatile attractants listed at the top are all distinguished from one another by *C. elegans* in cross-saturation assays. The volatile attractants are recognized predominantly by two types of chemosensory neurons, the AWA and AWC neurons, but other neurons also contribute weakly to the responses. AWA is most important at sensing one nanolitre of diacetyl, but AWC also appears to sense diacetyl at higher concentrations; other attractants might also be sensed by different neurons at different concentrations. The four water-soluble attractants sensed predominantly by the ASE neurons fall into four different classes in cross-saturation assays. From Bargmann & Horvitz (1991a) and Bargmann et al (1993).

in chemotactic responses (Table 1) (Bargmann & Horvitz 1991a, Bargmann et al 1993). None of these neurons overlap with the neurons that appear to be most important for repulsion, the ASH and ADL neurons (Thomas & Horvitz 1993, B. E. Kimmel & C. I. Bargmann, unpublished results). Interestingly, the ASH neurons sense both chemical repellents and light touches to the nose—an aversive mechanical stimulus (Kaplan & Horvitz 1993).

Since the anatomy of the nervous system is well described, potential synaptic partners of the sensory neurons can be identified (White et al 1986). ASH and ADL, which mediate avoidance, synapse onto neurons that are thought to control whether the animal moves forwards or backwards. By contrast, the neurons involved in chemotaxis synapse onto neurons that might initiate the turning behaviour needed to track to the peak of a gradient. The functions of these potential synaptic partners in chemotaxis can also be tested by laser killing.

Some of the neurons that mediate chemotaxis also regulate other sensory functions. One group of neurons regulates both chemotaxis and a developmental decision called dauer larva formation that is under chemical control (Bargmann & Horvitz 1991b). Several other neurons involved in chemotaxis also regulate egg-laying (E. Sawin & H. R. Horvitz, personal communication). It is unlikely that all of the sensory responses of *C. elegans* are known at this time; the properties of single chemosensory neurons will probably turn out to be more complex as further information is gathered.

Single neurons sense multiple classes of distinguishable odorants

Classes of odorants distinguished by *C. elegans* are defined by two different sorts of experiments (Bargmann et al 1993). In one, *C. elegans* attempts to find the point source of an attractant in the presence of high uniform levels of a second attractant (saturation assay); in the second, *C. elegans* is exposed to one odorant for long periods of time and then challenged with a second odorant in a chemotaxis assay (adaptation assay; see below for more details). Cross-saturation and cross-adaptation are the criteria classically used to divide odorants into classes in human psychophysics as well (Finger & Silver 1987). In *C. elegans*, both kinds of experiment give similar results, defining at least seven distinct classes of attractive odorants. One class includes isoamyl alcohol and other small alcohols; a second class, 2-butanone; a third class, benzaldehyde and some other aromatic compounds, and so forth.

In principle, the distinctions among odorant classes could be made by sorting information to different sensory neurons, just as repulsive and attractive information appear to be sorted at the level of the sensory neurons. However, this does not appear to be the case. An animal can distinguish between two odorants that appear to be sensed by one neuron (defined by laser killing experiments) (Table 1). For example, both benzaldehyde and 2-butanone are sensed almost entirely by the AWC neurons, yet the two attractants are clearly distinguished by *C. elegans*. Similarly, diacetyl and pyrazine are two different odorants that are both sensed mainly by the AWA neurons.

The disparity between the cellular and behavioural classes of odorants in *C. elegans* is strikingly similar to that observed in recordings obtained from single olfactory units in the vertebrate neuroepithelium. In vertebrates, single olfactory neurons usually respond to a broad spectrum of odorants that are easily distinguished by the animal (Sicard & Holley 1984).

Chemosensory mutants have defects in responses to specific odorants

The complexity of single sensory neurons in *C. elegans* is emphasized further by analysis of mutants with defective odorant responses (anosmic mutants). We have defined mutants based on their defective responses to various volatile odorants, including benzaldehyde and diacetyl. These mutants have complex patterns of deleted responses. Some genes are required for all normal chemotaxis responses, while others are only required for responses to subsets of odorants. Of the more specific genes, two general types can be identified. One type of gene appears to affect many or all of the functions of a single cell type. For example, mutations in *odr-1* or *odr-5* cause the same behavioural effects as are observed when the AWC neurons are killed with a laser, and *osm-9* mutants have the same behavioural defects as animals in which the ASH and AWA neurons have been killed (J. Thomas, personal communication, and P. Sengupta & C. I. Bargmann, unpublished results).

A second type of gene has a more complex set of functions that do not correspond to the functions of a single cell type. Mutations in *odr-2*, *odr-3*, and *odr-4* all eliminate some, but not all, of the functions of the AWC neurons and also eliminate some functions mediated by other neurons. Interestingly, whenever a mutation affects the response to one of a group of odorants defined by the behavioural classes (for example, isoamyl alcohol), it affects the response to all of the odorants in that behavioural class (all attractive alcohols). Thus these mutations respect the behavioural, but not the cellular, borders that define odorant responses.

One interpretation of the behavioural and genetic data is that single sensory neurons integrate information about disparate odorants, just as single bacterial cells can discriminate among multiple chemical attractants (Adler 1975). A second possibility is that the perception of odorant quality is obtained by comparing the activities of broadly tuned neurons, to arrive at a more specific perception of odorant class than can be derived from any single neuron.

A single odorant is detected by distinct sensory cells at high and low concentrations

C. elegans will often respond to an odorant across a 1000-fold concentration range or more. Is all of this response mediated by one system with a highly adjustable sensitivity, or do different systems detect different odorant concentrations? Both cellular and genetic data indicate that the latter explanation is true in *C. elegans*.

Some compounds, including benzaldehyde and 1-hexanol, are attractive to *C. elegans* at low concentrations but repulsive at high concentrations (Bargmann et al 1993). It appears that the attractive and repulsive qualities of each odorant are mediated by distinct sensory neurons. Attractive responses to the volatile odorant benzaldehyde are mediated almost entirely by the AWC neurons, but avoidance of benzaldehyde does not require these neurons. The response to hexanol may be explained by considering the responses of *C. elegans* to other alcohols. Butanol and pentanol are attractive, while heptanol and octanol are repulsive to *C. elegans*. Attractive alcohols are sensed primarily by AWC, and repulsive alcohols are sensed primarily by ADL. It is likely that hexanol, which is intermediate in size between the attractive and repulsive alcohols, is activating

ADL and thus avoidance at high concentrations and AWC and attraction at low concentrations.

Even for an attractant, several neurons may respond to different concentrations of the same compound. A point source of one microlitre of the odorant diacetyl (2,3-butanedione) is attractive to *C. elegans*, but less than one nanolitre of diacetyl is also attractive. Mutations in at least three different genes, *odr-3*, *odr-4* and *odr-7*, cause animals to be unable to respond to one nanolitre of diacetyl (Table 2). However, animals mutant for any one of these genes respond normally to 100 nanolitres of diacetyl. Even when animals are doubly mutant for two of these genes they are still attracted normally to 100 nanolitres of diacetyl.

Animals with mutations in the genes *odr-1* and *odr-2* appear to be normal for their responses to diacetyl at all concentrations (Table 2). Surprisingly, when animals are doubly mutant for one of these two genes and for one of the genes *odr-3*, *odr-4* and *odr-7*, they fail to respond to diacetyl at any concentration. These results indicate that there are two groups of genes that act redundantly to promote diacetyl responses at high concentrations. At least one gene from each category must be inactivated for diacetyl responses to be lost completely.

What are the redundant functions in sensing diacetyl? The simplest model that explains the genetic data is that two kinds of neurons can sense 100 nanolitres of diacetyl, but only one kind senses one nanolitre of diacetyl. Killing the AWA neurons has a similar effect on diacetyl responses as do mutations in *odr-3*, *odr-4* and *odr-7*. The activity of these genes could be required for the AWA neurons to sense one nanolitre of diacetyl normally. The genetic redundancy of the response to 100 nanolitres of diacetyl suggests that additional cells that are affected by *odr-1* and *odr-2* sense diacetyl at high concentrations.

TABLE 2 Genetic redundancy of responses to diacetyl

	Response to diacetyl	
Mutations	*100 nl*	*1 nl*
None	+	+
Class I (*odr-3, odr-4, odr-7*)	+	−
Class II (*odr-1, odr-2*)	+	+
Two Class I mutations	+	−
Two Class II mutations	+	+
One Class I + one Class II	−	−

Other experiments indicate that *odr-1* and *odr-2* affect the functions of the AWC neurons, indicating that the AWC neurons might affect responses to 100 nanolitres of diacetyl. This possibility has been confirmed by laser ablation experiments that indicate that the AWC neurons respond to 100 nanolitres of diacetyl, but are neither necessary nor sufficient for responses to one nanolitre of diacetyl.

Further analysis of mutants has shown that the responses to high and low amounts of diacetyl can be separated completely by genetic manipulation. Some mutants respond only to high concentrations of diacetyl; other mutant combinations respond only to low concentrations. Taken together, our results suggest that different sensory neurons are tuned to different concentrations of the odorant diacetyl.

An interesting and puzzling phenomenon in human olfaction is the observation that chemical structure correlates poorly with perceived odorant quality. One example of this disparity is perception of α-ionone, which smells of cedar at high concentrations but smells of violets at low concentrations. If different sensory neurons are activated by different concentrations of α-ionone, this could result in the observed changes in perceptual quality.

Olfactory responses can be altered by experience in *C. elegans*

Adaptation is a general function of sensory systems. Even unicellular organisms are able to attenuate their responses to a chemical stimulus after prolonged exposure to the stimulus. In fact, adaptation is a necessary element of chemotaxis by bacteria, which move up a gradient of attractant by constantly adapting to the ambient attractant concentrations (Goy et al 1977). Similarly, vertebrate and invertebrate eye photoreceptors adapt to high or low light levels to increase their useful range of activity. These adaptation events occur within seconds of exposure to the sensory stimulus, as does olfactory adaptation described in vertebrates (Finger & Silver 1987). We do not know if such short-term adaptation occurs in *C. elegans*, since the assays we have for olfaction take at least 30 minutes to complete. However, we have observed a longer-term change in olfactory responses after minutes to hours of exposure to odorant. When *C. elegans* is exposed to an odorant like benzaldehyde for two hours, it loses its ability to be attracted to benzaldehyde. Animals adapted to benzaldehyde are still attracted to other odorants. This adaptation is fully reversible and recovers over about three hours.

From laser killing experiments, we know that the AWC sensory neurons are necessary for both benzaldehyde and 2-butanone responses, yet after exposure to benzaldehyde only the benzaldehyde response is eliminated. Thus it appears that only some of the functions of the AWC neurons are affected by benzaldehyde adaptation. One model is that separate receptor molecules on the AWC neuron recognize benzaldehyde and 2-butanone, and that these receptors can be regulated separately so that only one is inactivated following benzaldehyde treatment.

We have identified mutants that fail to adapt normally to benzaldehyde and continue to respond even after prolonged treatment. Interestingly, these mutants do not have detectable defects in their normal chemotactic responses. Therefore, the process of adaptation affected in these mutants is not analogous to the adaptation over the course of seconds that is necessary for bacterial chemotaxis. Rather, the mutations only affect integration of chemical information over long periods of time (minutes to hours).

The simplest explanation for the sensory adaptation that we have observed is that limited changes occur within a neuron to alter its responses to particular odorants. A more extensive form of behavioural plasticity occurs when animals are starved and crowded. Water-soluble chemicals that are strong attractants to naïve animals are ignored by crowded, starved animals, while certain volatile odorants appear to become much more attractive. Another behavioural response, positive thermotaxis (the ability to migrate to a preferred temperature), is also suppressed when *C. elegans* is starved (Hedgecock & Russell 1975). In the case of odorant responses, the changes induced by crowding and starvation persist for hours after the worms are separated and fed. One possibility is that naïve worms tend to respond to sensory stimuli that will identify local sources of food, while crowding and starvation enhance the importance of sensory stimuli that are used to locate distant food supplies.

Conclusion

The advantage to studying olfaction in a simple organism like *C. elegans* is that it is possible to correlate the behaviour of the animal, the functions of single neurons and the properties of individual genes. Cloning the genes that affect odorant responses in *C. elegans* will be a first step in seeing whether the molecular mechanisms of olfaction are similar between nematodes and vertebrates.

References

Adler J 1975 Chemotaxis in bacteria. Annu Rev Biochem 44:341–356

Bargmann CI, Horvitz HR 1991a Chemosensory neurons with overlapping functions direct chemotaxis to multiple chemicals in *C. elegans*. Neuron 7:729–742

Bargmann CI, Horvitz HR 1991b Control of larval development by chemosensory neurons in *Caenorhabditis elegans*. Science 251:1243–1246

Bargmann CI, Thomas JH, Horvitz HR 1990 Chemosensory cell function in the behavior and development of *Caenorhabditis elegans*. Cold Spring Harbor Symp Quant Biol 55:529–538

Bargmann CI, Hartwieg E, Horvitz HR 1993 Odorant-selective genes and neurons mediate olfaction in *C. elegans*. Cell 74:515–528

Chiel HJ, Weiss KR, Kupfermann I 1988 An identified histaminergic neuron modulates feeding motor circuitry in *Aplysia*. J Neurosci 6:2427–2450

Culotti JG, Russell RL 1978 Osmotic avoidance defective mutants of the nematode Caenorhabditis elegans. Genetics 90:243–256

Dusenbery DB 1974 Analysis of chemotaxis in the nematode *Caenorhabditis elegans* by countercurrent separation. J Exp Zool 188:41–47

Finger TE, Silver WL (eds) 1987 Neurobiology of taste and smell. Wiley, New York

Goy MF, Springer MS, Adler J 1977 Sensory transduction in *Escherichia coli*: role of a protein methylation reaction in sensory adaptation. Proc Natl Acad Sci USA 74:4964–4968

Hedgecock EM, Russell RL 1975 Normal and mutant thermotaxis in the nematode *Caenorhabditis elegans*. Proc Natl Acad Sci USA 72:4061–4065

Kaplan JM, Horvitz HR 1993 A dual mechanosensory and chemosensory neuron in *C. elegans*. Proc Natl Acad Sci USA 90:2227–2231

Newsome WT, Britten KH, Salzman CD, Movshon JA 1990 Neuronal mechanisms of motion perception. Cold Spring Harbor Symp Quant Biol 55:697–705

Selverston AI, Miller JP 1980 Mechanisms underlying pattern generation in lobster stomatogastric ganglion as determined by selective inactivation of identified neurons. I. Pyloric system. J Neurophysiol 44:1102–1121

Sicard G, Holley A 1984 Receptor cell responses to odorants: similarities and differences among odorants. Brain Res 292:283–296

Thomas JH, Horvitz HR 1993 Nociceptive neurons of *C. elegans* with multiple reactivities. In preparation

Ward S 1973 Chemotaxis by the nematode *Caenorhabditis elegans*: identification of attractants and analysis of the response by use of mutants. Proc Natl Acad Sci USA 70:817–821

Ward S, Thomson N, White JG, Brenner S 1975 Electron microscopical reconstruction of the anterior sensory anatomy of the nematode *Caenorhabditis elegans*. J Comp Neurol 160:313–337

Ware RW, Clark D, Crossland K, Russell RL 1975 The nerve ring of the nematode *Caenorhabditis elegans*: sensory input and motor output. J Comp Neurol 162:71–110

White JG, Southgate E, Thomson JN, Brenner S 1986 The structure of the nervous system of the nematode *Caenorhabditis elegans*. Philos Trans R Soc Lond B Biol Sci 314:1–340

Wood WB 1988 The nematode *Caenorhabditis elegans*. Cold Spring Harbor Laboratory Press, Cold Spring Harbor, NY

DISCUSSION

Reed: In these relatively simple responses, do you think it might be possible to blind the worms by using membrane-permeable second messengers? Could you look at the effect of Li^+ which would at least disrupt their inositol trisphosphate metabolism?

Bargmann: There are a lot more experiments with extracellular pharmacological agents that we could do. We have tried some blocking experiments but

I don't feel that the results are significant. The problem is that if you don't see any effect, you don't know whether this might not just be because the drug didn't get into the worm.

Getchell: The natural food of *C. elegans* is bacteria and we know that many bacteria produce endotoxins which presumably cause a repellent response by the worm. Because the molecular pharmacology of many of these endotoxins is well known, might this be another approach that you could use to gain some insight into the molecular basis of these chemoreceptors?

Bargmann: We haven't looked at any compounds like that yet.

Ache: Are the repellent and attractant systems basically parallel detection systems addressing different motors?

Bargmann: Yes. We have identified some neurons involved in repulsion. Some large alcohols, such as 1-octanol, are repulsive rather than attractive. Octanol avoidance employs different neurons from those required for the attractive response that normally occurs for other alcohols. Jim Thomas (personal communication) has identified neurons involved in other avoidance responses, such as the avoidance of high osmotic strength. These are different from the ones used in avoidance or attraction of volatile odorants. Repellent responses are different from attraction in a number of ways: for instance, they don't adapt after any length of time.

Caprio: In the channel catfish, it appears that the experimenter can change whether a compound can act as an attractant or a repellent. Tina Valentincic, in my laboratory (unpublished results), showed that the same amino acid that previously released feeding activity could release alarm behaviour if the catfish was frightened by something as simple as waving one's hand over the top of the aquarium prior to administering the amino acid stimulus. In nature, it would be ill-advised for a fish to detect an amino acid and come swimming out, because the origin of the stimulus might not be a prey, but a predator. If you alarm a worm and then present a previously attractive chemical, will the worm then go in the opposite direction?

Bargmann: Worms are much simpler organisms than fish, but they do show some forms of behavioural plasticity—habituation and sensitization of responses, for example. Also, under starved conditions their pattern of what is attractive and what is repellent is changed completely. Some things that were previously attractive are shut off; other things that were repellent become attractive. But that seems to be a general change, not based on recognizing one compound and a change of neural response to it.

Lancet: Are there any classes of chemicals that the worm definitely does not react to?

Bargmann: I tested 120 volatile chemicals from Sigma: the worms were attracted to 50, they avoided 10 and 60 seemed to produce no response, but I didn't test every concentration of each odorant. As a general rule of thumb, the worm fails to react to about 50% of chemicals.

Lancet: If you apply our model to the worm, that it reacts to 50% of them is exceptionally good for the small number of receptors that it has, with their apparent affinities. So there is a puzzle here that we need to address.

You described the mutants that cannot respond to a given chemical at low concentrations but can at high concentrations; the same is true for almost all cases of specific anosmia. Amoore showed that for specific anosmics, it is not that the threshold is absent, but that the threshold is always higher than normal. This may mean that if the high affinity gene is missing, another can replace it, and if this one is missing, yet another takes over. This theoretical framework would be useful for the analysis of your mutants, too.

Margolis: Doron, can we follow up on your comment about the fact that the apparent affinities in this animal don't seem to match the model that you created? It would also be useful to know where those numbers came from and how we can analyse the animals response to different ligand concentrations in this situation.

Bargmann: Some of these molecules are much more attractive than others. These others might simply not be being recognized as well. Since we look for the response of an animal on a gradient, we don't know what concentration the animal is facing when it starts changing its behaviour. We need better ways of looking at behaviour, such as video tracking.

Lancet: But the concentrations cannot be higher than the source.

Bargmann: That's right. We know, for instance, when we are seeing a saturation or adaptation, that this defines at least the point at which that odorant must be detected. Odorant responses saturate at around 10^{-5} M odorant.

Lancet: What is the concentration of the source?

Bargmann: Typically, we use 1 μl of a 1:1000 dilution of an odorant in ethanol as the source.

Lancet: That's about 10^{-2} M.

Bargmann: Yes; but we use just 1 μl in a 100 ml plate; the detected concentrations must be much lower than that—probably in the nanomolar to micromolar range.

Lancet: It's worth remembering that we see high apparent affinity in many olfactory systems. Cells often respond with a much higher apparent affinity because of amplification. This should show the way to looking for amplification mechanisms.

Bargmann: Of course, the concentration of the chemical that the worm must detect to cause movement may not represent half-saturation of the receptors. That is why the saturation numbers are very solid, because we are putting the worm in a uniform field of odorant. We know what the concentration is, so we know that the worm's olfactory system is saturated at a particular concentration (about 10^{-5} M).

Buck: Do you know anything about the connections that these sensory neurons make with other neurons that might suggest how the information might be processed further?

Bargmann: Each neuron has a characteristic set of targets. The neurons that are particularly important in chemotaxis have a common set of targets. We know this from anatomy, so we don't know if those connections are excitatory, inhibitory, modulatory or silent. But, for some of the interneurons, the same sorts of laser killing experiments have been done as were done with sensory neurons. A subset of interneurons is involved in chemotaxis; those are the ones that are anatomically connected with the chemotaxis sensory neurons. The problem with these experiments is that the specificity that you see at the sensory level (which is really what tells you that you are looking at a sensory defect and not at something that's making the animal confused, sick or uncoordinated) disappears when you start hitting those second-order cells. So, at that point, the laser-operated animals start responding badly to everything. On the one hand we can say these neurons are required for chemotaxis, on the other hand so are the muscles—there could be a trivial explanation for the effect.

Siddiqi: What do you know about the orientation mechanism of the worms in these tests?

Bargmann: The worms reach an attractant by sensing with their heads and orienting in a straight line along a gradient. The best experiment to show this is an old experiment that Sam Ward did using an animal with a bent head (Ward 1973). If you place a normal worm in a gradient of an attractant, it will move to the attractant source in a fairly straight line. But if you take an animal that has a kink in its head, it will spiral in to the attractant.

Lindemann: I would like to ask about the *deg-1* gene. Canessa et al (1993) cloned the α subunit of the epithelial Na$^+$ channel from rat colon. This subunit is already a functional channel. Surprisingly, it has homology to the *deg-1* gene in *C. elegans*. Knowing that a protein similar to the epithelial Na$^+$ channel must be present in this worm, could this protein be involved in salt detection? You mentioned the NaCl attractive response: is it amiloride-blockable?

Bargmann: The amiloride experiment has not been done, but the NaCl response is still present, at least in part, in *deg-1* mutants. However, we have seen defects in chemotaxis to lysine in *deg-1* mutants. Despite its name, it's not correct to think of *deg-1* as involved in degeneration, because the degenerative mutants are rare gain-of-function mutants near the transmembrane domain of these channel-like proteins. Therefore, over-function or mis-function of the genes causes degeneration. Loss-of-function of the gene is what would define its normal effect.

DeSimone: You mentioned that cAMP itself is a chemoattractant: is that like the chemoattraction that is associated with the aggregation of slime moulds (Robertson & Cohen 1972)?

Bargmann: cAMP is not known to be produced and secreted by worms: it is more likely to be used by *C. elegans* to find a slime mould or bacterium that would be its food. *E. coli* is rather sloppy about producing cAMP, so cAMP could be a reasonable predictor of food sources.

DeSimone: Do the worms have an extracellular receptor for cAMP, something like that of slime moulds?

Bargmann: It's a behavioural response to cAMP and no one has looked for an equivalent of the *Dictyostelium* cAMP receptor in *C. elegans*.

Van Houten: To follow up on John's question, if worms are homing in on cAMP to find *Dictyostelium*, they would be better off homing in on folic acid, because that is the cue followed by *Dictyostelium* when grazing on bacteria (Newell et al 1990, Van Haastart 1991). Do worms also respond to folic acid?

Bargmann: Biotin was the only amino acid that was attractive to worms.

Breer: You showed that three genes affect the worm's perception of benzaldehyde, whereas most other odorants have one receptor. Is benzaldehyde of particular relevance to *C. elegans*?

Bargmann: The mutants described were originally identified in screens for animals that failed to detect benzaldehyde. Subsequent screens with other odorants have identified additional genes required for other responses. Benzaldehyde is, if anything, the one attractant for which I can think of no logical justification. Many of the odours *C. elegans* is attracted to are produced by bacteria, or fermenting fruit and vegetable material, which are things that worms naturally eat: why they should like almonds is beyond me. We've done screens with a number of different odorants: benzaldehyde, diacetyl, pyrazine and some water-soluble attractants; in each case we are getting a comparable number of genes out.

Reed: Do they also show chemoattraction to sugars?

Bargmann: No; the only response to sugars that is known is avoidance of high osmotic strength.

Hatt: Do you think that your adaptation time constants are mainly based on peripheral processes and not so much on central adaptation?

Bargmann: The reason we think that adaptation is peripheral is that after benzaldehyde adaptation, somehow you have left the animal able to respond to 2-butanone but not benzaldehyde, both of which are detected by the sensory neuron. However, there are ways in which central processing could lead to specific peripheral effects, just as Dr Siddiqi mentioned with *gust E*, where central cells can feed back onto peripheral mechanisms. From the genetics we've done, I don't think we can exclude that.

Hatt: Are the central cells coupled electrically, as has been shown for many other systems?

Bargmann: You can see what appear to be both gap junctions and chemical synapses in the morphology of the different neurons: these synapses are all defined anatomically, rather than functionally. They are very reproducible between different animals.

Reed: Could you comment on the dismal state of worm electrophysiology?

Bargmann: C. elegans electrophysiology has suffered from the fact that the neurons are only 2–3 μm in diameter and they are packed together under the cuticle. Two groups have started doing worm electrophysiology in the past year or so; Leon Avery at Dallas in collaboration with Jim Hudspeth, and Shawn Lockery in Terry Sejnowski's lab at the Salk Institute. Both of them have managed to get patches from worm neurons and show that they contain channels. Neither of the groups has got intracellular recording to work. For the kinds of wiring things that we would really like to look at, that's going to be a must.

Reed: At much cruder levels, has anybody actually measured the equivalent of an electroolfactogram?

Bargmann: Leon Avery has been working out that technology. He records from the entire head end because he's interested in the pharynx; he looks at the rather large changes that happen when the pharyngeal muscles contract. It's possible that by modifying this technology you might be able to look at the sensilla at the tip of the nose.

Lancet: You said that the worm can actually sense gradients, but it cannot compare concentrations across its body. The only way to explain this is by means of time-dependent comparisons of concentrations. Then you fall back to the bacterial-type mechanism in which you are asking for adaptation over very short time periods. Is anything known about that?

Bargmann: Worms do compare concentrations of attractants over time— this is known from some earlier work. But the simplest version of the bacterial model is not the only mechanism for *C. elegans* chemotaxis. Bacteria use only temporal mechanisms for moving along a chemical gradient. The best movement pattern they can generate is a biased random path towards the attractant. *C. elegans* can clearly orient itself straight up the gradient, which demands more than just temporal information. The worms may be integrating information about their head position with temporal information about attractant concentration, for example. More experiments are needed to look at that directly.

References

Canessa CM, Horisberger J-D, Rossier BC 1993 Epithelial sodium channel related to proteins involved in neurodegeneration. Nature 361:467–470

Newell PC, Europe-Finner GN, Liu G, Gammon B, Wood CA 1990 Signal transduction for chemotaxis in *Dictyostelium* amoebae. Semin Cell Biol 1:105–113

Robertson A, Cohen MH 1972 Control of developing fields. Annu Rev Biophys Bioeng 1:409–464

Van Haastert PJM 1991 Transmembrane signal transduction pathways in *Dictyostelium*. Adv Second Messenger Phosphoprotein Res 23:185–226

Ward S 1973 Chemotaxis by the nematode *Caenorhabditis elegans*: identification of atrractants and analysis of the response by the use of mutants. Proc Natl Acad Sci USA 70:817–821

Genetic and pathological taste variation: what can we learn from animal models and human disease?

Linda M. Bartoshuk

Yale University School of Medicine, Department of Surgery (Otolaryngology), 333 Cedar St, PO Box 208041, New Haven, CT 06520-8041, USA

Abstract. The study of patients with taste disorders (i.e. 'experiments of nature') suggests that the old tongue maps (e.g. sweet on the tip, bitter on the back) that often appear in textbooks are wrong. If they were correct, severing the taste nerves that innervate the front of the tongue would result in a loss of the ability to taste sweet, etc. This does not occur. Severing these nerves has little effect on everyday taste experience because taste nerves inhibit one another. Damaging one nerve abolishes its ability to inhibit others and the release-of-inhibition compensates for the damage. There is sometimes a clinical cost for this redundancy; release-of-inhibition can produce taste phantoms. Genetic variation in taste ability occurs across and within species. For example, about 25% of humans are relatively unresponsive to a variety of sweet and bitter compounds (non-tasters) while another 25% are unusually responsive (supertasters). Supertasters have about four times as many taste buds as non-tasters and have smaller and more densely packed fungiform papillae. Since there are pain fibres associated with taste buds, supertasters are unusually responsive to the oral burn of spices.

1993 The molecular basis of smell and taste transduction. Wiley, Chichester (Ciba Foundation Symposium 179) p 251–267

Taste pathology

The myth of the tongue map

Many textbooks that cover taste include a sketch of the tongue showing the distribution of sensitivity to the four basic tastes. Sweet is always placed on the tip of the tongue and bitter on the base, although the placements of salty and sour vary. These maps are wrong.

Collings (1974) discovered the error when she attempted to replicate the work of Hänig (1901). At the turn of the century, Hänig measured taste thresholds around the perimeter of the tongue and found small but reliable differences in the thresholds for his four stimuli (sucrose, NaCl, HCl and quinine sulphate). He argued that the variation in thresholds supported the existence of four discrete receptor mechanisms for the four basic tastes. Hänig summarized his study with an idealized sketch showing how sensitivity (the reciprocal of threshold) changed around the tongue. For example, the sensitivity for sweet was at a maximum on the tongue tip and at a minimum on the tongue base.

Edwin Boring, the great historian of psychology at Harvard, may have intended to clarify Hänig's presentation in his classic text (Boring 1942) when he went back to the data in Hänig's tables and actually calculated the reciprocals of the average thresholds to get sensitivity scores. For each stimulus, Boring divided the sensitivity scores by the maximum sensitivity achieved for that stimulus. This gave him functions that resembled Hänig's sketch. Unfortunately, Boring did not label the ordinate of his graph. Readers could assume that where sensitivity was at a minimum it was absent altogether—and so the tongue map was born.

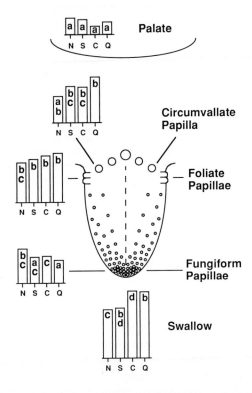

FIG. 1. Relative magnitude estimates of 0.32 M NaCl (N), 0.32 M sucrose (S), 10^{-2} M citric acid (C) and 1.8×10^{-4} M quinine hydrochloride (Q) applied to the indicated loci in the mouth ($n = 17$). Results of ANOVA and follow-up Newman-Keuls tests are indicated by letters on the bar graphs. For each stimulus, bars that do not share a letter are statistically significantly different from one another ($P < 0.05$).

When Collings attempted to replicate Hänig's experiment, she found some errors. For example, the threshold for bitter is actually lower on the front of the tongue than on the back. However, she supported Hänig's overall conclusions: there are variations in the thresholds across tongue loci, but they are actually quite small. All four basic tastes can be perceived on all loci where there are taste receptors. It is interesting to note that in many of the tongue maps that appear in modern texts, the authors rarely cite a source for the map. The tongue map has become an enduring scientific myth.

We routinely test various tongue loci in our studies on the spatial characteristics of taste. Figure 1 shows the results from one of these studies (Yanagisawa et al 1992). Note that perceived intensity varies as stimuli are moved across the tongue, but this in no way implies that taste qualities are localized on the tongue as suggested by the maps.

These tongue maps do not apply to other species, either, although they apply better to some than to humans. For example, the rat responds well to quinine on the back of the tongue but not on the front (Frank 1975).

Innervation of the anterior and posterior tongue

Although the exact innervation of the tongue is still not known (Tomita et al 1986, Catalanotto et al 1991), the tongue is divided into two portions according to the classic picture of its innervation. On the anterior portion of the tongue, the chorda tympani nerve (a branch of the facial (VII) nerve) innervates taste buds in fungiform papillae. These papillae resemble button mushrooms (hence their names) and are distributed most densely at the tip of the tongue, but can be found along the entire edge of the tongue all the way back to the foliate papillae. There are relatively few fungiform papillae in the centre of the tongue. Temperature, touch and pain are mediated by the trigeminal nerve (V) on the anterior tongue.

On the posterior portion of the tongue, taste, temperature, touch and pain are all mediated by the glossopharyngeal nerve (IX). Taste buds are found in the foliate and circumvallate papillae on the posterior tongue. The foliate papillae are on the edges at the base of the tongue. The skin is thinner there, so they appear redder than the surrounding tissue. The circumvallate papillae are relatively large circular structures on the back of the tongue. There is usually one larger papilla on the midline with 3–4 smaller papillae on either side arranged in an inverted V.

Anaesthesia of the taste system

If taste qualities were arranged chemotopically on the tongue such that sweet receptors were found on the front, then damage to the chorda tympani nerve would selectively impair one's ability to taste sweet. Not only does this not occur, damage to the chorda tympani often produces virtually no change in the subjective taste world of the patient.

Release-of-inhibition. In an effort to understand this puzzle, Catalanotto and I and our students have done a series of studies aimed at duplicating these clinical observations in normal individuals who have volunteered to have one or more taste nerves anaesthetized temporarily. In the first study, unilateral chorda tympani anaesthesia (Östrom et al 1985) actually caused some of the taste stimuli to taste more intense *after* the anaesthesia. The effect was not a large one and we might have failed to understand its significance had not Halpern & Nelson (1965) reported a phenomenon much like this in the rat. They recorded in the medulla, in an area that received input from both the anterior and posterior tongue. When they anaesthetized the chorda tympani nerve, the neural response to stimulation of the posterior tongue increased. They suggested that the chorda tympani input might normally inhibit the area in the medulla receiving input from the glossopharyngeal nerve. Thus when the chorda tympani nerve was anaesthetized, that inhibition was released and the responses from stimulation of the glossopharyngeal nerve increased.

To determine where the release occurred in human subjects, we 'painted' taste stimuli onto various areas on the tongue and palate after anaesthetizing the chorda tympani nerve unilaterally (Lehman 1991) with dental anaesthetic (which anaesthetized VII and V), or by anaesthetic used routinely by otolaryngologists to anaesthetize the ear drum for minor surgery (which anaesthetized only VII).

Anaesthesia of VII by either method caused some taste sensations at IX to increase, with the greatest increases on the contralateral side. Since the taste system projects ipsilaterally (Norgren 1990), an effect that crosses the midline must take place in the central nervous system. London et al (1990) identified descending pathways that cross the midline. These pathways descend both contra- and ipsilaterally; the contralateral pathway is the largest. Perhaps the inhibition implied by the anaesthesia experiment is mediated via these pathways.

Release-of-inhibition phantom. In another experiment, where we anaesthetized the chorda tympani via the ear canal (Yanagisawa et al 1992), some subjects developed a taste phantom that appeared to come from the contralateral rear of the tongue. Application of a topical anaesthetic to the area where the phantom appeared to be abolished the phantom. Since topical anaesthesia abolishes spontaneous activity, the phantom might result because release-of-inhibition allows the central nervous system to interpret spontaneous activity as a taste signal. This release-of-inhibition phantom could be related to clinical dysgeusia, because localized losses of taste are relatively common.

Nerve-stimulation phantom

Damage to taste structures sometimes produces abnormal stimulation of those structures and so produces another kind of taste phantom. EO experienced an intense salty taste following a mastoidectomy to remove infected bone around her ear. When she rinsed her mouth with a topical anaesthetic (0.5% dyclonine and 0.5% diphenhydramine in 0.9% saline), she experienced a dramatic *increase* in the intensity of her salty taste. When the anaesthetic wore off, her salty taste returned to its original intensity (Bartoshuk & Kveton 1991). We conclude that the neural responses from the glossopharyngeal nerve usually inhibit neural responses from the chorda tympani nerve. The anaesthetic rinse of EO's mouth abolished the input from IX and so removed its inhibition of the taste phantom coming from VII.

RY experienced a bitter phantom on the rear of her tongue apparently caused by damage to IX during a tonsillectomy. Anaesthesia of her mouth intensified that phantom, which suggests that VII also normally inhibits IX. When VII was anaesthetized, its inhibition of IX was abolished and the phantom originating from damage to IX was released from inhibition.

Damage to taste via herpes zoster oticus: more evidence for inhibition in the taste system

Herpes zoster oticus results when the virus that causes chicken pox reactivates and damages cranial nerves unilaterally. Dr Carl Pfaffmann lost all taste on his left side as the result of this disorder. He had no subjective awareness of any loss. As the taste nerves on the left began to regenerate, taste function on the unaffected right side began to decline, suggesting that with regeneration came the re-establishment of inhibition (Pfaffmann & Bartoshuk 1990).

Localization illusions: why people don't notice local areas of taste loss

Release-of-inhibition helps to explain why everyday tastes do not suddenly become weaker when a taste nerve is damaged. However, this still does not explain why people do not notice that one area in the mouth has become devoid of taste. The answer to this lies in the way in which taste sensations are localized perceptually. Some introspection shows that taste sensations are not localized to the receptors. If this were so, then taste sensations would seem to arise from the perimeter of the tongue and from a narrow band of tissue dividing the hard and soft palates. In fact, taste sensations seem to come from the entire mouth. This occurs because the brain uses the sense of touch to localize taste sensations (Todrank & Bartoshuk 1991).

Dr Pfaffmann's experience with herpes zoster oticus provides a clinical example: painting a taste solution from the damaged side to the intact side resulted in a sudden taste when the solution touched the midline. Painting from the intact side to the damaged side resulted in a taste sensation that crossed the midline and seemed to come from the denervated area.

Thus patients with localized taste damage but intact touch systems do not notice localized taste damage, because there is little change in taste intensity and because taste sensations seem to come from any areas touched in the mouth.

Genetic variation

History

In 1931, Fox reported a startling accidental discovery. He was synthesizing some phenylthiocarbamide (PTC) in his laboratory and some of it blew into the air. One of his colleagues commented on how bitter it was, yet Fox tasted nothing (Fox 1931). Family studies led to the conclusion that PTC tasting is produced by the dominant allele, T. Thus individuals with two recessive alleles, tt, are non-tasters and individuals with one dominant allele, Tt, as well as those with two dominant alleles, TT, are tasters.

In the early days of PTC research, thresholds were used to classify individuals as tasters and non-tasters. However, we now know that thresholds are not necessarily a valid measure of suprathreshold experience. The major barrier to using suprathreshold scaling to study the differences between tasters and non-tasters is a philosophical problem—there is no way to compare directly the experiences of different individuals. If we want to show that a compound is more bitter to one individual than to another, we must have a standard that we can reasonably believe tastes the same to both. In 1975, we first scaled the bitterness of PTC in tasters and non-tasters (Hall et al 1975) using NaCl as our standard. We found that a dilute caffeine solution was less bitter to non-tasters than to tasters. This was unexpected and stimulated our interest in PTC.

In subsequent studies, we used a chemical relative of PTC, 6-N-propylthiouracil (PROP), a medication used to suppress thyroid function. PROP lacks the sulphurous odour of PTC and, because PROP is a medication, safety limits can be set for its use (Fischer & Griffin 1964, Lawless 1980). We also began to evaluate an auditory standard (Marks et al 1988). Our studies revealed that the bitter tastes of saccharin (Bartoshuk 1979), KCl, sodium benzoate and potassium benzoate (Bartoshuk et al 1988) are greater for tasters than for non-tasters. Non-taster–taster differences for $CaCl_2$ and casein led us to look at some milk products; cheddar and Swiss cheeses turned out to be more bitter to tasters (Marino et al 1991). A study with 5–7 year olds showed that cheddar cheese was less palatable to tasters (Anliker et al 1991). Work with sweeteners showed that saccharin, sucrose and neohesperidin dihydrochalcone (a sweetener made from citrus fruit peels) were sweeter to tasters than to non-tasters (Gent & Bartoshuk 1983). This was a startling result, because early speculation about PTC/PROP assumed that only bitter tastes were involved.

Supertasters of PROP

Throughout these studies, there seemed to be a subset of tasters that gave very large responses to the bitterness of PROP. We began to refer to them as 'supertasters' and to wonder if they were homozygous for the dominant allele. In order to test this with family studies, we needed a very robust, reliable method of sorting subjects into non-tasters, medium tasters and supertasters. Thresholds let us sort subjects into non-tasters and tasters. We turned to suprathreshold scaling to try to sort tasters into medium and supertasters.

Suprathreshold PROP functions show considerable variability. If the most responsive tasters are TT tasters, then these individuals should be more responsive to the substances that we have previously shown to differ for non-tasters and tasters. To sort subjects into non-tasters, medium tasters and supertasters, we measured PROP thresholds and then asked subjects to taste a series of concentrations of NaCl and PROP and to assign each a 'magnitude estimate'. The bitterness of PROP divided by the saltiness of NaCl provided a measure of suprathreshold responsivity to PROP (see Fig. 2). In a large sample ($n = 150$), the third of the tasters most responsive to PROP were designated supertasters (PROP ratio > 1.2). The remaining tasters were designated medium tasters. Figure 2 shows the PROP and NaCl functions for the subjects who participated in the anatomical studies discussed below.

To test the validity of our suprathreshold sort, we asked subjects to taste a variety of stimuli. Figure 3 shows preliminary results from some bitter compounds (Bartoshuk et al 1992). Supertasters tended to perceive more bitterness from all of the compounds. Note that the bitter tastes are expressed

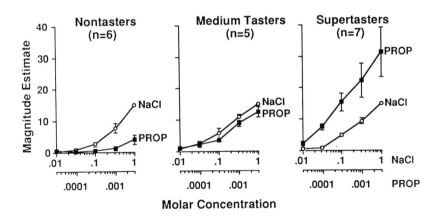

FIG. 2. Magnitude estimates (see text) of NaCl and PROP ($\pm$ SEM) for the subjects whose anatomical results are shown in Fig. 5. The PROP ratios were calculated from the magnitude estimates of the top two concentrations of NaCl (N) and PROP (P): PROP ratio = $(0.001P_1/0.32N_1 + 0.0032P_2/1.0N_2)/2$.

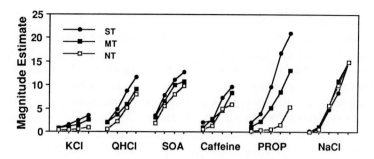

FIG. 3. Magnitude estimates of KCl, quinine hydrochloride (QHCl), sucrose octaacetate (SOA), caffeine, PROP and NaCl for non-tasters ($n=11$), medium tasters ($n=36$) and supertasters ($n=18$). Since the data were normalized to 1 M NaCl, the estimates of the bitter compounds are all expressed relative to 1 M NaCl. ANOVA (KCl, QHCl, SOA and caffeine) produced a main effect of PROP status ($F(2,62)=6.524$; $P<0.01$) and a significant interaction between PROP status and concentration ($F(6,186)=2.383$; $P<0.05$). Follow-up Newman-Keuls tests showed that the top two concentrations showed the effect of PROP status.

relative to the saltiness of 1 M NaCl, so this cannot simply reflect a tendency for supertasters to taste everything as more intense.

Supertasters and fungiform papillae. Miller & Reedy (1990) stained taste buds (with 0.5% methylene blue) so that they could count them in living human subjects. Tasters of PROP had more taste buds than non-tasters. Recently, working in collaboration with Reedy and Miller, we used their method with supertasters (Reedy et al 1993). Supertasters not only had the most taste buds,

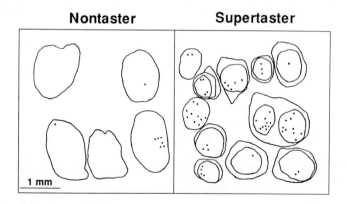

FIG. 4. Tracings of fungiform papillae from a non-taster and a supertaster. The small dots indicate taste pores (the taste pore is the opening to the taste bud). Note that the supertaster's fungiform papillae tend to be smaller and to be surrounded by rings. The area sampled was on the right tip of the tongue.

Genetic and pathological taste variation

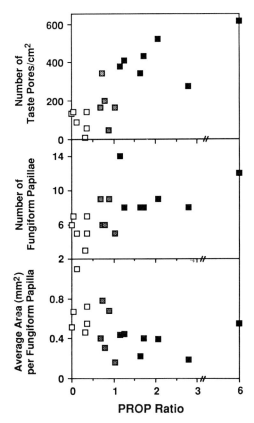

FIG. 5. PROP ratios (see Fig. 2) versus number of taste pores/cm^2, number of fungiform papillae and average area of fungiform papillae. Spearman rank correlation coefficients ρ were: $\rho = 0.79$ ($P < 0.01$), $\rho = 0.64$ ($P < 0.01$) and $\rho = 0.54$ ($P < 0.05$) for the data in the top, middle and bottom panels, respectively. Open squares represent non-tasters, grey squares represent medium tasters and black squares represent supertasters.

there were also anatomical differences between the fungiform papillae of non-tasters, medium tasters and supertasters. Figure 4 shows tracings taken from videotapes of the fungiform papillae of a non-taster and a supertaster. Note that the supertaster had more fungiform papillae, which were smaller, had more taste pores and had rings of tissue around them that were not seen on the non-taster's fungiform papillae.

Figure 5 shows plots of a measure of the perceived bitterness of PROP (see Fig. 2 for the calculation of the PROP ratio) versus various anatomical measures.

Among the 18 subjects shown in Fig. 5, the number of taste pores (i.e. taste buds) per cm^2 ranged from 11 to 611. Although there are differences in taste perception among these subjects, the differences are not as great as one might expect given the large anatomical differences. This suggests the possibility of lateral inhibition among taste buds in supertasters.

Supertasters and chilli peppers. Supertasters and medium tasters of PROP perceived greater burn from capsaicin (the active compound in chilli peppers) than non-tasters did (Karrer & Bartoshuk 1991). This is the first evidence that the burn of capsaicin shows genetic variation. Since super and medium tasters have more taste buds and since taste buds have nociceptive fibres associated with them (Whitehead et al 1985), super and medium tasters may have more nociceptive fibres and so experience greater burning from capsaicin. Tests on specific loci on the tongue (Karrer et al 1992) showed that variability in the burn associated with PROP status occurred only with areas innervated by VII and V. Under the tongue (innervated only by V) and on the back of the tongue (innervated only by IX), the burn of capsaicin is equivalent for non-tasters, medium tasters and supertasters.

Supertasters and ethanol. Because capsaicin was presented in 10% ethanol (to put it in solution) in the experiment just described, 10% ethanol was tested itself. The alcohol was perceived to be more bitter and more irritating on the tongue tips of supertasters. This alcohol effect is especially interesting because of the finding that alcoholism is associated with non-tasting (Pelchat & Danowski 1992). This suggests that super and medium tasters might be protected against alcoholism to some extent, because the alcohol is a less pleasant sensory stimulus to those individuals.

Genetic variation in bitter taste in other species. Some bitter compounds completely fail to cross adapt in human psychophysical studies (McBurney & Bartoshuk 1973), suggesting that some bitter receptor mechanisms are virtually independent of one another, while others may be stimulated by more than one bitter compound. Thus we should not be surprised to find genetic variation across bitter compounds.

Inbred strains of mice differ with regard to aversion to the bitter-tasting compound sucrose octaacetate (SOA). Mice of the SWR/J strain avoid SOA and so are designated as SOA tasters, while mice of other strains (e.g. C57BL/6J) do not avoid SOA and so are designated as SOA non-tasters (Lush 1981, Whitney & Harder 1986). Miller & Whitney (1989) showed that the SOA taster mice had many more taste buds in the vallate papilla than did SOA non-taster mice.

Conclusion

In the elegant mouse model and the genetic data from humans, we see that perceived taste intensities are greater in individuals with more receptors. Yet in animal studies and human pathology, we also see preservation of perceived intensity in the face of loss of input. These new quantitative insights about taste experience illuminate both basic mechanisms and clinical phenomena.

Acknowledgement

This work was supported by National Institutes of Health grant DC 00283.

References

Anliker JA, Bartoshuk LM, Ferris AM, Hooks LD 1991 Children's food preferences and genetic sensitivity to the bitter taste of PROP. Am J Clin Nutr 54:316–320

Bartoshuk LM 1979 Bitter taste of saccharin: related to the genetic ability to taste the bitter substance 6-*n*-propylthiouracil. Science 205:934–935

Bartoshuk LM, Kveton J 1991 Peripheral source of taste phantom (i.e. dysgeusia) demonstrated by topical anesthesia. Chem Senses 16:499–500(abstr)

Bartoshuk LM, Rifkin B, Marks LE, Hooper JE 1988 Bitterness of KCl and benzoate: related to PTC/PROP. Chem Senses 13:517–528

Bartoshuk LM, Fast K, Karrer TA, Marino S, Price RA, Reed DA 1992 PROP supertasters and the perception of sweetness and bitterness. Chem Senses 17:594 (abstr)

Boring EG 1942 Sensation and perception in the history of experimental psychology. Appleton-Century-Crofts, New York

Catalanotto FA, Lecadre Y, Devonshire F, Bartoshuk LM 1991 Foliate papillae taste perception in humans. Chem Senses 16:508(abstr)

Collings VB 1974 Human taste response as a function of locus of stimulation on the tongue and soft palate. Percept Psychophys 16:169–174

Fischer R, Griffin F 1964 Pharmacogenetic aspects of gustation. Drug Res 14:673–686

Fox AL 1931 Six in ten 'tasteblind' to bitter chemical. Science News Lett 9:249

Frank M 1975 Response patterns of rat glossopharyngeal taste neurons. In: Denton DA, Coghlan P (eds) Olfaction and taste. Academic Press, New York, p 588–618

Gent JF, Bartoshuk LM 1983 Sweetness of sucrose, neohesperidin dihydrochalcone, and saccharin is related to genetic ability to taste the bitter substance 6-*n*-propylthiouracil. Chem Senses 7:265–272

Hall MJ, Bartoshuk LM, Cain WS, Stevens JC 1975 PTC taste blindness and the taste of caffeine. Nature 253:442–443

Halpern BP, Nelson LM 1965 Bulbar gustatory responses to anterior and to posterior tongue stimulation in the rat. Am J Physiol 209:105–110

Hänig DP 1901 Zur Psychophysik des Geschmackssinnes. Philos Studien 17:576–623

Karrer T, Bartoshuk LM 1991 Capsaicin desensitization and recovery on the human tongue. Physiol Behav 49:757–764

Karrer T, Bartoshuk LM, Conner E, Fehrenbaker S, Grubin D, Snow D 1992 PROP status and its relationship to the perceived burn intensity of capsaicin at different tongue loci. Chem Senses 17:649(abstr)

Lawless HT 1980 A comparison of different methods used to assess sensitivity to the taste of phenylthiocarbamide (PTC). Chem Senses 5:247–256

Lehman C 1991 The effect of anesthesia of the chorda tympani nerve on taste perception in humans. PhD thesis, Yale University, New Haven, CT

London J, Halsell CB, Barry B, Donta TS 1990 An examination of the projection from the gustatory cortex to the NTS in the hamster. Chem Senses 15:609(abstr)

Lush IE 1981 The genetics of tasting in mice. I. Sucrose octaacetate. Genet Res 38:93–95

Marino S, Bartoshuk LM, Monaco J, Anliker JA, Reed D, Desnoyers S 1991 PTC/PROP and the tastes of milk products. Chem Senses 16:551(abstr)

Marks LE, Stevens JC, Bartoshuk LM, Gent JG, Rifkin B, Stone VK 1988 Magnitude matching: the measurement of taste and smell. Chem Senses 13:63–67

McBurney DH, Bartoshuk LM 1973 Interactions between stimuli with different taste qualities. Physiol Behav 10:1101–1106

Miller IJ, Reedy FE 1990 Variations in human taste bud density and taste intensity perception. Physiol Behav 47:1213–1219

Miller IJ, Whitney G 1989 Sucrose octaacetate-taster mice have more vallate taste buds than non-tasters. Neurosci Lett 360:271–275

Norgren R 1990 Gustatory system. In: Paxinos G (ed) The human nervous system. Academic Press, New York, p 845–861

Östrom KM, Catalanotto FA, Gent JF, Bartoshuk LM 1985 Effects of oral sensory field loss of taste scaling ability. Chem Senses 10:459(abstr)

Pelchat ML, Danowski S 1992 A possible genetic association between PROP tasting and alcoholism. Physiol Behav 51:1261–1266

Pfaffmann C, Bartoshuk LM 1990 Taste loss due to herpes zoster oticus: an update after 19 months. Chem Senses 15:547–658

Reedy FE, Bartoshuk LM, Miller IJ, Duffy VB, Lucchina L, Yanagisawa K 1993 Relationships among papillae, taste pores, and 6-n-propylthiouracil (PROP) suprathreshold taste sensitivity. Chem Senses, in press

Todrank J, Bartoshuk LM 1991 A taste illusion: taste sensation localized by touch. Physiol Behav 50:1027–1031

Tomita H, Ikeda M, Okuda Y 1986 Basis and practice of clinical taste examinations. Auris Nasus Larynx 13(suppl I):1–15

Whitehead MC, Beema CS, Kinsella BA 1985 Distribution of taste and general sensory nerve endings in fungiform papillae of the hamster. Am J Anat 173:185–201

Whitney G, Harder DB 1986 Single-locus control of sucrose octaacetate tasting among mice. Behav Genet 16:559–574

Yanagisawa K, Bartoshuk LM, Karrer TA et al 1992 Anesthesia of the chorda tympani nerve: insights into a source of dysgeusia. Chem Senses 17:724(abstr)

DISCUSSION

Lush: Can you be sure that the ability to taste PROP as well as PTC is a single-gene effect?

Bartoshuk: I am not a geneticist: I looked it up in a text book that described this as a simple Mendelian characteristic.

Bargmann: The sorts of ratio of non-tasters to medium tasters to supertasters is about 1:2:1. This ratio could arise from supertasters being homozygous for a single gene, as you suggest. But it could also arise if there were two genes involved, both dominant, with supertasters having both dominant alleles. So, unless the genetics have been done for PROP as well as for PTC, you can't be sure that it *is* a single gene effect.

Bartoshuk: They have been done, but I don't know if you would be happy with them: I'm very unhappy with the psychophysics in these genetic studies. I am a psychophysicist, not a geneticist and my task is to see that the psychophysics are done the best way they can be. My feeling is that the genetics should be re-done with better psychophysics.

Bargmann: You could use family studies to see if a second gene affected the supertasters, or if they were homozygous at one locus.

Bartoshuk: I wish we could talk somebody into doing that!

Lush: I do the PTC test regularly on a medical student class and it works very reliably. We do a threshold test and get a completely bimodal distribution with no overlap at all. I haven't thought of trying this with PROP.

Bartoshuk: Harry Lawless (1980) showed that PTC produces better separation between tasters and non-tasters than PROP.

Lush: PTC tasting seems to be fairly specific. I can test my students with quinine and sucrose octaacetate but there's no real relationship between their perception of these and their ability to taste PTC.

Bartoshuk: Were you measuring thresholds?

Lush: Yes.

Bartoshuk: You can't see these relationships clearly when you measure thresholds. These relationships are much easier to see, however, if you look at suprathreshold intensities. If you go back and do magnitude estimation or category scaling of the intensity of bitterness, you will find that the people who perceive PTC to be the most bitter are the people who also perceive sucrose, octaacetate and caffeine to be the most bitter. As a control, use something like a salt to normalize, to make sure you are not just getting people who like to use big numbers to describe their experiences.

Lush: Don't you think that what you're measuring by these tests is just the number of taste sensory cells people have?

Bartoshuk: Yes, I think it's really quite simple—the more receptors you have, the more intense the taste (Miller & Reedy 1990).

Lancet: I want to argue in favour of using thresholds to study taste. There is clearly more than one gene for bitter taste, because a lot of PTC non-tasters can taste other bitter substances—perhaps there are as many as 5–10 bitter genes to cope with the variety of chemicals that taste bitter. If you study bitter taste using suprathreshold intensities, you might confuse some details that the geneticist would like to know. In my threshold hypothesis, the highest affinity receptor for any given substance determines the threshold, in which case there is a higher probability that you are looking at one gene at a time if you measure thresholds. The only way to untangle the different bitter substances and the different genes that might be involved in their perception is to do careful threshold measurements and family and twin studies. This would enable us to count genes in parallel with Bob Margolskee's counting of genes the moment he clones them.

Bartoshuk: Another way to study classes of bitter is to use cross-adaptation. You can find classes of bitter that don't cross-adapt at all, for instance, urea and caffeine. This shows that there have to be multiple bitter receptor mechanisms. I never believed that PROP had any major significance—I thought it was just one bitter taste among many, until we saw this tremendous association between PROP intensity and number of receptors. I find it an amazing association; one we did not expect to see.

Reed: I would caution against attributing all these effects directly to receptor genes. For instance, when you look at the ability of individuals to perceive quinine as bitter, the quinine might be acting at the level of channels, which show allelic variation in the population. It is hard to tell whether these variations are due the binding proteins that represent receptors in the first components of some cascade. It may be that what PTC tasting is revealing is some developmental aspect of how many pores you have and nothing to do with receptor genes.

Lancet: When I referred to receptors, I didn't specify G protein-coupled seven-transmembrane-domain receptors: 'receptor' might also mean a channel protein that binds a bitter substance and undergoes a conformational change. It is also possible that this phenomenon is not occurring at the membrane protein level at all.

Margolis: It could be dangerous to assume an interpretation of the psychophysical results that relates to numbers and kinds of receptor molecules, because you would be ignoring everything that goes on beyond the taste bud.

Reed: Certainly, it is dangerous to go to flies or worms with a notion that each of these identified mutations is going to map to fundamental receptor systems.

DeSimone: Your studies emphasize the inhibitory events that occur during taste sensory processing. For some of the recent electrophysiological studies in rodents in the nucleus of the solitary tract and in higher brain taste areas, it seems that some responses are broadened while others may be sharpened through inhibitory neural circuitry (McPheeters at al 1990). Your nerve damage studies seem to suggest that mutual inhibition among the taste nerves is the norm.

Bartoshuk: There are several papers that show this inhibition (Halpern & Nelson 1965, Norgren & Pfaffmann 1975, Ninomiya & Funakoshi 1982, Ogawa & Hayama 1984, Sweazey & Smith 1987).

Getchell: Is there a similar way of classifying animals into non-, medium and supertasters for PROP or PTC? If so, this might be a way of studying the genetic mechanisms underlying this phenomenon.

Bartoshuk: The work hasn't been done for PROP in animals, but it has been done for sucrose octaacetate.

Lush: I usually work with strains of mice and there are certainly differences between them in their abilities to taste a large number of substances, judged by rejection and acceptance experiments.

Getchell: Has anyone ever investigated the size and the distribution of a variety of types of taste papillae on these animals' tongues?

Bartoshuk: Miller & Whitney (1989) did, for the sucrose octaacetate mouse model. That's why we looked in human, because they found an impressive association between the number of receptors in the circumvallate papillae and the magnitude of the response to sucrose octaacetate.

Lush: I'm not happy with that report. They took one strain which was a taster and another strain which was a non-taster, and found that the taster had slightly more papillae than the non-taster. This result could have been obtained by chance.

Margolis: That is true. However, with the development of the bilineal congenic lines of mice, differing only in presumably the locus that is involved in the ability to taste sucrose octaacetate, this result can be verified (Whitney et al 1989).

Lindemann: There is a disease called familial dysautonomia, where, among other defects, patients have no taste buds. In the literature, there are reports that on injection of an analogue of acetylcholine, taste returns within 5–10 min (Henkin & Kopin 1964). Have you ever tested this?

Bartoshuk: Patients with familial dysautonomia are very sick (Henkin 1967) and it's very difficult to make the ethical decision to do this kind of research with that population: I have chosen not to. I've read that observation, but I find it hard to believe. These are very ill patients who have never experienced taste! How can we tell what they are experiencing? Being a psychophysicist is interesting, because we are constantly asked to solve the philosophical question of comparing experience of one person with another. But this strikes me as a really difficult observation to make with any kind of certainty.

Caprio: In the taste illusion experiments where you paint a taste stimulus from the taste side of the tongue to the anaesthetized non-taste side and taste appears to follow the mechanical stimulus, what happens if you paint the taste solution from the rostral portion of the tongue to the rear of the tongue that does not contain taste buds?

Bartoshuk: You can cross the midline and you can go from one nerve field to another. The characteristics of that clearly do not follow anything that we expect to see in nerve fields.

Hwang: Are supertasters of PROP also supertasters of sweet as well?

Bartoshuk: No. The reason the taster/non-taster phenomenon was discovered originally by Fox in the 1930s was because the thresholds or suprathreshold intensities produce such a clearly bimodal distribution (Fox 1931). There are no other substances except the PTC-like compounds that produce this bimodal distribution.

Hwang: Have you any ideas why there are differences in the number of circumvallate papillae in different species of animals?

Bartoshuk: I have no idea. We don't even know what the circumvallate papillae are there for.

Margolskee: Do you also get a clear difference in the populations of super-tasters and non-tasters if you compare PROP taste to different types of bitter taste?

Bartoshuk: Yes; most bitters associate with PROP status. There is considerable variation between subjects, but, on average, bitter compounds are more bitter to supertasters than to non-tasters.

Lush: I think we should be very careful about the genetic explanation. What you ought to do is get your colleagues, find out which supertasters are married to non-tasters and test their children.

Bartoshuk: We haven't done this so far because these are expensive studies to do and I'm still hoping to find even better psychophysical ways to do this. While we were searching, we ended up looking at the anatomy, which is much better than looking at the ability to taste in defining non-, medium and supertasters.

Lancet: The question is: are these anatomical differences genetic?

Bartoshuk: Many family studies have been done to show that PTC non-tasting is a simple Mendelian recessive characteristic.

Lush: There are two kinds of variation: there is Mendelian variation, where you can classify people, and there is continuous variation. I think that both are occurring here. PROP/PTC tasting is a case of Mendelian segregation where you can identify a gene, but I think what you are now looking at is a non-classifiable phenomenon, i.e. quantitative variation, for example, in the number of papillae on the tongue. It is your classification that worries me, not your results.

Bartoshuk: We simply chose an empirical classification, based on the perceived bitterness of PROP. Having done this, we see that people classified as supertasters are living in a very different world from non-tasters; they're having very different experiences. I think it's a very interesting classification. If you ask whether we should draw the line where we do, I can't answer. I don't think we have any idea where to draw the line. The family studies might tell us that.

References

Fox AL 1931 Six in ten 'tasteblind' to bitter chemical. Science News Lett 9:249

Halpern BP, Nelson LM 1965 Bulbar gustatory responses to anterior and to posterior tongue stimulation in the rat. Am J Physiol 209:105–110

Henkin RI 1967 Sensory mechanisms in familial dysautonomia. In: Bosma JF (ed) Symposium on oral sensation and perception. Charles C. Thomas, Springfield, IL, p 341–349

Henkin RI, Kopin IJ 1964 Abnormalities of taste and smell thresholds in familial dysautonomia: improvement with methacholine. Life Sci 3:1319–1325

Lawless HT 1980 A comparison of different methods used to assess sensitivity to the taste of phenylthiocarbamide (PTC). Chem Senses 5:247–256

McPheeters M, Hettinger TP, Nuding SC, Savoy LD, Whitehead MC, Frank ME 1990 Taste-responsive neurons and their locations in the solitary nucleus of the hamster. Neuroscience 34:745–758

Miller IJ, Reedy FE 1990 Variations in human taste bud density and taste intensity perception. Physiol & Behav 47:1213–1219

Miller IJ, Whitney G 1989 Sucrose octaacetate-taster mice have more vallate taste buds than non-tasters. Neurosci Lett 360:271–275

Ninomiya Y, Funakoshi M 1982 Responsiveness of dog thalamic neurons to taste stimulation of various tongue regions. Physiol & Behav 29:741–745

Norgren R, Pfaffmann C 1975 The pontine taste area in the rat. Brain Res 91:99–117

Ogawa H, Hayama T 1984 Receptive fields of solitario-parabrachial relay neurons responsive to natural stimulation of the oral cavity in rats. Exp Brain Res 54:359–366

Sweazey RD, Smith DV 1987 Convergence onto hamster medullary taste neurons. Brain Res 408:173–184

Whitney G, Harder DB, Gannon KS 1989 The B6.SW bilineal congenic sucrose octaacetate (SOA)-taster mice. Behav Genet 19:409–416

General discussion III

Fesenko: I would like to make some comments and pose some questions for our general discussion. My first comment was stimulated by Heinz Breer's and Doron Lancet's papers (this volume: Breer 1993, Lancet 1993) and relates to the kinetic properties of olfactory receptors. How large can the rate constant of association (k_{on}) of olfactory receptors and stimuli be? Apparently, because of diffusional limitations, this constant will probably not exceed 10^8 $M^{-1}s^{-1}$. According to our data (Fesenko et al 1985), the value for k_{on} is about 10^6 $M^{-1}s^{-1}$. If, using these values, we calculate the time needed to reach the equilibrium state for stimulus–receptor binding at the normal physiological concentrations of odorant (10^{-9}–10^{-10} M), we obtain a time constant of no less than 1–2 min. However, we know that the generation of an electrical response by a receptor cell takes about 0.1 s. There is a discrepancy between these two values. It follows that an olfactory receptor cell reacts not to the total binding of odorant, but rather to the rate of binding, that is the first derivative of the process of binding. Such a situation can be accounted for if we introduce a process of rapid inactivation of the receptor–stimulus complex with a time constant of about 0.1 s. In relation to this, Kaissling (1975) introduced a similar process, referred to as 'early inactivation', on the basis of physiological investigations of sensory transduction in insect olfactory receptors. Two conclusions may be derived from the foregoing considerations. Firstly, k_{on} is a more important property of the receptor molecule than K_d. Secondly, a mechanism of rapid inactivation of the activated receptor must exist.

The only suggestion of this we have, so far, is the phosphorylation of receptors that was demonstrated by Heinz Breer (this volume: Breer 1993), which occurs in less than a second. Thus, the detection of a stimulus is a dynamic process which is not based on reaching a stable state. This is very important when we consider the role of K_d in determining sensitivity.

In fact, there is no direct correlation between K_d and sensitivity. The high sensitivity of the olfactory receptor cell is a consequence of signal amplification mechanisms that are sufficient to realize the physical limit of sensitivity—detection of a single molecular event. The problem is: how can this sensitivity be achieved and how can the signal be recognized within such a short time? Let us imagine, for example, that in this room there is one stimulus molecule and one very sensitive receptor cell containing one receptor molecule. The sensitivity is not a problem in this case. Rather, the problem is the collision of these two molecules. The time limit for this collision might be hundreds of years. Hence, to increase the probability of collision, we need to concentrate the receptor and the stimulus in a limited space. We must also choose a receptor with a maximal value for k_{on}.

We measured the sensitivity of the olfactory epithelium of the Black Sea skate to amino acids by means of electroolfactogram. We obtained a value of 10^{-7}–10^{-8} M, which at first glance appears to be in agreement with the expected K_d value. But when we cut the olfactory tract off from the brain and chose a nerve fibre which could selectively respond to serine, lysine or any other amino acid, we obtained very high sensitivity (10^{-14}–10^{-13} M), much higher than the anticipated K_d value (Brown & Fesenko 1981).

Lancet: I have no problem converting all our arguments about K_d to k_{on}. The receptor with the highest k_{on} determines threshold. Receptors differ in their values for k_{on} and therefore differ in their ability to respond to different odorants. The problem I see is that receptors in biology tend to have similar k_{on} values because, in most cases, they are diffusion controlled. The selectivity range that is afforded by k_{on} is extremely narrow, yet we may see selectivity of up to 10 orders of magnitude in terms of affinities and thresholds. So a likely explanation for this would be variation in k_{off}, leading to a range of K_d values. But I appreciate Dr Fesenko's comments on the fact that receptors don't always reach full saturation. In our models, we should all consider what happens at steady-state, at very low saturations. We need to address what goes on in this situation and how it is actually related to the k_{off}, because it is the k_{off}, as Paolo Pelosi has pointed out (p 183), that in most cases determines the distinction of many ligands by many receptors.

Bargmann: When you are talking about receptors that have a potential to recognize many different ligands, you are not talking about one molecule coming to one receptor, you are talking about a molecule competing with many other molecules and either overriding them or adding to signals that already exist. So, in some cases, the relative relationship may be as important or more important than the absolute affinity of a receptor for a ligand.

Fesenko: The second point on which I would like to focus our attention is the possibility of the induction of transcription in olfactory receptor cells by odorous stimuli. We have recently tried to check this in our laboratory (E. E. Fesenko, V. A. Ivanov, M. Y. Zemskova & V. I. Novoselov, unpublished results). We incubated tissue slices of rat olfactory epithelium with a mixture of odorants (camphor, citral, linalool, isoamylacetate and menthone—all at 1 mM) and assayed them for odorant-induced expression of the c-*fos* gene (looking for the mRNA), using a fragment of mouse c-*fos* kindly provided by Dr M. E. Greenberg. Exposure of olfactory epithelium to this mixture resulted in a sevenfold increase in c-*fos* transcription within 5 min, as measured by hybridization of cellular RNA to a radioactive c-*fos* probe. The activation of c-*fos* transcription was transient; it was maximal within 10–15 min of addition of odorants and decreased to basal levels within 30 min. Under identical conditions, no change in c-*fos* transcription was observed in rat lung cells. The kinetics of induction and repression of c-*fos* transcription seen here is very similar to those of PC12 cells in response to nicotine reported by Greenberg et al (1986). The stimulation of the c-*fos* gene by these odorants is specific in as much as this odorous stimulus does not specifically affect the expression of several other genes, including α-globulin, c-*src* (a proto-oncogene that encodes a protein tyrosine kinase) and c-*erbB* (a proto-oncogene that encodes the EGF receptor). Transcription of another nuclear proto-oncogene, c-*myc*, was also unaffected, at least within the first several hours of treatment with odorants.

Thus, these findings suggest that odorants may rapidly activate specific gene transcription in olfactory cells. The precise function of Fos is not known. According to an idea that comes from neurobiology, it acts as a 'third messenger' in signal transduction systems, where it would couple the short-term intracellular signals, elicited by a variety of extracellular stimuli, to long-term responses, by altering gene expression (Ohlsson & Pfiefer-Ohlsson 1987). In the olfactory system, the odorant-induced expression of c-*fos* might indicate the induction of other genes in the olfactory receptor cell. It would be interesting to compare expression maps of c-*fos* with Linda Buck's maps of olfactory receptor gene expression (this volume: Buck 1993). Does anyone know of any examples of the control of gene expression in olfactory epithelium by olfactory stimuli?

Van Houten: Some systems that we've talked about would be good for studying the possible induction of gene expression upon stimulation. In particular, John Carlson's systems are intriguing, because not only can he use the enhancer trap to identify genes that are involved in olfaction, he can also stimulate the cells to see whether gene expression is enhanced or reduced by stimulation (this volume: Carlson 1993).

Has anyone considered what second messengers might be doing in this system other than activating channels, such as activating transcription factors? Is anyone looking at long-term changes due to stimulation?

Getchell: Christine Gall is studying the expression of c-*fos* and c-*jun* upon odour stimulation in the mammalian olfactory system (Gall & Guthrie 1993).

Fesenko: The third problem, which Stuart Firestein has already mentioned (this volume: Firestein & Zufall 1993), is that of the noise in the cell membrane generated by random fluctuations of ionic channels. It is evident that for the cell to reach the physical limit of sensitivity (detection of a single molecular event), it must have a noise level that is less than the magnitude of the response to the minimal stimulus. In photoreception, for example, the signal:noise ratio is about three (Matthews & Baylor 1981).

According to binomial statistics, the noise generated by the channels is determined by the expression

$$\frac{1}{\sqrt{nP}}$$

where n is the total number of channels in the membrane and P is the probability that an individual channel is open at a given moment. For one open channel, at a given moment, there is 100% noise; for 10^4 channels, there is 1% noise. Thus, by increasing the number of channels in the membrane, we can diminish the level of noise. If we want low noise and minimal leakage, the membrane should contain a large number of channels of a very small (0.1 pS) unitary conductance. Thus, Nature has found a very elegant solution to the problem of cell noise. My question is whether it is the only solution to this problem, or is there an alternative way of obtaining a low level of noise not using binomial statistics, for example, by introducing a completely silent (not fluctuating) channel that could be opened by a stimulus?

Firestein: One possibility is that there is a pulsatile second messenger signal. Typically, frequency modulation signal systems are quieter than amplitude modulation systems, so if the second messenger system is working on a frequency basis, that might be an answer.

Reed: One other solution that is used in detection of decay events is coincidence. One way of silencing a system like this is to demand coincidence between, say, an inositol trisphosphate ($InsP_3$) and a cAMP signal.

Getchell: Are there any examples of this?

Bargmann: The Cdc2 kinase has to have both cyclin and a particular spectrum of phosphorylation and dephosphorylation to be activated (Nigg et al 1992).

Firestein: What about the $InsP_3$ system itself, which uses both diacylglycerol and Ca^{2+} as second messengers?

Reed: Long-term potentiation appears to be a perfect example: you must achieve both depolarization and a glutamate signal simultaneously.

Lancet: But the coincidence has to be at the single molecule level. If you have many molecules that activate cAMP generation, many molecules that activate the $InsP_3$ mechanism and then integration occurs downstream, the simultaneity principle may not work.

Reed: No, the system could be organized such that the amount of stimulation that you would get from cAMP is always subthreshold and activation of both pathways would be required.

Ronnett: There are several important points that need stating regarding our interpretation of experiments from different laboratories. A lot of the differences we have seen, especially regarding the modulation of the second messengers involved in olfaction, are probably due to the fact that we are all using different species. It is important to bear this in mind when comparing results. Although the same principles probably apply, the way each species modulates or 'fine-tunes' the second messenger signals may be quite different.

Secondly, when studying biological systems, it is often much easier to obtain results by disruption and subcellular fractionation. This has been especially true in olfaction, in which the ability to isolate the chemosensory cilia has permitted analysis of second messenger signalling. The more you disrupt a system, the more accessible the inside is to experimentation. That is why the cilia and synaptosomes have proved so useful. However, we must keep in mind that these are disrupted subcellular fractions which have been washed extensively and exposed to extraordinarily high Ca^{2+} levels. There have indeed been differences when comparing the results obtained in whole-cell systems with results obtained using cilia preparations. These approaches should be viewed as complementary, not mutually exclusive. We must remember that in preparation of subcellular fraction, regulatory proteins may be lost or uncoupled and this may result in distortion of the physiological signal.

A third point is that when we perform *in vitro* assays, we often modify conditions to optimize a response, which may distort the true biological signal. One of the best examples of this is when we modulate Ca^{2+}. When we fix one variable in a system, such as when we clamp Ca^{2+}, we remove the normal temporal fluxes of Ca^{2+} that occur. In the cell, Ca^{2+} concentrations can rise and fall very rapidly. This creates a far more complicated interplay than we can ever mimic in the test-tube, be it in a whole cell or in isolated cilia. We must take these factors into account when we start trying to apply what we find biochemically to how the system works physiologically. We are looking at windows of function and hopefully these windows will allow us eventually to develop an understanding of the true *in vivo* mechanisms.

Getchell: During the last several years we have begun to understand the role of perireceptor events in sensory transduction. As we have heard many times in this symposium, the factors that regulate the access of ligands to receptor molecules are of prime importance in this process. We have also come to appreciate the role that phase I (e.g. P_{450} cytochromes) and phase II biotransformation enzymes play in odorant inactivation. However, we are only just at the beginning of these experiments and I would urge caution in the creative interpretation of these results so that the models we develop do not get too far ahead of the experimental evidence.

We have also discussed mucus barriers. Returning to the original work of Reese (1965), I am struck by the observation that olfactory mucus has a fundamentally different organization from that of mucoid surfaces in other epithelial tissue (Lopez-Vidriero 1989). We know that it is an important barrier in terms of immunological function (Getchell & Getchell 1991, Mellert et al 1992).

There have also been fascinating discussions arising from the research of Hartwig Schmale on von Ebner's gland protein, Solomon Snyder's research on odorant-binding protein (Pevsner & Snyder 1990) and Paolo Pelosi's research (Pelosi & Maida 1990) on the characterization of transport proteins. These proteins have been shown to bind ligands but, again, I would urge caution in our extrapolation of these results to their having a transport function.

References

Breer H 1993 Second messenger signalling in olfaction. In: The molecular basis of smell and taste transduction. Wiley, Chichester (Ciba Found Symp 179) p 97–114

Brown HR, Fesenko EE 1981 Impulse reactions of single nerve fibers of olfactory tract of Black Sea skates. Dokl Akad Nauk USSR 259:1006–1009

Buck LB 1993 Receptor diversity and spatial patterning in the mammalian olfactory system. In: The molecular basis of smell and taste transduction. Wiley, Chichester (Ciba Found Symp 179) p 51–67

Carlson J 1993 Molecular genetics of *Drosophila* olfaction. In: The molecular basis of smell and taste transduction. Wiley, Chichester (Ciba Found Symp 179) p 150–166

Fesenko EE, Novoselov VI, Novikov IV 1985 Molecular mechanisms of olfactory reception. VI. Kinetic characteristics of camphor interaction with binding sites of rat olfactory epithelium. Biochim Biophys Acta 839: 268–275

Firestein S, Zufall F 1993 Membrane currents and mechanisms of olfactory transduction. In: The molecular basis of smell and taste transduction. Wiley, Chichester (Ciba Found Symp 179) p 115–130

Gall CM, Guthrie KM 1993 Odor induced c-*fos* mRNA expression reveals a functional unit in main olfactory bulb. In: Development, growth and senescence in the chemical senses (NIDCD monograph). US Dept Health & Human Services, NIH, p 119–128

Getchell ML, Getchell TV 1991 Immunohistochemical localization of components of the immune barrier in the olfactory mucosae of salamanders and rats. Anat Rec 231:358–374

Greenberg ME, Ziff EB, Greene LA 1986 Stimulation of neuronal acetylcholine receptors induces rapid gene transcription. Science 234:80–83

Kaissling K-E 1974 Sensory transduction in insect olfactory receptors. In: Jaenieke L (ed) Biochemistry of sensory functions. Springer-Verlag, Berlin, p 243–273

Lancet D, Ben-Arie N, Cohen S et al 1993 Olfactory receptors: transduction, diversity, human psychophysics and genome analysis. In: The molecular basis of smell and taste transduction. Wiley, Chichester (Ciba Found Symp 179) p 131–146

Lopez-Vidriero MT 1989 Mucus as a natural barrier. Respiration 55(suppl 1):28–32

Matthews G, Baylor DA 1981 The photocurrent and dark current of retinal rods. Curr Top Membr Transp 15:3–18

Mellert TK, Getchell ML, Sparks L, Getchell TV 1992 Characterization of the immune barrier in human olfactory mucosa. Otolaryngol Head Neck Surg 106:181–188

Nigg EA, Gallant P, Krek W 1992 Regulation of p34^{cdc2} protein kinase activity by phosphorylation and cyclase binding. In: Regulation of the eukaryotic cell cycle. Wiley, Chichester (Ciba Found Symp 170) p 72–96

Ohlsson RI, Pfiefe-Ohlsson SB 1987 Cancer genes, proto-oncogenes and development. Exp Cell Res 173:1–16

Pelosi P, Maida R 1990 Odorant-binding proteins in vertebrates and insects: similarities and possible common function. Chem Senses 15:205–215

Pevsner J, Snyder SH 1990 Odorant-binding protein: odorant transport function in the vertebrate nasal epithelium. Chem Senses 15:217–222

Reese TS 1965 Olfactory cilia in the frog. J Cell Biol 25:209–230

Summing-up

Frank L. Margolis

Laboratory of Chemosensory Neurobiology, Roche Institute of Molecular Biology, 340 Kingsland Street, Nutley, NJ 07110-1199, USA

I have been impressed with both the quality and the quantity of the science and with the intensity of the interactions that we have experienced during this symposium. Furthermore, I am very pleased to note that several unanticipated collaborations have been generated as a result of this meeting. In these past few days we have been able to learn of recent accomplishments in the field, identify and begin to resolve potential areas of disagreement and project future needs and directions.

One of the things that struck me most as I went back and re-read the proposal for this symposium that Tom Getchell and I put together two years ago, is the fact that several of the points that we raised as areas of total ignorance in need of exploration are now overflowing with almost more information that we had ever bargained for.

The exciting advances that we have been witness to these past few days would probably have seemed to be science fiction if viewed from the vantage point of the first Ciba Foundation Symposium on Taste and Smell in 1969. In addition to making one cautious about trying to predict the future, I think that this rapid progress illustrates the major role that the application of contemporary technologies and multidisciplinary approaches to diverse biological systems has had in enabling us to achieve our current level of understanding. Thus, we have seen in these few days the dramatic impact on this field of the application of molecular biology, patch-clamp physiology, genetic analysis, stopped-flow technology, contemporary anatomical and imaging techniques, and sensory psychophysics to systems as diverse as cell fractions, invertebrate organisms and humans. One reason that we have been able to ask questions and do experiments that could not have been done previously relates to the sensitivity of some of the new technologies. They have enabled us to identify molecules of extremely low abundance or infrequent events and then selectively amplify them or the responses that they engender, thus facilitating the generation of specific reagents to use as probes of function in the original system.

Lest this accolade give us an unwarranted arrogance, it is perhaps useful to consider a few of the questions for which we still don't have answers (but probably will quite shortly). Although it seems all but certain that the chemical senses function similarly to other biological ligand detection systems, we still don't really understand how the chemical senses actually identify and discriminate molecular signals.

To date, many putative olfactory receptor molecules have been cloned in several vertebrate, but not yet in invertebrate systems. All appear to be members of the G protein-coupled seven-transmembrane-domain protein superfamily. Why has it been so difficult to demonstrate olfactory receptor function of these molecules in reconstituted systems? Is this related to our lack of knowledge of the ligand specificity, to the need for specific post-translational modifications, to specific second messenger coupling requirements, or to structural demands related to ciliary organization? Is it a function of the level of expression and/or membrane insertion of proteins expressed *in vitro*? Will the answers to these questions facilitate our subsequent studies of taste receptor molecules?

How does genotype determine the observed sensory phenotype for both smell and taste? Is the chemosensory phenotype stable and invariant or can it change in response to experience? What are the molecular bases for anosmia and ageusia? We need additional genetic models to facilitate investigations of transduction and of the genetic mechanisms regulating the determination of cellular phenotype in the olfactory and gustatory systems in vertebrates and invertebrates.

It is evident from the comparative studies of taste transduction (e.g. amiloride) in the mudpuppy and in mammals that no single animal model will enable us to define taste transduction. What other models would be useful that are not yet being studied in taste or olfaction?

Although we have learned a great deal, we still need to know more about the detailed mechanisms by which components of secretory cells, mucus and saliva, influence ligand access to, and elimination from, receptor sites. Do the mechanisms of activation and termination of olfactory and taste transduction vary at different anatomical loci? What alterations in specificity do they manifest across the olfactory and vomeronasal mucosa and the Organ of Masera; what are the variations in the various taste bud loci on the tongue and in other areas of the oro-pharynx?

How are receptor distribution patterns determined? Do they alter during development and ageing? Are they constant across sex, strain, species? Are the olfactory receptor distribution patterns in any way analogous to visual cortex columns or whisker barrels? What are the genetic mechanisms that determine when and where and how much of the various sensory cell molecules are expressed? How do the various components interact and relate to specific sensory deficits and to information processing?

In most of our studies of transduction we are probably never studying a system truly at equilibrium, but more realistically, in kinetic flux. In part, this is a reflection of the way in which the biological systems are structurally organized. Thus, it is essential to consider the kinetics of events rather than equilibrium in our calculations and models. How does this influence our interpretations of physiological vs. biochemical measures of transduction?

It is essential, when interpreting data, that we keep in mind the fact that some of the techniques we use monitor summed population properties while others monitor events in individual cells, for example, polymerase chain reactions, electroolfactograms, patch-clamp measurements, stop-flow biochemistry, immunocytochemistry, *in situ* hybridization and other imaging techniques. This consideration can have a significant impact on the way in which we interpret various observations on transduction processes and measurements of changes during ontogeny and in response to lesions.

The questions raised in these concluding remarks point, indirectly, to a major need in olfactory and taste research. That is, the desirability of *in vitro* cell lines of olfactory and gustatory sensory cells, support cells and secretory cells that can be differentiated separately and in mixed populations to enable us to study the events, factors and mechanisms involved in the regulation of gene expression and the role of various cellular components in transduction, adaptation and cellular development.

These comments are, of necessity, brief, biased and incomplete. I anticipate that the speed with which knowledge is accruing will shortly make them appear to be quite naive as well. I truly hope so, as that will be solid confirmation that we have achieved our goal in convening this symposium and will give us impetus to plan for its successor.

Index of contributors

Non-participating co-authors are indicated by asterisks. Entries in bold type indicate papers; other entries refer to discussion contributions.

Indexes compiled by Liza Weinkove

Ache, B. W., 88, 90, 91, 92, 93, 109, 144, 147, 245
*Ahlers, C., **167**
*Akeson, R., **3**

Bargmann, C. I., 66, 74, 84, 142, 143, 162, 213, **235**, 245, 246, 247, 248, 249, 262, 263, 269, 271
Bartoshuk, L. M., 93, 145, 199, 212, 213, 215, **251**, 262, 263, 264, 265, 266
*Ben-Arie, N., **131**
*Bläker, M., **167**
Breer, H., 47, 66, 84, 85, 90, 94, **97**, 109, 110, 111, 112, 113, 114, 147, 148, 162, 183, 184, 248
Buck, L. B., 20, 23, 40, 41, **51**, 64, 65, 66, 73, 74, 75, 85, 88, 89, 92, 93, 94, 95, 144, 145, 147, 165, 197, 247

Caprio, J., 21, 22, 25, 48, 67, 88, 92, 113, 165, 245, 265
Carlson, J., 41, 85, 86, 94, **150**, 162, 163, 165, 166, 196
*Cohen, S., **131**
*Colbert, H. A., **235**

DeSimone, J. A., 40, 43, 46, 47, 48, 88, 92, 109, 111, 128, 144, 184, 210, 211, 212, 213, 214, 215, 216, **218**, 230, 231, 232, 233, 248, 264
*Dwyer, N., **235**

Fesenko, E. E., 42, 44, 45, 268, 269, 270
Firestein, S., 21, 24, 25, 45, 46, 47, 48, 64, 65, 66, 67, 85, 89, 91, 92, 93, 110, 111, 112, 113, **115**, 127, 129, 143, 144, 145, 148, 213, 216, 233, 271

*Gat, U., **131**
*Getchell, M. L., **27**
Getchell, T. V., 20, 21, 22, 25, **27**, 40, 41, 42, 43, 44, 46, 47, 48, 64, 86, 113, 144, 165, 180, 197, 211, 212, 232, 233, 245, 264, 270, 271, 272
*Graminski, G. F., **76**
*Grillo, M., **3**
*Gross-Isseroff, R., **131**

Hatt, H., 25, 44, 45, 84, 91, 112, 113, 128, 162, 212, 248, 249
*Heck, G. L., **218**
*Horn-Saban, S., **131**
Hwang, P. M., 20, 41, 43, 44, 112, 165, 181, 182, 197, 199, 231, 232, 265

*Jayawickreme, C. K., **76**

*Karne, S., **76**
*Khen, M., **131**
*Kimmel, B. E., **235**
Kinnamon, S. C., 64, 92, 110, 143, 184, 199, **201**, 210, 211, 212, 213, 214, 230, 234
*Kock, K., **167**
*Kudrycki, K., **3**
Kurahashi, T., 110, 113, 128, 213, 214

Lancet, D., 22, 23, 24, 41, 45, 46, 47, 48, 66, 67, 75, 85, 90, 92, 93, 95, 110, 113, 128, 129, **131**, 142, 143, 144, 145, 146, 148, 162, 165, 166, 181, 182, 183, 197, 200, 210, 214, 215, 231, 232, 245, 246, 249, 263, 264, 266, 269, 271
*Lehrach, H., **131**
Lerner, M. R., 47, **76**, 84, 85, 86, 87

Lindemann, B., 40, 42, 45, 85, 126, 143, 148, 198, 210, 211, 212, 213, 214, 229, 247, 265
Lush, I., 144, 262, 263, 264, 265, 266

*McClintock, T., **76**
*McKinnon, P. J., **186**
*McLaughlin, S. K., **186**
Margolis, F. L., **1**, **3**, 20, 21, 24, 25, 41, 43, 44, 45, 48, 73, 74, 75, 86, 88, 91, 92, 93, 94, 110, 142, 147, 148, 162, 181, 183, 184, 212, 231, 246, 264, 265, **274**
Margolskee, R. F., 73, 74, 86, 89, **186**, 196, 197, 198, 199, 200, 266

*Natochin, M., **131**
*North, M., **131**

Pelosi, P., 41, 44, 46, 145, 182, 183
*Potenza, M. N., **76**

Reed, R. R., 20, 23, 24, 47, 48, 65, **68**, 73, 74, 75, 85, 86, 87, 88, 89, 92, 94, 111, 113, 127, 129, 146, 147, 148, 181, 183, 197, 232, 244, 248, 249, 264, 271

*Robichon, A., **186**
Ronnett, G. V., 42, 90, 91, 111, 113, 129, 148, 200, 271

Schmale, H., 25, 41, 42, 45, 74, **167**, 180, 181, 182, 184, 211, 212
*Seidemann, E., **131**
*Sengupta, P., **235**
Siddiqi, O., 21, 92, 141, 144, 162, 163, 164, 165, 233, 247
*Spickofsky, N., **186**
*Spielman, A. I., **167**
*Stein-Izsak, C., **3**
*Su, Z., **27**

Teeter, J. H., 88, 91

Van Houten, J., 88, 89, 184, 248, 270

*Walker, N., **131**
*Wang, M. M., **68**

*Ye, Q., **218**

*Zufall, F., **115**

Subject Index

acetone, 152, 153, 157
acetophenone, 102
N-acetylglucosamine (GlcNAc), 42
  olfactory cilia glycocalyx, 35, 38
  secretory mucins, 33–34
acj6 mutation, 152–157, 162, 165
adaptation, *C. elegans*, 242–243, 249
adenylate cyclase
  melanophores, 77–78, 85
  olfactory transduction, 88–89, 90–91, 97–98
  taste transduction, 187, 195, 198
  type III, 68, 69, 70
β-adrenergic agonists, 42, 43–44
β-adrenergic receptor kinase 2 (βARK2)-like kinase, 102, 148–149
β_2-adrenergic agonists, 78–79, 80, 82
β_2-adrenergic receptors (β_2-AR), 78–79, 80, 82, 85–86, 92
age-dependent expression *see* developmental expression
alcoholism, 260
aldosterone, 211, 233
amiloride
  analogues, 210
  salty taste and, 213–214, 221
  sour taste and, 205–208
  see also Na^+ channels, amiloride-sensitive
amphid, 237
amyl acetate, 142, 162
anaesthesia, tongue, 253–254, 265
anion effect, salty taste, 213–214, 215, 219–221, 227, 233
anosmia
  C. elegans mutants, 239–240
  specific, 134, 136, 246
antenna, 92, 151–152
  odorant responses in *acj6* mutants, 154–157
  olfactory gene expression, 158–159

antenno-maxillary complex (AMC), 152, 158
Antheraea perny, 183–184
Antheraea polyphemus, 45, 183
aphrodisin, 178, 180, 182
L-arginine, 22
arginine vasopressin (vasopressin), 207, 210–212, 233
β-arrestin, 148–149
attractants, *C. elegans*, 237, 238, 239–240, 248–250
avoidance behaviour, *C. elegans*, 237, 240–241, 245

B50/GAP-43, 20–21
baculovirus/Sf9 cell system, 105, 147
behavioural responses
  C. elegans, 235–237, 238
  Drosophila, 152, 163, 165
benzaldehyde
  C. elegans responses, 238, 239, 240, 241, 242, 248
  Drosophila responses, 152, 153, 154–157, 163
BG protein, 183
bilin-binding protein, 170
biotin, 238, 248
biotransformation enzymes, 133–134, 144, 272
bitter compounds
  block of apical K^+ channels, 203–205
  VEG-P binding, 177, 181, 182
bitter taste
  genetic variation, 256–260, 262–266
  role of VEG-P, 177
  tongue map, 251–253
  transduction, 186, 187, 194, 195, 198, 199–200
bland gene, 163
blindness, congenital, 196–197

bombesin, 80, 81, 82
  receptor-expressing melanophores, 80, 81, 82
bombykol, 183
Bombyx mori, 183
bovine serum albumin (BSA), 44, 45
Bowman's glands, 29, 35, 43
8-bromocyclic AMP, 207
butanol, 152, 153, 156, 157, 241
2-butanone, 238, 239, 242
butyl acetate, 164

Caenorhabditis elegans, 137, 235–250
  chemosensory mutants, 239–240, 246, 247
  chemosensory neuron functions, 237–239
  chemotaxis *see* chemotaxis, *C. elegans*
  electrophysiology, 249
  odorant concentration effects, 241–242, 246–247
  sensory adaptation, 242–243, 249
  single neuron responses, 239
caffeine, 256, 258, 263–264
calcium (Ca^{2+}) channels
  olfactory neurons, 116
  taste cells, 202–203, 208
calcium ions (Ca^{2+}), 213, 272
  bitter taste transduction, 187, 194
  olfactory mucus, 40
  olfactory transduction, 116, 118–120, 126–127, 128
cAMP *see* cyclic AMP
capsaicin, 260
catfish, 61, 91, 92–93, 120, 245
Cdc2 kinase, 271
chemosensory cells, gut, 215, 218
chemosensory neurons, *C. elegans*, 236, 237–242, 247
chemotaxis, *C. elegans*, 235–237, 245, 247, 248–249
  adaptation, 243–244
  concentration effects, 241
  neurons regulating, 237–239
chilli peppers, 260
chloride (Cl^-) channels, taste cells, 233, 234
chloride ions (Cl^-)
  C. elegans responses, 238
  Na^+ taste response and, 213, 220, 221, 226, 227, 233–234

chorda tympani nerve, 253
  damage/anaesthesia, 253–254
  responses to Na^+ salts, 220–221, 222, 224–225
cilia, olfactory, 31, 68, 115
  ion channels, 118–119
  isolated, 98, 110–111, 271–272
  microchemical organization of glycocalyx, 35–37, 38
  signal transduction, 97–105
circumvallate papillae
  gustducin/transducin expression, 191, 193, 197, 200
  innervation, 253
  sour taste transduction, 208
  sucrose octaacetate mouse model, 265
  VEG-P expression, 169, 173
citraldimethylacetal, 102
citralva, 98, 102
citric acid
  ionic current generation, 203, 204, 205, 207, 210, 212
  tongue sensitivity, 252
citronellal, 102
citronellylacetate, 102
clonal exclusion, 134
cyclic AMP (cAMP), 271
  C. elegans responses, 238, 248
  melanophores, 77–78, 79
  olfactory transduction, 68, 100, 101, 109, 115–116
    ion channel activation, 118–120, 121, 123, 129
    kinetics, 98, 99, 111, 113–114
    odorant specificity, 90–91, 101, 102, 113
    signal termination, 101–103, 104, 148–149
  taste transduction, 186–187, 194–195, 198, 199
cyclic GMP (cGMP), 103–105, 112–113, 118
cyclic nucleotide (cAMP)-gated ion channels, olfactory (OcNC), 97–98, 101, 115–130
  gene, 69, 70
  kinetics, 112, 120–121
  membrane current generation, 118–120, 123, 126–127
cytochrome P_{450}, 93, 133–134, 144

Datura stramonium agglutinin (DSA), 33–34, 37, 38, 42
dauer larva formation, 239
deg-1 gene, 247
denatonium, 182, 187, 198
deoxyglucose, radiolabelled ([³H]deoxyglucose), 163
desensitization
  olfactory receptors, 101–103, 104, 148–149
  receptors expressed in melanophores, 84
developmental expression
  Olf-1-binding activity, 11, 13
  olfactory-specific genes, 22–24
  OR genes, 22–24, 66, 94, 159
diacetyl, 238, 239, 240, 241–242, 248
diacylglycerol (DAG)
  melanophores, 79–80, 87
  olfactory transduction, 100
diaminophenylindole (DAPI), 71
digital imaging, 82
DNA-binding proteins
  olfactory neurons, 15–17, 69–72, 73
  see also Olf-1
DNase footprint analysis, 5–9
dopamine 2 receptors, 80, 86
dopamine 3 receptors, 80, 87
double-pulse experiments, 123–125
Drosophila, 92, 137, 150–166
  gust E expression, 21, 163
  odorant receptor affinity, 141–142
  salt sensitivity, 233
dysautonomia, familial, 265

E12, 73
east gene, 163
EC_{50} values, 84, 127–128
egg-laying, *C. elegans*, 238
electroantennogram (EAG), 152, 154–157, 162, 164–165
electroolfactogram (EOG), 22, 24, 249, 269
electropalpogram (EPG), 152, 153, 156
electrophoretic mobility shift assay (EMSA), 5, 9
endothelin C-specific receptor, 87
enhancer trap screen, 151, 157–160
erythropoietin receptor, mouse, 74
ethyl acetate, 141–142, 162, 164, 165
  response in *acj6* mutants, 153, 154–156, 157

ethyl alcohol (alcohol), 142, 260
ethylvanillin, 102
eugenol, 102

facial (VII) nerve, 253, 254, 255
fatty acids, VEG-P affinity, 173, 174, 178, 182
fenoterol, 79, 80
50.06 gene, 70
foliate papillae
  gustducin/transducin expression, 191, 193
  innervation, 253
  VEG-P expression, 169, 173
c-fos gene, 269–270
fungiform papillae
  gustducin/transducin expression, 197, 200
  innervation, 253
  PROP supertasters, 258–260
  sour taste transduction, 205–208
furfurylmercaptan, 102

G_{14}, 198
G protein-coupled receptors
  new tool for investigating, 76–87
  olfactory system, 52–53, 89–90, 92, 132, 133, 275
G proteins
  α subunits *see* Gα subunits
  βγ subunits, 90
  feedback loops, 111–112
  γ subunits, 89
  olfactory transduction, 68, 89–90, 98, 100, 133
  taste transduction, 186–200
  see also G_{olf}; gustducin; transducins; *other specific proteins*
Gα subunits, 111, 187
  cloning from taste tissue, 189
  olfactory neurons, 89–90
  taste transduction, 187–189
  see also gustducin; transducins
GAP-43, 20–21
GAP-43 gene, 17
gap junctions, 219
gastrin-releasing peptide (18-27), 80, 81
gastrointestinal tract, chemosensory cells, 215, 218
gene clusters, OR genes, 74–75, 134–135

gene regulation
  olfactory neuron-specific genes, 68–75
  OMP, 4, 15–17, 23–25, 69–70
  OR genes, 54, 72, 74
genetic polymorphism, olfactory receptors, 136–137
genetic screen, Drosophila mutants, 151, 152–157
genetic variation, taste, 256–260, 262–266
geraniol, 102
G_i, 85
glossopharyngeal (IX) nerve, 253, 254, 255
glycocalyx, olfactory cilia, 35–37, 38
glycoconjugates, secretory, 30–31, 33–35
G_o, 85, 89
goldfish, 22, 25
G_{olf}, 68, 69, 89–90, 187
  gene regulation, 70
G_q, 85, 89, 198
G_s, 85, 187, 195
GTPase-activating proteins (GAPs), 111–112
gust E gene, 21, 163, 248
gustducin (α gustducin), 187, 189
  cloning from taste tissue, 189
  role in taste transduction, 193–195, 198
  sequence comparisons, 188, 189
  tissue-specific expression, 190–191, 192, 193, 197–198, 200

H^+ see protons
hedione, 102
herpes zoster oticus, 255
hexamethylene amiloride, 210
1-hexanol, 182, 241
HLH proteins, 71, 73
hyperosmia, specific, 134, 136

IBMX, 199, 200
immunocytochemistry
  G proteins, 191, 193, 197–198
  Na^+ channels, 230–232
  VEG-Ps, 169, 173
in situ hybridization
  G proteins, 191, 193, 200
  OR genes, 55–61, 64, 105–107
  VEG-Ps, 169, 173
inhibition
  olfactory receptor neurons, 109–110
  taste, 215, 254, 255–256, 264

inositol 1,4,5-trisphosphate (InsP$_3$), 80, 165, 271
  bitter taste transduction, 187, 194
  olfactory transduction, 88–89, 98–101, 105, 109
  odorant specificity, 90–91, 101, 102, 113
  receptors, 35, 42, 91
ion channels
  inositol 1,4,5-trisphosphate-gated, 99, 101
  noise generation, 270–271
  olfactory cyclic nucleotide-gated see cyclic nucleotide-gated ion channels, olfactory
  olfactory neurons, 100, 101, 133
  taste cells, 202–203, 208
  see also K^+ channels; Na^+ channels
ion pathways in taste buds, 218–234
  apical membrane channels and Na^+ effects, 221–225
  paracellular, 219–221, 226, 227
  transcellular, 219, 226
α-ionone, 242
iridophores, 77, 84
isoamyl acetate, 164
isoamyl alcohol, 238, 239–240
2-isobutyl-3-methoxypyrazine, 44, 182, 183
isoeugenol, 102
isomenthone, 98, 99
isovaleric acid, 91, 102

jump response, Drosophila, 152, 156
c-jun gene, 270

K^+ channels
  olfactory neurons, 88, 116, 118, 128
  sour taste transduction, 202–205, 208, 212–213
  sweet taste transduction, 186
kinases, protein, 100, 101, 102–103, 148–149

β-lactoglobulin, 170, 172
lacZ reporter gene, 11–15, 20, 24, 157, 158
lectinoprobes, 33–35, 38, 42
lilial, 102, 148
Limax flavus agglutinin (LFA), 33, 34, 38, 42

Subject index

lingual epithelium, electrical potential, 219–220, 221
lipase, lingual, 178
lipocalins, 170, 182
  functions, 177–178
  gene structure, 172
  sequence similarities, 170, 171
lithium chloride (LiCl), 212
litorin, 80, 81
lobster, 88–90, 91, 101, 109
loose-patch recordings, 202, 203
lozenge mutation, 162
lyral, 102, 148
lysine, 238

macroglomerulus, 163
malvolio mutation, 164
maxillary palp, 152
  odorant responses in *acj6* mutants, 152, 153, 156
  olfactory gene expression, 159
mec-4 gene, 248
mechanosensory responses, *acj6* mutants, 156, 157
melanocyte-stimulating hormone (MSH), 77, 78
melanophores, 77–86
melatonin, 77, 78, 85
membrane currents, olfactory transduction, 115–130
metaproterenol, 79, 80
N-methyl-D-glucamine, 207
Mg^{2+}, olfactory transduction and, 119, 120
moths, 44, 45, 46, 162–163, 183–184
mouse urinary proteins (MUP), 172, 177, 181, 182
mucins
  membrane-bound, 30, 31, 35
  secretory, 30–31, 33–35
mucociliary complex, 31
mucous domains, olfactory epithelium, 27–50
mucus
  barrier, 31, 47–48, 272
  olfactory *see* olfactory mucus
  taste pore, 167
mudpuppy (*Necturus maculosus*), 110, 201–217
mushroom bodies, 159
MyoD, 69, 71

N-glycosidic sugar residues, 30, 31, 42
Na^+ channels
  α subunit, 248
  amiloride-sensitive, 205–208, 210, 213–214, 230–232
  antibodies, 230–231
  olfactory neurons, 110, 116, 118
  rat brain type II, 17
  taste bud nerve fibres, 232
  taste cells, 202–203
    ion selectivity, 212
    membrane localization, 214, 229–232
    role in sour taste, 205–208, 210, 213
    salty taste transduction, 213–214, 221–228, 230–232
    species-dependent expression, 214
Necturus maculosus (mudpuppy), 110, 201–217
nerve-stimulation taste phantom, 255
neuromedin B, 80, 81
nitric oxide (NO), 103–105, 112
nitric oxide (NO) synthase, 112–113
L-N^G-nitroarginine, 103
NMDA (N-methyl-D-aspartate) receptors, 119, 121

O-glycosidic sugar residues, 30–31, 42
odorant-binding protein II (OBP II), 170, 171
odorant-binding proteins (OBPs), 29, 43–44, 272
  ligand binding, 173, 181–183
  localization, 40–41
  putative function, 44, 45, 46, 47
  VEG-Ps and, 169–170
odorant receptors *see* olfactory receptors
odorants
  C. elegans responses, 236, 237, 238, 239–243, 245–247, 248–249
  concentration effects, 92, 127–128, 240–242, 246–247
  inactivation, 134, 144
  induction of transcription by, 269–270
  melanophore responses, 85
  membrane currents induced by, 121–125, 127–128
  receptor affinity, 133–134, 141–144
  responses in *Drosophila* mutants, 151, 152–157, 162

odorants (*cont.*)
  second messenger pathway specificity, 90–91, 101, 102, 113
  second messenger signalling, 97–114
  sensing by *Drosophila*, 151–152
  spectrum of receptor responses, 92–93, 147–148
  transport to receptors, 29, 44–47
odr mutations, 240, 241–242
Olf-1, 4, 5–7, 16, 70–72, 73–74, 162
  age-dependent expression, 11, 13
  bulbectomized rats, 11, 14, 73
  localization, 71
Olf-1-binding site, 3–26, 70–72, 73–74
  biological function, 11–17, 72
  identification, 5–7
  sequence analysis, 8, 9
  tissue specificity, 7–9, 12
olf E gene, 163
olfactomedin, 28
olfactory bulb, organization, 62, 65–66, 67
olfactory bulbectomy, 11, 14, 24–25, 73
olfactory cilia *see* cilia, olfactory
olfactory cyclic nucleotide-gated ion channels (OcNC) *see* cyclic nucleotide-gated ion channels, olfactory
olfactory epithelium
  mucous domains, 27–50
  spatial expression of OR genes, 54–62, 64–67, 105–107
olfactory genes
  identification in *Drosophila*, 157–160, 162, 163–164
  regulation of specific, 68–75
  see also olfactory receptor genes
olfactory marker protein (OMP), 4, 11, 17, 21, 22
  regenerating cells, 24–25
  see also OMP gene
olfactory mucus, 28–30, 40–48, 272
  chemical composition, 28, 40–42
  ligand translocation, 29–30, 44–47
  microchemical organization, 28, 33–35
  microstructural organization, 28, 29, 31–33, 42–43
  pH regulation, 48
  physical state/viscosity, 28, 42
  regulation, 29
  secretory glycoconjugates, 30–31

olfactory receptor (OR) genes, 133
  chromosomal localization, 74–75, 134–135
  Drosophila, 164, 165
  evolution, 137–138, 142
  expression during development, 22–24, 66, 94, 159
  expression within zones, 61, 64–65
  expression zones, 59–62, 64, 65–67
  multigene family, 52–54
  pseudogenes, 75
  regulation of expression, 54, 72, 74
  sequence diversity, 53–54, 89
  spatial expression, 54–62, 64–67, 105–107
  variations in humans, 134, 136–137
olfactory receptor neurons
  axotomy studies, 22
  chemical specificity, 21, 22, 23, 25
  death, 25–26
  differentiation, 20–22, 69
  immature, 11, 16, 20–22
  inhibitory responses, 109–110
  membrane currents, 121–125
  odorant-induced transcription, 269–270
  Olf-1-binding activity, 11, 13, 14
  OMP expression, 4, 11, 17, 21–22, 24–25
  organization of axonal projection, 62, 65–66, 67
  receptor expression, 55–59, 64–65, 88–89, 92, 94
  regenerating, 11, 24–25
  signal transduction pathways *see* olfactory transduction
olfactory receptors (OR), 51–67, 88, 105–107, 131–146, 275
  affinity for odorants, 133–134, 141–144
  expression of cloned, 85, 105, 147
  expression in individual neurons, 55–59, 64–65, 88–89, 92, 94
  expression on olfactory axons, 94
  genetic polymorphism, 136–137
  integration of responses over time, 46–47
  kinetic properties, 268–269
  phosphorylation, 103, 148–149, 268
  spectrum of odorant responses, 92–93, 134, 147–148
  testis, 74, 94

Subject index

olfactory thresholds, 143, 145–146
  genetic basis, 134, 142, 263
  individual differences, 136–137
olfactory transduction, 88–92, 133
  membrane currents, 115–130
  second messengers, 97–114
OMP *see* olfactory marker protein
OMP gene
  ablation studies, 20
  Olf-1-binding site *see* Olf-1-binding site
  regulation, 4, 15–17, 23–25, 69–70
  upstream binding element (UBE), 7–9, 12, 16
Oncorhynchus mykiss (rainbow trout), 22, 67
opsins, photoreceptor, 137, 193–194
osm-9 mutation, 240
osmotropotaxis, 163
otal mutation, 159, 165

paracellular pathways, taste buds, 219–221, 226, 227
patch-clamp recordings, 117, 202, 203–205
perireceptor events, 272
  olfaction, 27–28, 44–47
  taste, 167–185
phantom taste, 254–255
phenamil, 210
phenylethylalcohol (phenylethanol), 102, 182
phenylethylamine, 102
phenylthiocarbamide (PTC), 256, 262, 263–264, 266
pheromaxein, 178, 180
pheromone-binding protein (PBP), 44, 45, 46, 47, 183–184
pheromones, 163
  binding to PBPs, 183–184
  salivary gland proteins and, 178
  transporters, 29, 44, 45, 46–47
phosphatases, protein, 103
phosphodiesterase (PDE)
  inhibitors, 195, 199–200
  melanophores, 85
  olfactory cilia/neurons, 111, 113
  taste transduction, 194, 198, 199–200
phospholipase C, 80, 90, 100
phosphorylation, protein
  melanophores, 77–78

olfactory transduction, 101–103, 104, 148–149, 268
polymerase chain reaction (PCR), 24, 52–53, 95, 189
potassium chloride (KCl), 258
potassium ion channels *see* K^+ channels
prealbumin, tear, 173, 177
presynaptic influences, 25
proline-rich proteins (PRPs), 168, 181
PROP *see* 6-*N*-propylthiouracil
propionic acid, 152, 153, 156, 157
6-*N*-propylthiouracil (PROP), 199, 256–260, 262–266
  supertasters, 257–260, 262, 265
  tasters/non-tasters, 256
prostaglandin D synthetase (PGD), 172, 184
protein kinase inhibitors, 101, 103, 104
protein kinases, 100, 101, 102–103, 148–149
protein phosphatases, 103
protein phosphorylation *see* phosphorylation, protein
protons (H^+)
  block of apical K^+ channels, 202–205
  Na^+ channel permeability, 205–207, 210, 213
  sour taste transduction, 201–202, 207–208
psychophysics
  human olfaction, 134, 142, 144–146
  human taste, 256–260, 262–266
PTC (phenylthiocarbamide), 256, 262, 263–264, 266
ptg^{3D18} mutation, 156
pyrazine, 98, 238, 239, 248
pyrrolidine, 102

quinine (hydrochloride), 182, 199, 252, 258, 263, 264

rainbow trout (*Oncorhynchus mykiss*), 22, 67
rdgB mutation, 159, 165–166
release-of-inhibition, taste system, 254, 255–256
repellants, *C. elegans*, 236–237, 241, 245
reproductive tract, olfactory gene expression in *Drosophila*, 159

retinol-binding protein (RBP), 170, 173, 178
rhodopsin, 193–194

salamander, 25, 31–33, 91, 121, 126, 142, 143
salbutamol, 79, 80
saliva, 43, 167–168, 184
  von Ebner's, 178, 184
salivary glands, 168–169
  see also von Ebner's glands
salty taste, 186, 205, 215–216
  anion effect, 213–214, 215, 219–221, 227, 233
  Na^+-restricted rats, 210–211
  sour taste overlap, 212, 214
  tongue map, 251–253
  transduction, 213–215, 218–234
    apical membrane channels and, 221–225, 230
    paracellular pathways, 219–221, 226, 227
    vasopressin and, 210–211
  see also sodium chloride
scalloped mutation, 164
second messengers
  olfaction, 97–114
    alternative pathways, 98–101, 109
    kinetics, 98, 99, 122–125
    odorant-specific pathways, 90–91, 101, 102, 113
    role of cGMP, 103–105, 112–113
    signal termination, 101–103, 104
  see also cyclic AMP; cyclic GMP; inositol 1,4,5-trisphosphate
sensilla, *Drosophila*, 151–152, 158, 162
sexual dimorphism, olfactory gene expression, 158–159
sialic acid
  olfactory cilia glycocalyx, 35, 38
  secretory mucins, 33–34
signal transduction *see* olfactory transduction; taste transduction
smellblind mutation, 159
sodium acetate (NaAc), 220–221, 222, 226, 227
sodium chloride (NaCl), 205, 226, 227, 233
  C. elegans response, 248
  chorda tympani responses, 220–221, 222

PROP supertasters, 257, 258
  tongue sensitivity, 252
  see also salty taste
sodium gluconate (NaGlu), 221, 223, 224–225, 226, 227
sodium ion channels see Na^+ channels
sodium ions (Na^+)
  C. elegans responses, 238
  electrochemical concentration effects, 221–225, 228
sodium (Na^+) salts, taste cell responses, 220–221, 222, 226, 227, 233
sour taste, 186
  salty taste overlap, 212, 214
  tongue map, 251–253
  transduction, 201–217
    amiloride-sensitive Na^+ channels and, 205–207
    apical K^+ channels and, 202–205
    hypothesis, 207–208
starvation, *C. elegans*, 243, 246
stopped-flow method, 111, 129
sucrose, tongue sensitivity, 252
sucrose octaacetate (SOA), 258, 263
  mouse model, 260, 264–265
superoxide dismutase, 112
sustentacular cells, 29, 35
sweet taste, 214
  PROP tasters, 256
  tongue map, 251–253
  transduction mechanism, 186–187, 195, 198

tannic acid, 181
taste
  genetic variation, 256–260, 262–266
  intravascular stimulation, 219
  localization illusions, 255–256
  pathological variation, 251–256
  perireceptor events, 167–185
  phantom, 254–255
  thresholds, PROP supertasters, 257, 263
  tongue map myth, 251–253
  see also specific modalities
taste buds, 215
  gustducin/transducin expression, 191, 192, 193, 197–198
  innervation, 253
  ion pathways, 218–234
  PROP supertasters, 258–260
  sour taste transduction, 202, 205–207

taste cells
  apical membrane channels and Na^+ effects, 221–225
  membrane potential, 220, 232–233
  peptidergic innervation, 212
  sour taste transduction, 202–208
taste papillae *see* circumvallate papillae; foliate papillae; fungiform papillae
taste transduction, 275
  bitter *see* bitter taste, transduction
  G proteins, 186–200
  ion pathways, 218–234
  salty, 213–215, 218–234
  sour, 201–217
  sweet, 186–187, 195, 198
tear prealbumin, 173, 177
TEF-1, 164
terbutaline, 79, 80
testis, 41, 73
  olfactory receptors, 74, 94
12-*O*-tetradecanoylphorbol-13-acetate (TPA), 79–80
tetraethylammonium (TEA), 203, 205
threshold hypothesis, olfactory, 134, 263
tight junctions, 219, 220
tongue
  anaesthesia, 253–254, 265
  innervation, 253
  map, 251–253
  VEG-P expression, 169, 173
transcription factors, 164
  olfactory neuron-specific, 4, 15–17, 69–72, 73
  see also Olf-1
transducins, 111, 187
  cone α ($\alpha_{t\text{-cone}}$), 189
    cloning from taste tissue, 189
    role in taste transduction, 193–195
    sequence comparisons, 188, 189
    tissue-specific expression, 190–191, 192, 193, 197–198
  knockout mutants, 196
  rod α ($\alpha_{t\text{-rod}}$), 189
    cloning from taste tissue, 189
    role in taste transduction, 193–195
    sequence comparisons, 188, 189
    tissue-specific expression, 190–191, 192, 197–198
transgenic mice, *OMP* gene-expressing, 4, 11–15, 20, 24
transposase, 157
triethylamine, 102
trigeminal (V) nerve, 253, 254
trimethylthiazole, 238

UDP glucuronosyltransferase (UGT), 133–134, 144
upstream binding element (UBE), *OMP* gene, 7–9, 12, 16
urea, 199, 263–264

vallate papillae *see* circumvallate papillae
vasopressin (arginine vasopressin), 207, 210–212, 233
VEG-P *see* von Ebner's gland proteins
video imaging, digital, 82
voltage-clamp studies, 91, 117, 220–221, 222, 228
vomeromodulin, 29, 73–74, 180
vomeronasal organ, 43, 64, 73
von Ebner's gland proteins (VEG-P), 41, 44, 169–170, 272
  gene structure, 172
  ligand binding, 173, 174, 180, 181–183, 184
  localization studies, 169, 173
  molecular cloning, 170–172
  purification, 173
  putative functions, 174–178
  rat versus human, 170
  species-specific expression, 174, 176, 181
von Ebner's glands (VEGs), 41, 168–170, 178

water
  electroantennogram response, 154–156
  taste, 199
whole-cell recordings, 117–118, 202, 203, 207

yeast genetic system, 70–71